The True Creator of Everything

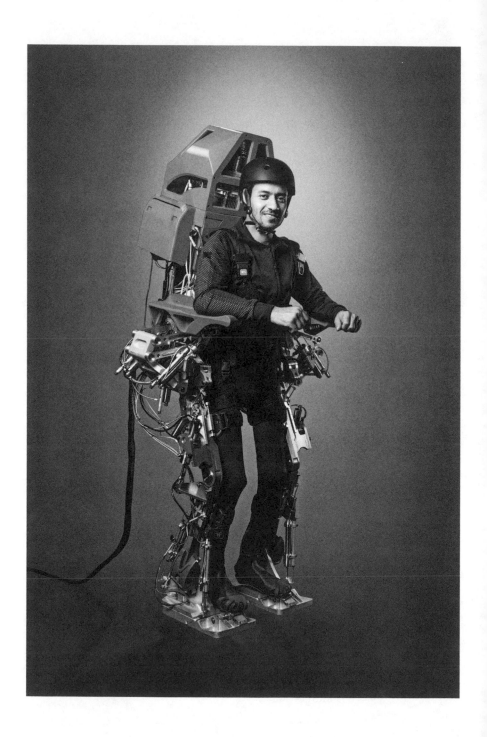

THE TRUE CREATOR
OF EVERYTHING

HOW THE HUMAN BRAIN SHAPED THE
UNIVERSE AS WE KNOW IT

Miguel Nicolelis

Yale UNIVERSITY PRESS   NEW HAVEN AND LONDON

Yale University Press books may be purchased in quantity for educational, business, or promotional use. For information, please e-mail sales.press@yale.edu (U.S. office) or sales@yaleup.co.uk (U.K. office).

Frontispiece: Juliano Pinto inside the brain-controlled robotic exoskeleton built by the Walk Again Project. (Courtesy of the Alberto Santos Dumont Association for Research Support [AASDAP])

Excerpt at the end of Chapter 12 from *THE ROCK: A Pageant Play*, Book by T. S. Eliot. Copyright © 1934 by Houghton Mifflin Harcourt Publishing Company, renewed 1962 by T. S. Eliot. Reprinted by permission of Houghton Mifflin Harcourt Publishing Company. All rights reserved.

Set in Scala type by Newgen North America.
Printed in the United States of America.

Library of Congress Control Number: 2019943685
ISBN 978-0-300-24463-2 (hardcover : alk. paper)
A catalogue record for this book is available from the British Library.

This paper meets the requirements of ANSI/NISO z39.48-1992 (Permanence of Paper).

10 9 8 7 6 5 4 3 2 1

To the teachers who introduced me to the different faces
of the True Creator of Everything:
Juarez Aranha Ricardo
César Timo-Iaria
John Chapin
Rick Lin
Jon Kaas
and
Ronald Cicurel

Contents

PREFACE

When Brazil was officially announced as the host of the 2014 FIFA Soccer World Cup in 2007, I came up with an idea to introduce a global audience to the edge of modern brain research and show how much it has to offer for the betterment of human lives. After five years of planning, I approached the president of Brazil and the secretary-general of FIFA to propose running a scientific demonstration during the opening ceremony of the upcoming World Cup. The central goal of this event would be to highlight the fact that, thanks to new technological developments and major insights on the basic operation of the human brain, neuroscientists were getting close to achieving a magnificent feat: restoring mobility to millions of people worldwide paralyzed by serious spinal cord injuries.

To those in charge of the World Cup's opening ceremony, I proposed having a young Brazilian, completely paralyzed from the chest down by a spinal cord injury, deliver the symbolic opening kickoff of the World Cup. In reply, the event organizers immediately posed to me the question anybody confronted with such an outrageous plan would ask: how will a paraplegic deliver such a kick? My answer bewildered them even more: by using a lower-limb robotic exoskeleton directly controlled by his or her brain, I said nonchalantly.

To my total surprise, the organizers agreed.

The easy part was done. Now came the tough part: actually pulling it off.

To do this, I created the Walk Again Project, an international nonprofit scientific consortium. In a matter of months, dozens of engineers, neuroscientists, roboticists, computer scientists, physicians, rehabilitation personnel, and a large variety of technicians from twenty-five countries joined us. The next eighteen months were the craziest of my life, and possibly in the lives of everyone involved in the project. By November 2013, eight strong-minded Brazilian paraplegic patients had volunteered to take part in the Walk Again Project. Every day for the next six months, these patients practiced a unique routine: first they imagined moving their legs to walk. Then, using a brain-machine interface that allowed their electrical brain activity to be decoded and transmitted to a lower-limb robotic exoskeleton that encased their paralyzed limbs, they used their motor thoughts to move the machine's legs at will.

And so, on the chilly austral winter afternoon of June 12, 2014, at precisely 3:33 p.m. Brasília time, Juliano Pinto (see frontispiece), one of the recruits of the Walk Again Project, made a final effort to straighten his torso. Inside his brand-new robotic cockpit, he stood tensely at the edge of the soccer stadium's immaculate grass field. Closely monitored by a crowd of sixty-five thousand fans, not to mention an estimated global audience of 1.2 billion people, Juliano waited for his moment to make history.

When the moment of truth came on that wintry afternoon, I was standing, together with twenty-four of the Walk Again Project's team members, just a few feet behind Juliano. The ceremonial soccer ball was placed in front of Juliano's right foot. To give an idea of how the exoskeleton worked, we had placed two long strips of LEDs from the edges of Juliano's helmet to the lower part of the exoskeleton's legs. When Juliano turned on the exoskeleton, the LEDs began to rhythmically flash an intense blue light.

We were ready to launch!

Releasing all the energy, anguish, and hope of one who knows intimately what it is to be confined to a wheelchair for almost a decade, Juliano under-took a movement that, just six months earlier, he never imagined he would be able to make again. As his brain generated the electrical signals that contained the needed motor instructions, the exoskeleton's computer translated Juliano's mental desire to move into a coordinated robotic leg movement sequence. At that moment, the flashing blue of the LED strips was replaced by fast sequences of intense green and yellow light pulses flowing from the top of Juliano's helmet, throughout the exoskeleton's frame, all the way to his feet.

Time seemed to slow down. In the span of an unforgettable split second, Juliano's body weight first shifted to the left, as a result of the now symbiotic movement of his own torso and the exoskeleton's balancing system. Next, gently moved by the robotic metal casing, his right leg began to sway, cocking back an unmistakable Brazilian kick. Reaching the zenith of the sway, his body moved forward, ready to unleash the most improbable of all kicks.

And then, as Juliano's right foot impacted the soccer ball, making it gently roll down to the edge of the redwood platform where he now stood tall and whole again, he let loose a loud, almost guttural, and hard-earned scream, throwing his clenched right fist up into the gray Brazilian sky to celebrate his *gol*. A sense that something quite magical had just happened overwhelmed all of us. We all ran to Juliano to embrace him in the most PhD-rich huddle ever to celebrate a score in a World Cup soccer match. Amid hugs and kisses, in a mixture of his and our own tears, Juliano shouted something that captured the profound and unexpected essence of what had just happened: "I felt the ball! I felt the ball!"

There were more surprises ahead. Throughout the Walk Again Project, our clinical protocol required that the patients have a neurological exam routinely. This was considered a simple academic exercise, given that their clinical state had not changed during all those years they were completely paralyzed and could not feel any part of their bodies below the level of their injuries. We did not expect to observe any change in their neurological status at all. But then one of our female patients told one of our physicians that, during a weekend out on the beach, she had, for the first time in fourteen years, felt the intense heat of the sun on her legs. We began to suspect that something unusual was happening.

Spinal cord injuries are assessed according to a scale developed by the American Spinal Cord Injury Association (ASIA). Seven of our patients were classified as ASIA A, meaning they exhibited complete paraplegia and no tactile sensitivity below the spinal cord lesion. The eighth patient was classified as ASIA B, meaning a complete paraplegia but with some preserved sensation below the injury level.

By August 2014, the clinical data we collected had us scratching our heads: after eight months of training, our patients were showing clear signs of clinical improvement—voluntary motor control and tactile sensitivity in their legs was reemerging, and their ability to control bowel and bladder functions was improving too.

Taken by surprise, we repeated the entire series of neurological exams three months later to rule out the possibility that those were simply transient clinical fluctuations. By early 2015, the data told us something we could hardly believe. Not only was the clinical recovery in all patients robust, but their motor, sensory, and visceral functions had all improved even further. The patients had regained the ability to voluntarily contract multiple hip and leg muscles; at least three were able to produce voluntary compound movements of their legs when suspended in the air. One of these patients could literally "walk again" in the air.

In parallel, when all patients' data generated by several tests of somatic sensitivity were averaged, they had become more sensitive to pain and had an improved ability to discriminate between tactile stimuli delivered many body segments below the original level of their spinal cord injuries. Pressure and vibration detection also increased significantly.

Ultimately, by the end of 2015, thanks to this unprecedented neurological recovery, all seven patients who had remained engaged in our training (one patient had to drop out of the study by the end of 2014) had to be reclassified to the ASIA C level, meaning that they were now considered as being only partially paralyzed! Juliano Pinto, for example, could now experience some crude tactile sensation when his toes and feet were touched!

But this was not the end. By 2016, two of our original patients, including Juliano, had improved enough to take advantage of a neurorehabilitation tool known as noninvasive functional electrical stimulation. Before they enrolled in our protocol, this stimulation technique, in which small electrical currents are applied to the skin's surface to help improve muscle contractions, would have been useless for these patients. Now, two of them were able to begin walking, using only a simple walker, while applying 30–40 percent of their body weight on the ground. By late 2017, these patients had been able to produce close to five thousand steps using this minimal apparatus.

Further clinical analysis revealed that our female patients could now feel abdominal contractions indicating that their monthly period was about to happen. One of the women enrolled in the project recovered so much visceral function and also tactile sensation in the perineum region that she decided to become pregnant again. Nine months later, after experiencing the baby's kicks and uterus contractions around the time she was due, she delivered a healthy baby boy.

In addition to the unexpected partial clinical recovery, we discovered that our neurorehabilitation protocol had also been able to reshape our patients' own sense of self. As a result, their brains had assimilated an artificial tool, the robotic exoskeleton, as a true extension of their biological bodies!

In all our minds a major question emerged: what mechanisms or brain properties could have accounted for this radical reshaping of the patients' self and triggered such remarkable and unprecedented neurological improvements?

Although many may have identified in Juliano's kick that afternoon the ultimate poster image of the cyborg age or an endorsement of the transhumanistic movement, I took a diametrically different reading. Where many saw the triumph of a hybrid and seamless communion between man and machine, I detected yet another clear demonstration of the unsurpassable and truly inspiring, adaptive power that the human brain is capable of unleashing, over and over again, throughout humankind's history, whenever it has to confront itself with never before encountered world contingencies.

To fully justify this interpretation, along with my contention that the human brain embodies a type of organic computing device without parallel in any machine ever built, including the most pervasive and successful of them all, the digital computer, I soon realized we were going to need a completely new theory in modern neuroscience: one that, at long last, recognizes how the human brain evolved, over millions of years, to become the True Creator of Everything.

ACKNOWLEDGMENTS

Although *The True Creator of Everything* project took about five years to execute, its content is based, among other things, on about thirty years of extensive theoretical, basic, and clinical brain research conducted since I moved to the United States from Brazil in the winter of 1989, first as a postdoctoral fellow in the lab of John Chapin and, for the past twenty-five years, as faculty at the Department of Neurobiology at Duke University. Therefore, I would like to thank all the undergraduate and graduate students, postdoctoral fellows, technical and administrative staff, and other collaborators in the U.S. and abroad who were once part of the Nicolelis Lab at the Duke University Center for Neuroengineering for all that I learned from our conversations and collaborations, and from the hundreds of experiments we conducted together during this period. I also would like to thank my colleagues and collaborators at both the AASDAP Neurorehabilitation Lab in São Paulo, Brazil, the world headquarters of the Walk Again Project, and the Edmond and Lily Safra International Institute of Neuroscience (ELS-IIN), in Macaíba, Brazil, for these past seventeen years of insightful intellectual exchanges and the great adventure of building, from scratch, a major scientific program in my beloved tropics.

I am deeply indebted to my literary agent and New Yorker friend, James Levine, for the continual support, total commitment, and deep friendship he bestowed on my projects and me over the past ten years. Without Jim's calm demeanor and decisive backing, none of my literary projects would have seen the light of the day, let alone this *True Creator of Everything*. For all the battles, the tough times, and the great times, thank you very much, Jim. I am also very grateful to Elizabeth Fisher and all the other friends from Levine, Greenberg,

Rostan Literary Agency for all their tremendous support to make this book successful around the world.

I would like to thank profusely Jean Thomson Black, my editor at Yale University Press, for supporting this project since the very first minute we discussed it over the phone and for following it up until its completion with utmost professionalism and enthusiasm. Likewise, I would also like to thank Michael Deneen and everyone at Yale University Press for providing the best possible support for the completion of this project. I also thank Robin DuBlanc for her outstanding copyediting and many insightful suggestions. I also would like to thank my good friend and gifted artist Custódio Rosa for producing some of the most fundamental illustrations of this book and for his constant kindness in always being ready to discuss minor improvements and details. A fellow fan of my beloved Palmeiras soccer club, Custódio was always available to offer his time to this project. To you, Custódio, my *alviverde*, thanks.

Nobody read this book more often and thoroughly than Susan Halkiotis, my long-term assistant and one of my best friends. As with all my projects over the past seventeen years, since she joined my lab at Duke, Susan was enthusiastically on board with this literary adventure from the first second, and never let it go until all was finished and done to her satisfaction. Over multiple versions and revisions during the past five years, Susan was always my first reader and the first to offer fundamental insights on how to improve the way to communicate my ideas and the book's overall message. There are no further words I can offer to thank Susan for all the professional excellence, brotherly love, and total support she dedicated to this project, to my lab, to me, and to all members of my family. Susan, it has been a total privilege, a delight, and above all, extreme great fun to be able to work with you for these two decades. A major big hug to you!

From the Brazilian side, Neiva Paraschiva has also been a reader of this book even before it existed. For the past forty years, we have collaborated in all sorts of projects, and since 2003, when I decided to create AASDAP and the ELS-IIN in Brazil, Neiva has always been there to provide encouragement, friendship, moral and intellectual support, and a big dose of tough love and reality-checking. Above all, Neiva has always been the person who encouraged me to dream at the limit of my imagination and to translate these dreams into practice. Without her unwavering determination and support, it is very likely that *The True Creator* would not have made it to print. Neiva, a kiss, a hug, and many thanks for everything.

For the past fourteen years, my entire worldview and scientific core have been revolutionized by my interactions with the Swiss-Egyptian mathematician, philosopher, and author Ronald Cicurel. I used to call him my best friend, but lately I realized that, without ever telling me, my sainted mother must have traveled to Egypt incognito in her teenage years, because it is absolutely clear that Ronald can only be described as my lost brother; the brightest intellectual and humanistic sun to shed light over the past decades of my adult life. Never before meeting Ronald in Lausanne in November 2005 had I the privilege to meet someone so intellectually gifted but, at the same time, so willing to share his vast and unique cultural and scientific knowledge and profound humanism. Thinking more clearly as I write, Ronald is more than a brother—he is the greatest teacher I have ever had. Without his wisdom, pointed critique and commentary, invaluable contributions, and unlimited kindness in sharing his time to read every line of this manuscript many times over, *The True Creator of Everything* would never have materialized. For that, and for the innumerous life lessons, a big hug to you, my brother. See you at the Palais Oriental, my friend.

Finally, I would like to thank my sons Pedro, Rafael, and Daniel for supporting my scientific adventure for the past thirty years and being there to talk to me when I needed to be reminded what this adventure was all about. For you big guys, a big kiss from Dad.

In the beginning,
The True Creator of Everything proclaimed:
Let there be light!
After a brief silence,
He then decreed:
And let it be
$E = mc^2$

THE TRUE CREATOR
OF EVERYTHING

1 • In the Beginning . . .

In the beginning, there was only a primate brain. And from the depths of that highly convoluted, 86-billion neuronal mesh, sculptured through a blind evolutionary walk and multiple mental big bangs over the span of millions of years, the human mind emerged. Unbounded, unconstrained, expanding quickly like some sort of biological plasma, it soon was welded into a continuum, spinning off a combustive mixture of bipedal walking, manual dexterity, tool making, oral and written language, elaborate social entanglements, abstract thinking, introspection, consciousness, and free will. From that same mental cauldron, the most comprehensive notion of space and time ever conceived by organic matter blossomed, serving as the ideal scaffolding for the genesis of a deluge of emergent mental abstractions, the true holy organic tablets of humankind. Soon these mental constructs began dictating the essence of the human condition and civilization: from our egotistical sense of self to our deepest beliefs to elaborate economic systems and political structures, all the way to our unique neuronal reconstructions of what is out there surrounding us all. From humble neuronal electromagnetic storms emerged the magnificent sculptor of our material reality, the virtuoso composer and sole architect of our epic and tragic history; the most insightful investigator of the deepest mysteries of nature; the restless seeker of the elusive truth of our origins; the master illusionist; the unorthodox mystic; the artist of many talents; the lyric poet that lent its unmistakable neurobiological rhymes to every thought, utterance, mythological conceit, cave painting, religious credo, written record, scientific theory, erected monument, exploratory voyage, gruesome genocide, and epic conquest, as well as every gesture of love and every dream and hallucination ever conceived by any hominid that roamed this imperfect blue sphere we call home.

And then, roughly one hundred thousand years since its explosive rise, the True Creator looked back at its almost miraculous accomplishments and saw, to its own astonishment, that it had created a whole new universe.

The True Creator of Everything is a story about the works of the human brain and its unique central position in the cosmology of the human universe. By human universe I mean the immense collection of knowledge, perceptions, myths, beliefs and religious views, scientific and philosophical theories, culture, moral and ethical traditions, intellectual and physical feats, technologies, art, and every other by-product that has emerged from the workings of the human brain. In brief, the human universe is all that defines, for the good or the bad, our legacy as a species. This is not, however, a history book, nor a comprehensive compendium of what neuroscience knows, or thinks it knows, about how the human brain does its tricks. Rather, this is a science book that intends to present the brain in a completely new framework. The core of the book's narrative introduces the details of a new theory on how the human brain, working in isolation or as part of large networks of other brains, accomplishes its amazing feats. I call this new theoretical framework the relativistic brain theory.

When I began planning this book, I tried to build my central argument by focusing on the scientific field where I have spent most of my professional life: brain research. Soon, however, I realized that such a choice was way too narrow-minded. I needed to broaden the scope of my intellectual journey and venture into fields that neuroscientists rarely visit these days—disciplines such as philosophy, art, archaeology, paleontology, the history of computational machines, quantum mechanics, linguistics, mathematics, robotics, and cosmology.

After months of reading, amid mounting frustration that I had not yet found the true beginning of my narrative, I came in contact, almost by accident, with the glorious book *The Story of Art*, by the distinguished German-British historian E. H. Gombrich. Worrying about my writer's block, my mother, a well-known Brazilian novelist, had given me the book as a present on Christmas Eve 2015. Arriving home late that night, I decided to read a bit before sleeping. Instead, the first few sentences jolted me awake. There it was! Written in plain black ink on white glossy paper: the initial thread of my own story. I would not close the book until early the next morning.

This is what Gombrich had written: "There is no such thing as Art. There are only artists. Once these were men who took colored earth and roughed out the forms of a bison on the wall of a cave; today some buy their paints, and design posters for hoarding; they did and do many other things."

Unexpectedly, I had found an ally. Someone who could see that without a human brain, this particular primate brain of ours, shaped and molded through a unique evolutionary process that very likely will never, ever reoccur, anywhere in the vast cosmos that engulfs us, there is no such thing as art because artistic manifestations are all by-products of inquisitive and relentless human minds eager to project to the outside world images from their own internal neuronal universes.

This may seem like a tiny issue, a meaningless semantic twist of the way we always see things. But placing the human brain at the center of the human universe has profound implications for the way we look at our lives and decide what kind of future our kin should inherit. Indeed, with a few small word replacements, Gombrich's remarks could open any other book describing the products of the human mind—for example, a book about physics. Our physical theories are so successful at describing natural phenomena happening at multiple spatial scales that most of us, including the scientists involved in working on a daily basis in these domains, tend to forget what key constructs in physics, such as mass or charge, really mean. As my good friend Marcelo Gleiser, a Brazilian theoretical physicist working at Dartmouth College, wrote in his wonderful book *The Island of Knowledge*, "Mass and charge do not exist *per se*: they only exist as part of a narrative we humans construct to describe the naturalistic world."

Both Marcelo and I came up with the same representation of what the human universe means: if another intelligent being, say the famous Mr. Spock from Vulcan, arrived on Earth and could, by some miracle, communicate effectively with us, we would very likely find that the explanations and theories, not to mention basic concepts and constructs, that he would employ to explain his species' cosmological view of the universe would be totally different from ours (figure 1.1). And why should we expect otherwise? After all, Mr. Spock's brain would be totally different from ours because it would represent a product of an evolutionary process and a cultural history that took place on Vulcan, not on Earth. From my point of view, neither description would be the most or least accurate: they would simply represent the best approximation two different types of organic intelligence had been able to construct from what was offered to them by the cosmos. At the limit, whatever exists out there in this 13.8-billion-year-old universe (a human estimate, mind you), from our brain's own point of view—and, I hazard to say, for any alien's brain too—the cosmos

Figure 1.1. Braincentric cosmology: the human brain's description of the universe—in this case through the use of mathematics—is very likely to be distinct from the one created by an alien's central nervous system. (Image credit to Custódio Rosa.)

is a mass of potential information, waiting for an intelligent observer to extract knowledge out of it and, almost in the same breath, stamp meaning on it all.

Giving meaning to things—creating knowledge—that is a domain in which the True Creator excels. Knowledge allows us to adapt to the ever-changing environment and maintain our ability to continue sucking up even more potential information from the cosmic soup. Protons, quarks, galaxies, stars, planets, rocks, trees, fish, cats, birds: it doesn't really matter what we call them. (Mr. Spock would certainly say that he had better names.) From our human brain's own point of view, those are all different ways to describe the raw information provided to us by the cosmos. Our brains baptize all these objects with both names and, for operational expediency, meaning, but their original content is always the same: potential information.

Before you begin to think that someone must have put something funny in the water that Brazilian neurobiologists and physicists drink when they are growing up in São Paulo or Rio de Janeiro, let me make this point clearer. Most of the time we talk of physics as if it were some sort of universal entity, with a life of its own, like the Art with a capital A that Gombrich referred to. Yet physics per se does not exist at all. What truly exists is the collection of human mental constructs that provide the best and most accurate account, to date, of the natural world that exists out there. Physics, like mathematics or any other accumulated body of scientific knowledge, is defined by the reverberations and echoes of the electromagnetic brainstorms that once crisscrossed the visionary brains of people called Thales, Pythagoras, Euclid, Archimedes, Diophantus, Al-Khwarizmi, Omar Khayyam, Copernicus, Kepler, Galileo, Newton, Maxwell, Bohr, Curie, Rutherford, Einstein, Heisenberg, Schrödinger, and Stueckelberg, among so many more.

By the same token, Gombrich's definition of art comprehends the dazzling collection of mental images generated by human brains that, for the past tens of thousands of years, have been carved, etched, sculptured, painted, or recorded in order to register inner memories, feelings, desires, cosmological views, beliefs, or premonitions into a variety of media (beginning with their own bodies, then using stone, bones, wood, rock, cave walls, metal, canvas, marble, paper, chapel ceilings and windows, videotape, CD-ROMs, DVDs, semiconductor memory, or cloud storage). That collection includes creations ranging from the anonymous and magnificent Upper Paleolithic cave wall paintings of Altamira and Lascaux to all the Botticellis, Michelangelos, da Vincis, Caravaggios, Vermeers, Rembrandts, Turners, Monets, Cézannes, van Goghs, Gauguins, and Picassos—just to name a few of the artists who translated their intangible brainstorms into colorful epic allegories of what it means to be human.

Using the same reasoning, our best and most accurate description of the universe is nothing but a distinguished and elaborate tale of mental derivatives, such as mathematics and logic, which usually go by the names of their creators: Kepler's laws of planetary motion, Galileo's astronomical observations, Newton's laws of motion, Maxwell's equations for electromagnetism, Einstein's special and general relativity, Heisenberg's uncertainty principle, or Schrödinger's equations of quantum mechanics.

Before any physicist jumps out of his chair, this view, rather than demeaning the stunning discoveries and achievements of the physicist brotherhood, simply adds to them by verifying that above all, physicists are also talented neuroscientists, capable of reaching the innermost workings of the human mind (even if most of them usually try to deny the interference of their consciousness in the process of scientific inquiry). But this notion also means that the search for the holy grail of physics, the theory of everything, cannot succeed without the incorporation of a comprehensive theory of the human mind. And although most traditional physicists tend to object adamantly to the idea that the intrinsic physiology of the human mind has anything do with the formulations of the main theories in the field, which they assume to be independent of humans' subjectivity, I hope to show in this book that some of the most enigmatic natural phenomena, including primordial concepts such

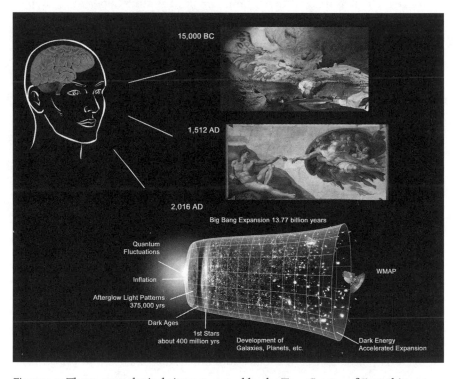

Figure 1.2. Three cosmological views proposed by the True Creator of Everything at different moments in time: The "Hall of the Bulls" in the Lascaux cave painted by our Upper Paleolithic ancestors; Michelangelo's Sistine Chapel; the latest description of the origins of the universe according to NASA.

as space and time, cannot be fully understood unless a human observer—and the human brain—is brought to the foreground.

From that point on, it's off to the races.

According to the most accepted human description of events, barely four hundred thousand years after the singular explosive event that gave rise to the cosmos, light finally escaped, traveling across the universe until it encountered someone or something who could reconstruct its epic journey and attempt to give some meaning to it all. On the surface of a little bluish rock, made by the fusion of intergalactic dust about 5 billion years ago, while orbiting a mediocre yellow star, itself lost in an undistinguished corner of an average galaxy, that primordial light encountered beings who longed to understand it and who, using all their evolution-endowed mental faculties and tools, began in earnest to re-create, inside their minds, the path whence that stream of potential information came and what it possibly meant. The three cosmological views depicted in figure 1.2 offer a tiny glimpse of the enormity of this epic, collective human act of mental creation. And whether one looks at NASA's latest visual description of the known universe, Michelangelo's frescoes, or the painted walls of the Lascaux cave, there is no way to avoid feeling temporarily breathless, humbled and, above all, deeply moved by all the splendorous magnificence that this True Creator of ours has accomplished, in so little time.

2 • The True Creator of Everything Makes Its Evolutionary Entrance

By the time the bison, alerted by the high-pitched whistle coming from the bushes, raised his bulky black head from the grass, his fate was already decided.

Without being able to see much at first glance because of the dense fog that embraced the valley, the mighty bull experienced a sickening sense of terror when flames erupted in sync with a barrage of wild grunts and screams from the dense foliage right in front of him. Following this first moment of perplexed hesitation, he turned his massive body and prepared to run, both from the fire and from the horde of bipedal creatures emerging from the bushes and rushing toward him. At this chaotic transition from immobilizing fear to the overwhelming desire to flee, the bull felt the first piercing impact on his back. The ensuing pain was sharp and deep, but before he could realize that his legs could no longer answer the urgent commands issued by his aroused brain, several other similar impacts followed, one after another, in a matter of seconds, sealing his fate. All he could do now was yield to the weakness that began to overtake his body and simply crash to the ground.

The wild grunts drew closer and closer until, inexplicably, they began to recede, even though the bull could now see that he was surrounded by a large pack of jubilant hunters, each of them clad in multiple layers of tanned animal skins, each one holding a menacing stone blade built by agile and precise prehensile hands. The receding sound of their voices did not mean the hunters were going away at all. Rather the opposite; they would be around for millennia to come. The only thing fading rapidly that morning was the bull's ability to stay alert. He was now experiencing his final seconds on Earth, still stunned by the swiftness with which his life had come to an end.

And even though it would provide no consolation for him, the scene that had just taken place would almost certainly be immortalized in a cave painting: to honor his memory and sacrifice, to educate further hunters on the tactics employed this morning, and perhaps to depict a belief in a mystic realm to which the bull would now pass to continue his existence after having fallen prey to the ingenuity of a world-shattering new way of life that it could not grasp. Indeed, in his final moments of awareness, the magnificent animal had no way of knowing that his downfall had been planned carefully, way ahead of time, and then put into practice, flawlessly, by the most powerful, most creative and effective and, in some instances, most deadly parallel organic computer ever shaped by the blind paths of natural selection: a human brainet.

The reconstruction of such a prehistoric hunting scene, fictitious as it is, captures some of the key neurobiological attributes resulting from a convoluted evolutionary process that began when our primordial kin diverged from the common ancestor we shared with the modern chimpanzee about 6 million years ago. Altogether, this process endowed our species with unprecedented mental capabilities. Still today many doubts remain in defining the precise causal chain of events that precipitated the emergence of such extraordinary neurological adaptations. My goal here, therefore, is not to get lost in the details but to recover, using large brush strokes, some of the essential transformations and potential neurobiological mechanisms that allowed the brain of modern *Homo sapiens* to emerge and take over the whole planet. More specifically, my objective is to describe how such an organic computer—the way I like to describe the human brain—achieved its modern configuration and, in the process, acquired the means to generate a series of essential human behaviors, which turned out to be fundamental for the ascent of the True Creator of Everything as the center of the human universe.

Historically, the first factor that caught the attention of paleontologists and anthropologists as the potential cause underlying the increasing complexity of human behavior over evolutionary time was the growing size of our brains. Such a process, known as encephalization, began about 2.5 million years ago (figure 2.1). Until then, the brain of the first walking hominids, such as the *Australopithecus afarensis* individual known as Lucy, had a brain volume of roughly 400 cubic centimeters, similar to that of the modern chimpanzee and gorilla. By 2.5 million years ago, however, *Homo habilis*, a tool-making hunter, had a brain whose volume, about 650 cubic centimeters, was already more than 50 percent larger than Lucy's.

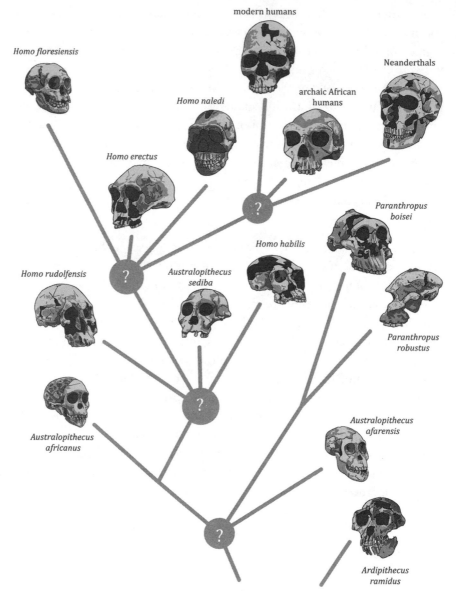

Figure 2.1. A speculative family tree for hominid species. Question marks indicate places where paleoanthropologists are not sure of the way branching occurred. (Courtesy of John Hawks. Originally published in Lee Berger and John Hawks, *Almost Human: The Astonishing Tale of Homo Naledi and the Discovery That Changed Our Human Story* [New York: National Geographic, 2017].)

Two million years would pass before a second phase of accelerating brain growth took place. It started around five hundred thousand years ago and continued for the next three hundred thousand years. During this period, the brain of *Homo erectus*, the next main actor in our evolutionary play, reached a peak of 1,200 cubic centimeters. From two hundred thousand to thirty thousand years ago, the brain volume of our human relatives reached a pinnacle with the Neanderthals, at around 1,600 cubic centimeters. However, by the time our own species appeared, the brains of men had been reduced to about 1,270 cubic centimeters, while the brain volume of women reached about 1,130 cubic centimeters. A key factor that has to be considered when we look at these numbers is that, at the end of this 2.5-million-year history, the brains of our human lineage had grown much more than the rest of our bodies had. That meant that the threefold increase that took place to produce the modern human brain generated a central nervous system that is about nine times bigger than would be expected in another mammal with our body weight.

When one tries to pinpoint what accounts for this extraordinary threefold increase in brain size from *Australopithecus afarensis* to *Homo sapiens*, we notice that most of the growth, already normalized by the equivalent change in body weight, is accounted for by a tremendous increase in the volume of the neocortex, the convoluted slab of neuronal tissue that defines the outermost layer of our brains. This is particularly relevant because the neocortex is known to mediate all of our most advanced cognitive capacities, the mental stuff that truly defines the essence of what it is to be human. In most primates, the neocortex amounts to around 50 percent of brain volume. In humans, however, the neocortex represents almost 80 percent of the total volume of the central nervous system.

Any theory that aims at explaining the explosive brain growth experienced by our human lineage has to cope with the paradox that our brain tissue consumes a lot of energy. Therefore, as our ancestors evolved large brains they had to strive much more to find the caloric resources to support the maintenance of their energy-hungry central nervous systems. Indeed, even though a human brain accounts for about 2 percent of human body weight, it consumes roughly 20 percent of all energy generated by each of us. That entails either eating a lot more food, which would require animals to spend more time exposed to predators, as our bison was, or alternatively, to change their diet in order to consume more caloric meals. Such an energetic surplus began to materialize when hominids shifted their earlier primate-like diet of foliage and fruits to incorporate a readily available food supply that could generate a much bigger energy yield per ingested volume. That food was fat- and protein-rich animal meat. Things improved even more when hominids learned to control fire and discovered the art of cooking. By cooking animal meat and highly energetic vegetables, these hominids improved the ease with which they could digest their meals and hence were able to extract more energy from them. Such a shift in diet was paralleled by—and may even have driven—a very significant evolutionary adaptation: a considerable reduction in the intestine's (especially the colon) size and complexity. Since large and complex intestines consume a lot of energy to work, such a visceral reduction produced extra energy savings that could now be directed to support the operation of larger brains.

Accounting for the energy sources needed to maintain larger brains, however, does not explain why such disproportionally large nervous systems emerged in the first place. Following a few failed attempts, a plausible and appealing hypothesis to account for the increase in primate and human brain size began to materialize in the 1980s, when Richard Byrne and Andrew Witten argued that brain size in apes and humans grew as a function of the increasing complexity of their societies. Named the Machiavellian theory of intelligence, this theory proposes that for ape and human social groups to survive and prosper, individuals have to deal with the complexity of the fluid dynamics underlying their social relationships. Acquiring and properly interpreting and using such social knowledge is essential to recognizing friends and collaborators as well as potential threats. According to Byrne and Witten, therefore, the great challenge involved in handling large amounts of social information required that apes, and particularly humans, develop bigger brains.

Put in other words, the Machiavellian theory of intelligence proposes that bigger brains are needed to develop a brain-based social map of the group to which one belongs and interacts on a daily basis. As such, this proposition converges with the idea that bigger brains like ours can develop a mental construct known as a theory of mind. Generally, such a cognitive skill bestows on us the ability not only to recognize that other members of our social group have their own particular internal mental states but also that we can continually hypothesize on what those states might be as we interact with them. That is, a theory of mind capability allows us to think about what other people are thinking, either about us or about others in our social group. Evidently, to take advantage of such a tremendous capability, one has to assume that our big brains also endowed us with self-recognition, self-awareness, and the establishment of the brain's own point of view.

In the 1990s, Robin Dunbar, a British anthropologist and evolutionary psychologist at the University of Oxford, introduced a new way to provide experimental support for the Machiavellian theory. First, instead of looking at whole brain size, he focused on the neocortex. Although the rest of the brain plays important physiological roles, when we approach skills like tool making, language, the establishment of a sense of self, a theory of mind, and many other mental attributes, it is the neocortex at which we must look.

Dunbar decided to test this theory using the only parameter of social complexity he could easily put his hands on in a very quantitative way: the size of primate social groups. The stunning result obtained by Dunbar's intelligent guess is plotted in monkeys and apes in figure 2.2. As one can easily see, the logarithm of group size for several primate species can be fitted along a straight line as a function of the logarithm of their corresponding neocortex ratio. As a result, this graph enables us to easily estimate the ideal size of a species' social

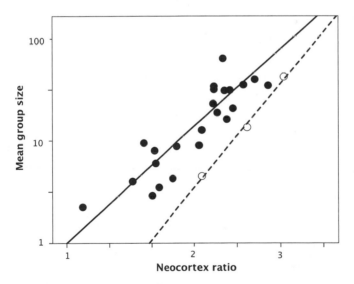

Figure 2.2. Correlation between mean group size and neocortical ratio for different anthropoid primates (monkeys in black dots and apes in open dots). (From R. I. Dunbar and S. Shultz, "Evolution in the Social Brain," *Science* 317, no. 5843 [2007]: 1344–47. Reprinted with permission from AAAS. Reproduced with permission. Originally published in L. Barrett, J. Lycett, R. Dunbar, *Human Evolutionary Psychology* [Basingstoke, UK: Palgrave-Macmillan, 2002].)

group from its neocortex ratio. In honor of his discovery, any estimation of animal group size obtained through this curve became known as the Dunbar's number for a given species. Thus, in the case of chimpanzees the Dunbar's number equals fifty, meaning that this ape's cortex is compatible with handling the social complexity generated by a group of about fifty individuals.

According to Dunbar's social brain hypothesis, as his theory became known, our overgrown cortex endows us with the mental skills to handle a close social group formed by about 150 other human beings, an estimate that matches well with data on modern hunter-gatherers as well as archaeological data indicating the populations of the earliest Neolithic farming villages in the Middle East.

Further support for Dunbar's claim that there seems to be a cap on the level of social complexity we can handle by interpersonal contact alone, without any external or artificial mechanisms of social control, can be demonstrated when human groups began to exceed 150–200 individuals. That is best illustrated in companies whose number of employees grows beyond Dunbar's threshold. Above that plateau, there is a growing need for the introduction of managers, supervisors, and administrative procedures simply to keep track of what goes on.

As a matter of curiosity, figure 2.3 shows Dunbar's estimates for social group size calculated for most of our key ancestors, based on the reconstruction of their relative neocortical ratio from fossil skulls. We can simply look at this graph to see how much social impact was produced by brain growth over the past 4 million years.

But how do primate social groups maintain the integration of such a large number of individuals? In nonhuman primates, grooming seems to be the main behavior employed for maintaining the cohesiveness of social relationships. The idea that grooming plays such an important social function is supported by the finding that primates spend 10 to 20 percent of their time dedicated to this activity. That endogenous opiates—so called endorphins—are liberated during grooming in monkeys likely explains in part how efficiently the collective exploitation of the exquisite sense of touch of primates creates the enduring bonding conditions needed to maintain the cohesiveness of their societies; groomed animals tend to relax and show much lower levels of stress.

Unlike our primate relatives, we do not spend much time grooming each other as a way to maintain the harmony of our social groups. Dunbar estimates that it would take 30 to 40 percent of the entire day to maintain a 150-person social group by grooming alone. Instead, as Dunbar argues, we

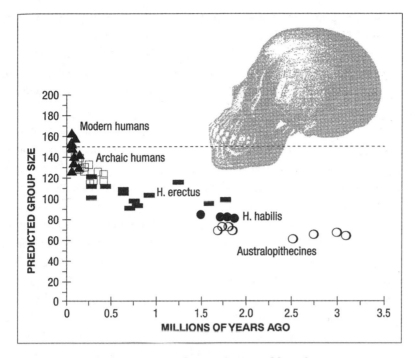

Figure 2.3. Estimated group size for populations of five of our ancestor hominids—australopithecines, *Homo habilis, Homo erectus*, archaic humans (including Neanderthals), and *Homo sapiens*—represented as a function of their estimated fossil age. (Reproduced with permission. Originally published in Robin Dunbar, *Grooming, Gossip, and the Evolution of Language* [London: Faber and Faber, 1996].)

may have resorted to the use of language to fulfill the same goal that grooming does in other species.

Complemented by hand gestures, grunts, and whistles, language may have endowed early humans with a very efficient medium capable of holding large groups together. Indeed, Robin Dunbar offers a very nice example of the impact of language as a human social bonding tool. Studying the content of conversations across many different social groups in modern England, he found that no matter who is talking, about two-thirds of the conversation revolves around our social lives. In other words, according to Dunbar's research, gossiping seems to be our favorite modern subject, suggesting that it likely served as the main mechanism through which the first members of our species, hundreds of thousands of years ago, were able to establish and maintain proper functioning of large social groups.

Despite the apparent elegant simplicity of Dunbar's arguments, animal evolution rarely follows a simple linear cascade of events like his theory suggests. Instead, many causal loops seem to interact with one another so that many traits tend to coevolve as a result of a particular selective pressure and even influence the evolution of each other. As other authors have argued since Dunbar proposed the social brain hypothesis in the 1990s, such a complex nonlinear causal chain is likely to have influenced the relationship between neocortex growth and the increased complexity of social behavior. For starters, one can say that the growth of the brain and the emergence of language not only enabled but were also required or perhaps even driven by the increasing complexity of human social behavior.

In this context, for the past twenty years another view has been proposed regarding what combination of factors may have driven human evolution and the process of encephalization. For example, Joseph Henrich, a professor in the Department of Human Evolutionary Biology at Harvard University, strongly argues that human culture played a central role in driving human evolution, and likely the growth of our brains. In his book *The Secret of Our Success: How Culture Is Driving Human Evolution, Domesticating Our Species, and Making Us Smarter*, Henrich describes in detail his theory of how the transmission of "practices, procedures, techniques, heuristics, tools, motivations, values, and beliefs" through generations made us "cultural animals" par excellence. By learning from one another, combining the knowledge accumulated, and transmitting it to our social groups and then to future generations, human culture not only provided better means for survival but, eventually, created a novel selective pressure that favored those individuals more capable of learning and assimilating such a cultural endowment. According to this view, human evolution was profoundly influenced by what Henrich calls culture-gene coevolution, the reciprocal and recurrent interaction between culture and genes. Such a process unfolds primarily because the dynamic human interactions in a social group generate cultural products as an emergent property of parallel interaction among the many individual human brains that constitute the group. Henrich defines this process of group learning, refinement, and transmission of knowledge as a product created by the group's "collective brain." I refer to this as the central function of human brainets, the main mechanism through which the human universe was shaped.

According to Henrich's view, the evolutionary success experienced by *Homo sapiens* depends much more on our ability to take advantage of our collective brains than on the power of our individual nervous systems. Such a hypothesis would partially explain, for instance, why small-brain hominids, whose fossils were found in the Island of Flores in Indonesia, were likely able to learn to use fire to cook or produce stone tools despite carrying a brain whose volume was equivalent to our *Australopithecus* cousins. Culture formation and transmission by brainets would have compensated for the small individual brains of *Homo floresiensis,* suggesting that brain size is not the only variable one should consider in evaluating the evolution of human cognitive skills.

Although I agree with most of Henrich's arguments, it is evident that the unique neuroanatomical and neurophysiological properties of our individual brains are essential to allow optimal brainets to be formed and endow human social groups with the ability to generate and transmit knowledge (See chapter 7).

The implication of the theory of culture-gene coevolution, as well as my own caveat about it, can be clearly illustrated by a major outcome of human evolution: our exquisite ability to create new tools. When our ancestors began walking upright about 4 million years ago, they significantly increased the spatial range of their daily roaming expeditions in search of food and shelter. In due time, this stunning biological innovation would allow African hominids to spread, first along the African coast and its interior and then throughout the entire world. As such, the first few waves of human colonization of the entire world, and the roots of what we know today as the process of globalization, were carried out on bare feet by African immigrants looking for better living conditions. Someone should remind modern politicians that without these epic migrant journeys, the world we know today would not have been possible.

But walking upright did much more than just increase human roaming range. It freed both the arms and hands of our ancestors to enact a variety of other motor behaviors, some of which required fine and precise coordinated movements using their opposable thumbs and the other fingers. Combined with the selective enhancement of frontoparietal cortical circuits, bipedalism gave us the opportunity to use our hands to produce tools.

For tool making to happen, however, our ancestors had to acquire the mental skill to seek and establish causal relationships in the surrounding world. For instance, a predecessor of ours might have thrown a piece of flint at a rock wall. Observing that some of the resulting splinters were capable of cutting all sorts of things, this hominid might have decided to begin to intentionally break rocks against each other to make more and better cutting tools. And as soon as this pioneer innovator succeeded in getting meat out of a carcass faster and more efficiently with his new tool, other members of the social group likely took notice and began to observe carefully how the innovator had produced the new instrument. The combination of the ability to produce insights followed by the dissemination of new knowledge among a social group defines a key neurological attribute that differentiates our species from other primates.

The phenomenon through which the same brain structures in multiple observers are concurrently activated by the motor acts of an individual is commonly referred to as motor resonance. If the observers begin to reproduce the motor behaviors they have observed, one talks about motor contagion. When the contagion manifests itself very rapidly, the phenomenon is known as mimicry. Particular circuits in the primate brain (see chapter 7) play a key role in the genesis of motor resonance, which triggers either contagion or mimicry in rhesus monkeys, chimpanzees, and humans. But comparative anatomical and physiological studies of such a cortical circuit in these three primates reveal important differences, both in terms of connectivity and patterns of activation during resonance. These findings are fundamental because they highlight how the evolutionary process may have influenced first the connectivity between different cortical regions of the temporal, parietal, and frontal lobes, and then the patterns of function activation circuitry, leading to the establishment of distinct brainets in different types of primates.

As a rule, rhesus monkeys rely less on social interactions to learn new skills than chimpanzees, which in turn exhibit less complex social learned actions than humans. That means that examples of motor skills learned by social interactions—by motor resonance and contagion—in rhesus monkeys are rare. Wild chimpanzees, on the other hand, are capable of expressing contagion of skills, such as communication gestures and tool making. In contrast to monkeys and chimpanzees, human beings excel in their capacity to employ motor resonance and contagion to spread new insights throughout their social groups, either locally through hand gestures and language, or at a distance by relying on a huge variety of communication media and technologies developed by their "collective brain."

There are two possible ways in which an observer can focus her contagion of a new motor act: emulation or imitation. Although emulation describes the act of simply focusing on copying the end goal of an observed motor act, imitation expands this focus to include the reproduction or copying of the entire process needed to achieve a particular goal. Interestingly, when all behavioral evidence available is analyzed, the consensus is that rhesus monkeys primarily emulate rather than imitate, whereas in chimpanzees imitation is more common. Indeed, chimpanzees are capable of observing, acquiring, copying, and transmitting new processes to execute motor behaviors to cospecific members of their groups, a trait that suggests the capacity to develop and sustain a prototype motor culture among these apes.

Yet, despite their clear capacity to imitate, chimpanzees do that much less often than humans. Essentially, this means that chimpanzees still tend to focus more on emulating—copying the end result of a motor behavior—while humans are much better imitators—focusing primarily on reproducing the process through which one achieves a motor goal. Moreover, thanks to the dramatic enhancement in communication provided by language, humans are much better at teaching new skills to others. In other words, in humans an insight spreads rather quickly and efficiently by gossiping.

Once an insight is produced by an individual or small team of collaborators, through a mechanism I discuss in chapters 7 and 11, motor resonance and contagion will ensure that this insight spreads and contaminates (almost like a virus) many individuals in a given social group. That mental recruitment accounts for the establishment of a tool-making human brainet that can improve the method, accumulate the knowledge, and distribute it to future generations.

However, the first hunting tool invented, the art of knapping, which constitutes the basis of the first man-made industrial revolution and tool making in general, evolved by an incremental process of discovery, enhancement, and additive complexity. Although millions of years were needed for the primitive rock hand axes made by our early ancestors to morph into the sharp spears that allowed *Homo sapiens* hunters to seize large prey, tool making and tool use became inseparable from any description of what it means to be human. Indeed, even though other animals, including chimpanzees, produce rudimentary tools, their artifacts do not show the same pattern of additive complexity that ours do. Nor do these animals ever exhibit the unique human capability of acquiring, accumulating, and transmitting such knowledge from one generation to another, for hundreds, thousands, or even millions of years.

Thus, because the mental skill needed to generate knowledge emerged in a species that loves to both cooperate and brag, innovative approaches for knapping rocks were able to spread, triggering a revolution in human life. From there on, to be successful and have a definitive impact, tool-making insight and skills had to be complemented by the bragging initiated by the close-knit skillful knap masters, otherwise the newly acquired knowledge would inevitably die, unheard of, as prisoners left in the solitary confinement of the masters' own individual brains.

The accumulation, refinement, and transmission of knowledge by human brainets may also have accounted for the most essential weapon for attaining success against large prey in a prehistoric hunt. I am referring to the human ability to plan and coordinate the activities of a large group of hunters. This huge task involved not only the ability to communicate effectively with all the individuals that formed the hunting pack at each moment in time, but also a series of subtler mental tasks that enabled each member of the team, as well as its leaders, to recognize what other individuals were thinking about the whole enterprise and the role assigned to them, what they could or could not endure mentally and physically during such a stressful event. At the same time, language was the main vehicle used to disseminate the tenets of new mythologies to entire human communities.

The emergence of language, tool making, theory of mind, and social smartness each offers a clue to what drove the tremendous growth in the brain's neocortex observed over the last 2.5 million years of our evolution. At the same time, the occurrence of so many evolutionary innovations raises a major puzzle: how could all these abilities be combined into a single fluid mind?

Steven Mithen, a professor of archeology at the University of Reading, England, has written extensively about this question and proposed a very interesting hypothesis about how a holistic mind, capable of cognitive fluidity like ours, may have emerged from the fusion of specific mental skills. Heavily influenced by the theory of multiple intelligences proposed by Howard Gardner, Mithen identifies three generic phases through which the amalgamation of the human mind might have taken place. According to Mithen's thinking, initially the mind of our early hominid ancestors was "a domain of general intelligence—a suite of general purpose learning- and decision-making rules." As time passed, our ancestors acquired new individual intelligences, such as tool making, language, and a theory of mind, but their brains were not able to integrate these modules; instead, according to Mithen's analogy, their brains would have operated like a sophisticated Swiss Army knife, which has multiple tools whose individual functions cannot be integrated. In the final stage in Mithen's model, the individual modules coalesced or fused into a single cohesive functional entity, giving rise to the modern human mind. At this point, information and knowledge acquired by each module can be exchanged freely, leading to the emergence of new mental derivatives and cognitive skills that endow the human mind with fluidity, creativity, intuition, and the possibility of generating insights and innovations that could not be produced by any of the individual modules in isolation.

Although other archeologists tend to criticize Mithen's theory and analogies, I find it interesting, at least as a starting point, to connect what is known about the details of the anatomical evolution of the human cortex since we diverged from our common ancestor with chimpanzees with the details of what we know about the functioning of the human mind today, after the supposed process of fusing intelligences took place. Mithen does not provide any neurobiological mechanism to account for this fusion of intelligences. This is totally justifiable because most inferences made about the evolution of the human mind rely solely on the analysis of endocasts made from the interior of fossilized skulls, which are rarely recovered intact, and indeed are often quite fragmentary and incomplete. That doesn't mean the fossils are useless: reconstructions of these skulls have allowed estimates of brain volume to be obtained for each of our ancestors, and endocasts often enable us to see the impressions made by brain tissue on the inner surface of the skull. Taken together, one can make some intelligent guesses about the shape and volume of different parts of the neocortex. Overall, this comparative endocast analysis reveals that brain shape has undergone significant modifications from *Australopithecus afarensis*, *Homo habilis*, *Homo erectus*, *Homo neanderthalensis*, and finally *Homo sapiens*.

Another way to evaluate what has happened during the evolution of the human brain is to compare its anatomy to the brains of other primates, such as the rhesus monkey and chimpanzee. The modern chimpanzee has also evolved since our species separated 6 million years ago, so although we cannot assume that it is identical to that of our common ancestor, it nonetheless offers a useful benchmark for comparison. Comparisons of these sorts have been made for many decades by neuroanatomists interested in brain evolution. The advent of modern brain-imaging techniques has provided many more details about where the tremendous expansion of the human neocortex, compared to our closest cousin's, has taken place.

In general, the neocortex consists of two major domains: gray and white matter. Gray matter contains large clusters of the main cell types that define a brain: neurons and the different type of cells, known as glia, that support them. White matter, on the other hand, is formed by the packing of the vast quantity of nerve bundles. These nerve bundles account for the extensive connectivity that exists between the different cortical areas that define the four cortical lobes—frontal, parietal, temporal, and occipital—in each cerebral hemisphere, left and right. The nerve bundles also provide the bulky link between the left and right cerebral hemispheres, via what are known as the callosal projections, and offer the neural highways through which the cortex receives and sends messages to what are known as subcortical structures, such as the spinal cord. Gray and white matter can be distinguished clearly in the neocortex. The former comprises a six-layer slab of neurons. These neurons, which are the neocortex proper, sit on top of a dense and thick block of white matter.

In the late 1990s, John Allman, a neuroscientist working at Caltech who spent a great deal of his career working on the central subject of mammalian brain evolution, demonstrated what has become a classic finding about the relationship between gray and white matter. Allman found that if one plots the volume of cortical gray matter versus the volume of its related white matter, considering a very large number of mammalian species, including many primates and humans, one finds a very clear power relationship.

White Matter Volume $=$ (Volume Gray Matter)$^{4/3}$

The exponent of this equation—4/3—indicates that as the cortex grows, white matter volume accumulates much more quickly (figure 2.4). When one examines similar data related to primates, and focuses on what exactly changed in the human neocortex to make it different from our primate relatives (chimpanzees and rhesus monkeys), we find that most of the cortical growth observed in humans was in the frontal lobe, particularly at its most extreme anterior portion (directly behind the forehead), called the prefrontal cortex, followed by the expansion of so-called association cortical areas in the posterior parietal lobe and temporal lobe.

In the case of the frontal lobe, when compared to rhesus monkeys, humans exhibit a thirtyfold increase in tissue volume. Interestingly, as Allman predicted, the largest share of this frontal lobe growth is represented by white matter hyperscaling. This resulted in the dramatic enhancement of the connectivity between the vastly expanded prefrontal cortex and the premotor and motor areas of the frontal lobe to several other parts of the brain, including other cortical areas in the parietal and temporal lobes as well as subcortical regions.

This unique volumetric explosion of the human frontal white matter, as well as the concurrent growth of parietal and temporal associations, means that a much larger proportion of the human neocortex became devoted to high-order conceptual and abstract thinking, the kind of stuff high cognitive

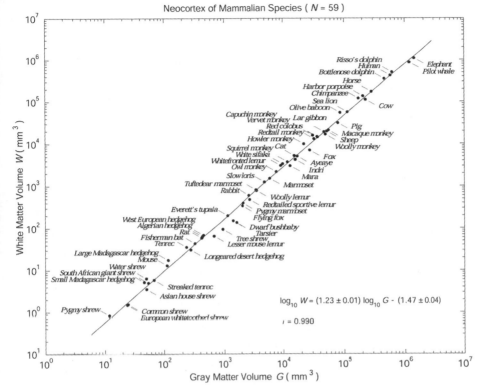

Figure 2.4. Relationship between cortical gray and white matter in 59 animal species. Note that both graph scales are logarithmic. (Reproduced with permission. Originally published in K. Zhang and T. J. Sejnowski, "A Universal Scaling Law between Gray Matter and White Matter of Cerebral Cortex," *Proceedings of the National Academy of Sciences U S A* 97, no. 10 [2000]: 5621–26. Copyright [2000] National Academy of Sciences, U.S.A.)

skills are made of. It is no coincidence, therefore, that neuroscientists seeking the neocortical circuitry underlying language, tool making, the clear definition of a sense of self, social smartness, and the theory of mind—all the attributes that emerged during the past 4 million years of hominid evolution—found them to reside in fronto-parietal-temporal cortical regions and their linking axonal highways. It is distributed in there, I would argue, that they identified the organic computing substrate that likely gave rise to the True Creator of Everything.

Taking all this information together, my own conclusion is pretty straightforward: to enable the evolution of large, complicated social groups capable of complex behavior, not to mention the acquisition of a culture that we could transmit through the millennia, we clearly needed more neurons. But as important as the sheer volume of neurons might be to achieve our present cognitive abilities, it is likely that the unique cabling of our brains was the main driving force behind the emergence of our species' exquisite mental skills.

Optimally wired up inside to be able to hyperconnect outside: that seems to be the motto behind the evolutionary history of brain growth.

Nevertheless, the motto is not enough to explain how all these human attributes could have fused to create the holistic and fluid mind of *Homo sapiens*, nor where we might look for a neurophysiological mechanism that could have enabled our enlarged neocortex to establish larger and more stable social groups. In modern neuroscience the first question is known as the binding problem. For the past thirty years, the binding problem has been a hot topic of discussion, particularly among those, like the distinguished German neuroscientist Wolf Singer, who study the visual system because the most classic theoretical framework in visual physiology, the Hubel and Wiesel model, named after the Nobel laureates David Hubel and Torsten Wiesel, whose work in the visual system revolutionized the whole field of systems neuroscience and led to a theoretical model that remains, after more than fifty years, the central dogma of visual physiology, cannot provide an adequate answer to it.

The second question, on the other hand, is essential to understanding why we humans have succeeded in building the kind of creative and enduring social groups that shaped the entire human universe. How does the neocortex fuse its parts into a single continuous—or analog—computing device, and how does it ultimately allow the activity of thousands, millions, or even billions of individual brains to be synchronized into functional brainets?

3 • Information and the Brain
A Bit of Shannon, a Handful of Gödel

The jolly young crowd filling the bucolic promenade that lined the Swiss border of Lake Leman along the village of Clarens, during a hot and humid summer afternoon in 2015, seemed to move according to the rhythm of yet another matinee at the outdoor stage of the Montreux Jazz Festival. Having enjoyed our lunch at Le Palais Oriental, just a few hundred meters away, my best friend, the Swiss-Egyptian mathematician and philosopher Ronald Cicurel, and I had decided to take a midafternoon walk to explore another major component of the theory we had been working on together. While we walked side by side, throwing ideas at each other in the midst of debating one of our favorite topics during that eventful Swiss summer (that is, the sequence of events that had allowed living organisms to emerge on Earth a few billion years ago and then evolve and succeed in their never-ceasing resistance against the relentless trend for entropy to increase all over the universe), we suddenly stumbled in front of the weird-looking tree (figure 3.1).

As we stood there, petrified, in the middle of the walkway, with all our attention directed to that twisted tree, an idea suddenly came to my mind. "Living is all about dissipating energy in order to embed information into organic matter," I said out of the blue, and then repeated the sentence a few times to make sure I would not forget it.

Taken aback by the thought, which immediately started to resonate inside his own brain, Ronald turned to stare at the tree once again, as if he were seeking a final reassurance. After an instant of silent contemplation, he smiled broadly, although he looked a bit more agitated than usual. He pointed to a nearby bench, inviting me to sit. "I think this is it!" he finally decreed.

Figure 3.1. Ronald Cicurel poses for a photo with the famous tree at the Leman lakefront in Montreux, Switzerland, after a major theoretic breakthrough. (Courtesy of Miguel A. Nicolelis.)

That was when I knew that we had finally found the thread we had sought the entire summer, walking by the same promenade every afternoon, observing the lake, the cranes, the ducks, and geese, disturbing busy pedestrians with our awkward habit of conducting thought experiments on public thoroughfares and, as destiny would have it, making our fateful acquaintance with that strange tree that clearly nobody else but us seemed to consider so important.

That morning, prior to our usual daily meeting, I had by chance focused my attention on the different branch and leaf patterns formed by the trees one can find on that Swiss lakeshore. Recalling the typical expansive canopy of tropical trees in Brazil, which I have admired since my childhood, I could now appreciate how differences in latitude had influenced the shape of tree leaves and the overall three-dimensional configuration of plants in different parts of our planet. I remember thinking what a magnificent adaptive mechanism nature had come up with to optimize the biological solar panels trees employ to collect as much energy from the sun as possible, depending on where on the Earth's surface they happen to sink their roots. That thought made me recall an almost forgotten lecture my high-school botany teacher had delivered forty years ago. It had to do with dendrochronology, one of the teacher's obsessions in 1977. According to Mrs. Zulmira, the great Leonardo da Vinci was the first to recognize that every year trees grow an extra ring of wood whose width reflects the climate conditions they endured during that season (some trees can, under certain climatic conditions, produce more than one ring per year). Armed with this knowledge, the American scientist and inventor Alexander Twining proposed that by synchronizing the ring patterns of a large number of trees it would be possible to establish the past climate conditions of any location on Earth. Properly moist years would yield wider rings, whereas periods affected by droughts would lead to the production of very thin circular wood additions.

Taking this idea further, Charles Babbage, the pioneer of modern age computing, came up with the insight of characterizing the age and the past climate conditions observed in geological strata by analyzing the rings of fossilized trees found embedded in it. Although Babbage proposed this procedure in the 1830s, dendrochronology became accepted as a real scientific field only thanks to the work and perseverance of the American astronomer Andrew Ellicott Douglass, who discovered a correlation between tree rings and cycles of solar spots as he built, over a three-decade period, a continuous sample of tree rings that dated back to 700 AD. Using this unique biological record of time, archeologists were able to pinpoint, for example, the precise years in which some Aztec ruins were built in the present U.S. Southwest. Today, dendrochronology allows scientists to reconstruct the occurrence of volcanic eruptions, hurricanes, glacial events, and precipitation that happened way back in Earth's past.

Altogether, therefore, tree rings illustrate pretty well how organic matter can be physically embedded with information that represents a detailed record of climate as well as geological and even astrophysical events that took place during the lifetime of a living organism.

Besides paying my dendrochronology debt to Mrs. Zulmira, I did not make much more of the two apparently uncorrelated observations—the exquisite shape of the leaves of the trees in the Montreux garden and the potential time records they carried inside their wooden core. Instead, I went back to my drawing—a hobby I rekindled while researching the material for this book—until it was time to join Ronald at Le Palais Orientale.

A few hours later, sitting next to Ronald on that garden bench, it all came back to me. The difference was that, by now, there was a clear logical and causal connection linking the tree's solar panels and its time-marking rings. "That is it, Ronald. Solar energy is dissipated in the form of information embedded in the organic matter that shapes the tree trunk. That is the key, energy being transformed into physically embedded information in order to maximize the local reduction in entropy needed for the tree to make it to the next day, so it can collect more energy, embed more information in its very flesh, and continue to resist annihilation!"

During the summer of 2015, Ronald and I had plunged deeply into the idea of using thermodynamics as a potential unifying framework to seamlessly link the evolution of the entire universe with the processes that led to the generation and evolution of life on Earth. During these discussions, it did not take much time for us to converge on the potential consequences of describing life and organisms as true evolutionary experiments aimed at reaching an optimal way to convert energy into embedded information as the ultimate strategy through which life can defeat, albeit briefly, the heinousness of terminal dissipation into a state we colloquially know as death.

Although many other authors for the past hundred years had discussed the mixing of concepts such as energy, information, and entropy in the context of living organisms, we now believed that we had come up with something a bit different on our walk that afternoon. For starters, our discovery required the introduction of a new definition of information, one that more closely reflected the basic operation of living systems and offered a contrast with the more well-known version of the term introduced by Claude Shannon in the electrical engineering context of studying messages transmitted through noise channels in artificial devices. Moreover, as we thought more thoroughly about what we had stumbled on that afternoon, it became evident we had uncovered yet another unique idea—one that equated organisms and even their cellular and subcellular components to a new class of computing devices, organic computers, which I had previously named in a 2013 paper, written under completely different circumstances.

Organic computers, which differ from the mechanical, electronic, digital, or quantum computers engineers build, emerge as a result of the process of natural evolution. Their main feature is that they utilize their very organic structure and the laws of physics and chemistry for acquiring, processing, and

storing information. This fundamental property means that organic computers rely primarily on analog computing to perform their tasks, although elements of digital computing can be used in several important cases. (Analog
computing relies on the continuous variation of a given physical parameter—
such as electricity, mechanical displacement, or fluid flow—to compute. The
slide rule is one of the simplest examples of an analog computer. Analog computers were very prevalent before the introduction of digital logic and digital
computers at the end of the 1940s.)

Given that thermodynamics was our point of departure, from the beginning of our collaboration Ronald and I had been deeply influenced by the
work of the Belgian-Russian chemist and Nobel laureate Ilya Prigogine and
his thermodynamic-based view of life. In one of his now classic books, *Order out of Chaos*, written in collaboration with Isabelle Stengers, Prigogine describes his theory, which concerns the thermodynamics of complex chemical
reactions, and the immediate consequences of his work, which led him to
develop a radical new definition of life in great detail. At its heart, Prigogine's
theory, which deals with what are now known as self-organizing chemical reactions, provides a way to understand how living systems can arise out of
nonliving matter.

At the core of Prigogine's thinking is the notion of thermodynamic equilibrium. A system is in equilibrium when there are no overall flows of energy
or matter within a system, or between a system and its surroundings. If for
whatever reason energy gradients appear, creating regions with more and less
energy, the system spontaneously dissipates the energy excess from the former to the latter. To understand this, imagine a little bit of water resting inside
a teakettle left at room temperature. Under these conditions, the water is in
thermal equilibrium, and no macroscopic changes of any sort can be identified, since the water remains peacefully in its liquid state. Now, if we decide to
heat up the water in order to prepare some tea, as the temperature rises and
gets closer and closer to the boiling threshold, the water moves farther and
farther from its liquid equilibrium state until it undergoes a phase transition
and becomes water vapor.

According to Prigogine, organisms from bacteria to trees and humans are
open systems that can survive only by maintaining themselves under far from
equilibrium conditions. That means that living requires a continuous process
of exchanging energy, matter, and information within the organism itself and
between it and its surrounding environment to maintain chemical and thermal gradients that develop within cells, within the organisms as a whole, and

between the organisms and their external environment. This struggle lasts for the entire life of an organism. The moment it fails to sustain these far-from-equilibrium conditions, the organism is irrevocably condemned to die and decay.

Energy dissipation is a phenomenon that we encounter in everyday life. For instance, when you turn the key of your car and the engine begins to work, some of the energy generated by the combustion of gasoline goes into moving the car, but a significant amount of it is dissipated in the form of heat, which is not readily used to generate further work. That's what dissipation is: the transmutation of one form of energy that can do lots of work into energy that can do less. Large structures that emerge in the natural world also result from processes that dissipate large amounts of energy. Hurricanes are a good example. The huge spinning whitish blob we see on satellite images results from a self-organization process of clouds and wind formed by the dissipation of large amounts of energy generated as huge volumes of hot and humid air, originating near the equator, ascend from the ocean surface to higher altitudes. As this hot and moist air climbs up the atmosphere, it generates a zone of lower pressure below it, which is soon filled with colder air rushing in from surrounding high-pressure regions. This air gets heated and moist and ascends. Upon reaching the cooler high altitudes, the water in the air condenses, forming clouds that begin to rotate, dragged by the violent winds that develop due to the high-speed circulation of hot and cold air. The hurricane structure and movement that we see result from the self-organizing process of energy dissipation generated by this weather mechanism that, in some extreme cases, can only be described as a ferocious climate bomb.

Prigogine and his collaborators found that there are chemical reactions that develop self-organizing structures in a laboratory petri dish that are not unlike those found in hurricanes. For example, varying the quantities of certain reagents, changing external conditions, such as temperature, or the introduction of a catalyst could drive the emergence of totally unexpected rhythmic oscillations in the reaction products. These patterns became known as chemical clocks. They also found that elaborate spatial structures, such as the segregation of different molecules in different regions of the reaction vessel, could emerge. In short, the random collisions of reagents could create order, driven by the dissipation of energy by the system.

Prigogine derived two major principles from his observations. The first, criticality, indicates that there is a sudden moment in which further addition of a small amount of reagent or a small temperature increment changes dramatically the way the chemical reaction organizes itself in time and/or space. Interestingly, at the end of the nineteenth century, the French mathematician Henri Poincaré had observed the same phenomenon mathematically during his studies of nonlinear differential equations: above a certain point the behavior of the equation can no longer be predicted precisely; from that point on the system behaves in a chaotic way, and the totality of the numerical values yielded by the equation defines a mathematical macrostructure known as a strange attractor. The second fundamental concept is known as synchronization. That refers to how, at certain far-from-equilibrium conditions, the molecules of the reagents seem to be "talking" to each other, so that highly elaborated temporal or spatial patterns can self-organize. Both these concepts are key to the definition of single brains and networks of synchronized brains (brainets) as organic computers (see chapter 7).

From these observations, jumping from chemical reactions to a theory on how living organisms operate was just the next logical step. And Prigogine took it with great gusto. To see how, let's now return to our Swiss tree on the lakefront promenade of Montreaux and align Prigogine's theory with our own ideas.

With deep roots established on the shore of Lake Leman for a long time, that tree took advantage of its extensive biological solar panels to absorb sunlight and carbon dioxide from the surrounding environment. It was able to capture a fraction of solar energy thanks to the existence of the light-absorbing pigment chlorophyll in the chloroplasts of leaf cells. Using sunlight, carbon dioxide, and water, chloroplasts carry out photosynthesis. Thanks to this latter process, plants are able to harness some of the energy from sunlight to maintain and expand the state of disequilibrium that existed in the seed from which the plant grew by adding and maintaining layers of organic tissue to its structure.

Plants capture sunlight, animals eat plants, and we eat plants and animals. In summary, life is about eating what the sun gives us; some get it firsthand, whereas others get their fair share of sunshine secondhand. What Ronald and I added to this idea is the notion that as the dissipative structure—the tree, in this case—self-organizes, it takes advantage of this process to physically embed information into the very organic matter of which it is made. As a tree grows, for example, information about the surrounding climate, water availability, sunspots dynamics, and many other variables are embedded in the rings that the plant adds to its three-dimensional structure every year. As such, a tree is able to perform all basic operations required of an organic computer according to our criterion. And even though trees may not be able to access the "memories" they deposit in their rings directly, external observers like us found a way to do it.

Put in a more formal way, what Ronald and I propose is this:

In an open, living system, energy dissipation allows information to be physically embedded into organic matter.

According to our view, this process is not equal in all living forms. We just saw that, in the case of trees, information embedded in the rings cannot be—as far as I can tell—recalled by the plant itself. In other words, the plant by itself cannot access this information to compute, let's say, the number of sunspots of a previous season. But in animals that have brains, information that was embedded in neuronal tissue not only can be recalled continuously but also is used to guide future actions and behaviors. In this case, the process of dissipating energy into embedded information underlies the fundamental phenomenon known as learning and is responsible for laying down memories in the animal brain. Furthermore, because in brains this process of information embedding involves direct modifications of neuronal tissue (that is, by physically changing morphological features of the synapses that connect two neurons), one can say that information exhibits "causal efficiency" on the nervous system. This means that the process of physically embedding information changes the physical configuration—and hence the function properties—of a neuronal circuit. This is the basis of a powerful neurophysiological property known as neuronal plasticity (see chapter 4).

Information embedding in animal brains represents a major jump from tree rings. Nonetheless, an even more impressive outcome takes place when we consider the human brain. In addition to being responsible for continuously laying down our memories, which exhibit an extraordinary and unique lifetime or long-term range, and for mediating learning and plasticity, in the human brain the process of energy dissipation accounts for the emergence of a much more precious and rare commodity: knowledge.

Energy dissipating into knowledge!

For me, that can be considered the pinnacle, the most transformative outcome to emerge from a thermodynamic description of life.

At this point, we need to introduce a very important thermodynamic concept: entropy. Entropy can be defined in multiple ways. One is to describe it as a measurement of the level of molecular disorder or randomness within a given macroscopic system. Another way of defining entropy is as the number of microstates a particular system, such as gas, can assume while still exhibiting the same macroscopic behavior. Suppose you enter a very large and empty hotel ballroom carrying a small birthday balloon full of helium gas. Because the volume of the balloon is small, the helium molecules are tightly packed next to each other, having a relatively low level of molecular disorder because they cannot expand much due to the small volume they occupy inside the balloon. Likewise, the number of microstates is relatively small: although each helium atom could swap places with any other one, still yielding the same macrostate of a small, helium-filled balloon, they are still constrained by the balloon itself from occupying all the other locations in the ballroom. By either description the helium is said to be in a low entropy state. Now, by the time you reach the center of the ballroom, you decide to pop the balloon and let the helium escape. The helium that was initially confined to a small space—the balloon volume—now spreads all over the much larger ballroom, increasing significantly its level of molecular disorder and the uncertainty in determining the precise location of each molecule inside the room. This uncertainty characterizes a state of high entropy.

The illustrious Austrian physicist Ludwig Boltzmann, one of the founding fathers of thermodynamics, was the first to come up with a way to describe this concept in a more quantitative way by creating a statistical formulation of the entropy of natural substances, like gases. According to his formulation,

$$E = k \times \log n$$

where E is the entropy, k is called the Boltzmann constant, and n is the total number of microstates in a system.

According to the original formulation of the second law of thermodynamics, made by William Thompson in 1852, the total entropy of an isolated closed system tends to increase over time. This law applies to the entire universe overall, but it does not preclude the emergence of "local pools of resistance" on the part of living organisms to delay ultimate disintegration and randomness. This guerrilla-style resistance of the living was nicely portrayed by another distinguished Austrian, the Nobel laureate physicist Erwin Schrödinger, one of the giants of quantum physics, who in his book *What Is Life?* proposed that living is a continuous struggle to generate and maintain true islands of reduced entropy, which we call organisms. In his own words: "The essential thing in metabolism is that the organism succeeds in freeing itself from all the entropy it cannot help producing while we are alive."

In *The Vital Question* Nick Lane, a British biochemist from the University College of London, elucidates even further the relationship between entropy and life, stating, "The bottom line is that, to drive growth and reproduction—living—some reaction must continuously release heat into the surroundings, making them more disordered." He continues: "In our own case, we pay for our continued existence by releasing heat from the unceasing reaction that is respiration. We are continuously burning food in oxygen, releasing heat into the environment. That heat loss is no waste—it is strictly necessary for life to exist. The greater the heat loss, the greater the possible complexity."

In Prigogine's terms, one can say that the greater the energy dissipation produced by an organism, the greater the complexity it can attain!

Since the late 1940s, the concepts of entropy and information have been intimately related, thanks to the work of the American mathematician and electrical engineer Claude Shannon who, as a thirty-two-year-old employee of Bell Telephone Laboratories, in 1948 published a seminal seventy-nine-page manuscript in the company's technical journal. In "A Mathematical Theory of Communication" Shannon described the first quantitative theory of information ever formulated. The same paper also became immortalized as the theoretical cradle from which one of the most influential human-idealized mathematical measurements of the twentieth century was born: the bit, a unit for measuring information.

Years prior to his revolutionary paper, back in 1937, Shannon, then an MIT master's student, had shown that one needs only two numbers, 0 and 1, and the logic that derives from using them—named Boolean logic in honor of its creator, George Boole—to reproduce any logical or numeric relationships in an electrical circuit. This incredible theoretical insight launched the era of digital circuit design, which coupled with both the invention of the transistor, in the same Bell Labs, and Alan Turing's initial theoretical formulation of an idealized computing machine, made digital computers possible, an event that dramatically changed the way humanity has lived for the past eight decades.

With his 1948 paper, Shannon offered a statistical description of information, much as his predecessors had quantified energy, entropy, and other thermodynamic concepts in the preceding century. Shannon's major interest was what he called "the fundamental problem of communication": "reproducing at one point either exactly or approximately a message selected at another point." In Shannon's approach to information, there was no role for context, semantics, or even meaning; these were all unnecessary complications for the narrow communication problem he set out to solve.

In *The Information: A History, a Theory, a Flood,* James Gleick nicely summarizes Shannon's key conclusions about his world-shattering probabilistic view of information. Three of these apply directly to the present discussion. They are:

1. Information is really a measurement of uncertainty, which can be measured by simply counting the number of possible messages. If only one message can be transmitted in a channel, there is no uncertainty about it, so there is no information.
2. Information is about surprise. The more common a symbol transmitted in a channel is, the less information the channel broadcasts.
3. Conceptually, information equals entropy, the key thermodynamics concept used by Schrödinger and Prigogine, to describe how energy dissipation gives rise to life from nonliving matter.

We will return to the broad consequences of the last shocking statement in a moment, but before we do that, it is important to show how Shannon's statistical view of information was described in an equation. In this mathematical formulation, Shannon's entropy (H) represents the minimum number of bits needed to accurately encode a sequence of symbols, each of which has a particular probability of occurrence. In a simplified notation, this formula is:

$$H(X) = \sum_{i=1}^{n} p_i \log_2 p_i$$

where p_i represents the probability of occurrence of each one of the symbols transmitted by the channel. The unit of H is given in bits of information.

For example, if a channel transmits only one 0 or one 1, with an equal probability of 50 percent for each of these two symbols, it needs one bit to accurately encode and transmit this message. On the other hand, if the channel always broadcasts 1s—meaning that the probability of the occurrence of this symbol is 100 percent—the H value equals 0: there is no information transmitted, since there is no surprise at all at the content of the transmission. Now, if this long string is made of 1 million independent bits (each having the same equal probability of having either a 0 or 1), this channel will transmit 1 million bits of information.

Shannon's definition basically means that the more random a sequence of symbols is—meaning the more surprising it is—the more information it contains. Just as popping a balloon enabled helium to go from a state of low thermodynamic entropy to a high one, the amount of information required to describe the location of every helium atom went up as well because of the increase in the uncertainty of its location in the much larger ballroom.

Thus, after Shannon, entropy began to be defined as the amount of additional information needed to define the precise physical state of a system, given its thermodynamic specification. From then on, entropy could also be considered as a measurement of our lack of information about such a system.

Proof of the success of Shannon's information breakthrough is that his concept rapidly crossed the boundaries for which it had been precisely tailored to reach a multitude of other disciplines, redefining many, sometimes in a radical way. For example, the discovery that long sequences of four basic nucleotides enabled DNA strands to encode all information needed to replicate organisms from one generation to another brought Shannon information to genetics and molecular biology. With the emergence of the genetic code, a growing consensus began to take shape. Roughly, this consensus proposes that everything we know in the universe can be encoded and decoded in bits according to Shannon's innovative and disruptive digital description of information. Indeed, in his paper "Information, Physics, Quantum: The Search for Links," John Archibald Wheeler, one of the greatest physicists of the last century, defended the proposition that "information gave rise to everything, every particle, every field of force, even the space-time continuum itself." He described this process with the expression "It from Bit," which immediately became very catchy.

Having made a major detour that took us to the distant shores of thermodynamics and the birth of the information age, we can now return to our beloved Swiss tree in the Montreux promenade to clarify what Ronald and I really meant. Basically, we proposed that living systems dissipate energy to self-organize and embed information into their organic matter to create the islands of reduced entropy that valiantly try to put the brakes on, even if only on a humble tiny scale, the drive toward inexorable randomness and nothingness to which the universe seems to be evolving. Although part of this information could be described by Shannon's classic formulation, we propose that the vast majority of it dissipates through a process that leads to the physical embedding of a different type of information in the organic tissue. Ronald and I decided to call it Gödelian information in honor of the greatest logician of the twentieth century, Kurt Gödel, who demonstrated the inherent limitations of formal systems, which are expressed by Shannon information. For now, it will be enough to contrast Shannon and Gödelian information as a way to set the stage for the continuation of this narrative.

For starters, rather than being binary and digital, Gödelian information is continuous or analog, given that its embedding in organic tissue is fueled by the process of energy dissipation in organisms. As such, Gödelian information cannot be digitalized or discretized and treated as binary bits of information flowing in a noisy communication channel. The more complex an organism gets, the more Gödelian information is laid out, embedded in the organic matter that forms it.

A series of examples may help clarify some of the main differences between Shannon and Gödelian information. Through the process of translation that takes part in ribosomes, individual amino acids are serially concatenated in order to generate the linear sequence of a given protein. As energy dissipates during translation, Gödelian information is embedded in this protein linear chain. To fully appreciate what this information encodes, however, the original linear amino acid chain that defines the protein needs to fold to assume its final three-dimensional configuration, also known as tertiary structure. By the same token, multiple folded protein subunits need to interact with one another to form the so-called quaternary structure of a protein complex, such as the one of hemoglobin, the oxygen-carrying protein found inside the red cells of our blood. Only when such a quaternary structure is formed can hemoglobin bind to oxygen and perform its main job.

Although linear protein chains assume their tertiary configuration very rapidly when placed in an appropriate medium, trying to predict this final folding from the protein's original linear chain using a digital computation algorithm is a daunting task. In our terminology, the Gödelian information embedded in the linear chain of a protein manifests itself directly (that is, it computes) by the physical folding process that generates the three-dimensional structure of the protein. The same process when approached in terms of digital logic can be considered either as nontractable or even totally noncomputable, meaning that no prediction of the final three-dimensional structure of the protein can be made solely based on its original linear amino acid chain. That is why we refer to Gödelian information as analog rather than digital. It cannot be reduced to a digital description, given that its full manifestation depends on a continuous—or analog—process of biological structure modification, dictated by the laws of physics and chemistry, not by an algorithm running in a digital computer.

Consider now a second and much more complex example. Suppose a recently wedded couple enjoys their first honeymoon breakfast on a hotel balcony, facing the Aegean Sea, on the Greek island of Santorini. As a typical Greek rose-fingered dawn unfolds in classic Homeric splendor, their hands touch and they exchange a brief passionate kiss. Fast-forward fifty years into the future. On the date of what would be the couple's fiftieth wedding anniversary, the only living witness of that first morning, the widow, returns to the same Santorini hotel balcony and orders the same Greek breakfast at dawn. As soon as she tastes her lonely meal, although half a century has elapsed, she again vividly experiences the same profound feeling of affection produced by the hand caress and kiss she shared with her beloved groom. And even though the skies are clouded this time around, and there is no wind, at that instant she feels almost transported to that original Santorini sunrise and experiences, once again, the sweetness of an early morning Aegean breeze brushing her hair as she touches the love of her life. For all intents and purposes, the widow is now experiencing the same sensations she felt half a century ago.

According to our view, in reality what she is experiencing is the overt manifestation of Gödelian information that had been originally imprinted in her memories and there remained for those fifty years until it was recalled abruptly when she first tasted the same Greek dish. And now, no matter how much she tries to talk about what she is experiencing, she will never be able to fully put into words those feelings of remembrance, tenderness, love, and loss. That is because even though Gödelian information can be partially projected into Shannon information and transmitted in the form of oral or written language, it cannot be fully expressed in those reduced digital terms.

This latter example shows two interesting properties. First, during the couple's honeymoon breakfast a series of sensory signals (taste, visual, auditory, tactile) were translated primarily into Shannon information, to the brains of the two interacting individuals. Once these multimodal messages reached their brains, they and their mutual relationships and potential causal-effect associations were compared with each brain's frame of reference, which was shaped by all their previous life experiences (figure 3.2). The result of this comparison was then readily embedded in their cortices as continuous Gödelian information. That indicates that human brains are continuously transforming Shannon information sampled from the outside by our peripheral array of sensory organs (eyes, ears, tongue, skin) into long-term mnemonic records of Gödelian information. Conversely, when triggered by similar sensory stimuli, like the taste of a meal previously savored in the same environment, Gödelian information records stored decades ago can be readily converted, at least partially, into streams of Shannon information for communication purposes. The portion that cannot undergo this conversion from Shannon to Gödelian cannot be verbalized and is experienced as one's emotions and feelings. Therefore, when we refer to this very human way of experiencing long-

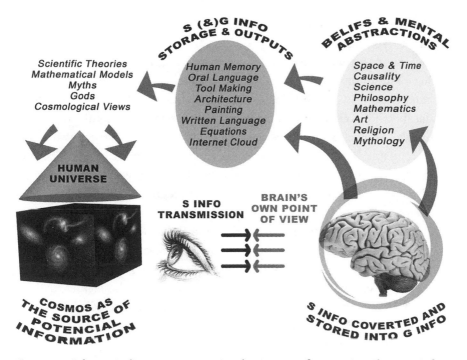

Figure 3.2. Schematic diagram representing the process of converting Shannon information into Gödelian information, the generation of mental abstractions, and the construction of the human universe as our attempt to represent the cosmos. (Image credit to Custódio Rosa.)

held memories, there is no stream of Shannon information, no mathematical algorithm, and no digital computer, no form of artificial intelligence trick that can get close to reproducing or mimicking what each of us really feels inside our heads. Basically, Shannon information alone is insufficient to comprehensively describe what brains are capable of storing, experiencing, and fully expressing. Thus, as Ronald proposes, if entropy is defined as the amount of additional information needed to specify the exact physical state of a system, Gödelian information is the entropy of the brain; that is, the extra chuck of information, not accounted for by Shannon information, that is necessary to fully describe the type of brain-embedded information that makes us human. Therefore, the existence of Gödelian information defines one of the key reasons why digital computers will never be able to reproduce the intrinsic works and wonders of the human brain; digital computers dissipate energy into heat and innocuous electromagnetic fields, while animal and in particular human brains use energy dissipation to accumulate Gödelian information into neuronal tissue (see chapter 6).

One phenomenon produced by the human brain I find particularly interesting, phantom limb sensation, may further explain the differences between Shannon and Gödelian information because it clearly illustrates that human brains, but not digital computers, can deal with potentially conflicting or ambiguous messages. Suppose a subject who had his right leg amputated is resting in a hospital bed and cannot see his limbs because his body is fully covered with a bedsheet. Now, the orthopedic surgeon who amputated this patient's leg comes to his bedside and informs him that, regrettably, his limb had to be severed a couple of hours ago due to untreatable gangrene. Even though the patient now knows the truth, he experiences a profound contradiction because he can still feel the presence of his right leg underneath the bedsheet, thanks to a clear manifestation of a phantom limb sensation, a well-known impression that affects close to 90 percent of amputees. In all these cases, patients report feeling very clear and detailed tactile sensation, including pain, and even movements of the amputated limb long after (in fact, months or even years after) the amputation surgery took place.

Because our hypothetical patient still feels the vivid presence of the amputated leg under the bedsheet, he insists to the surgeon that his limb was not amputated at all. This must be a mistake or worse, a scam worthy of a malpractice lawsuit! Stunned by this confrontational reply, the now mildly irritated surgeon, in a tasteless act, shows the amputated leg to the patient to convince him that the surgery really took place. And yet, despite seeing and identifying the amputated leg as belonging to him, the patient continues to experience and describe to the doctor the sensation that the leg is attached to his body. He can feel his foot moving even as they speak, although there is no movement in the amputated limb held by the surgeon.

This sad scene illustrates that the human mind can handle cases in which provability (not having a leg any more) and feeling (not being able to deny the sensation of still having a leg) diverge and coexist in the same brain. Conversely, a digital computer would not be able to cope with this ambiguity at all. Instead, it would halt its operation because digital logic cannot deal with the "fuzziness" of the situation. For a digital computer running on Shannon information, the leg is either attached (0) to the patient's body or it was amputated (1). There is no possible state in between. For a human brain, running on Gödelian information, however, both states can coexist and be dealt with to the point that the patient can report feeling an annoying itch on a leg that no longer exists.

As we will see, classic models of brain operation, like the one proposed by David Hubel and Torsten Wiesel in the 1960s, cannot account for the phenomenon of phantom limb sensation at all because they are basically using Shannon information to describe brain functions. Ronald and I believe that the phantom limb sensation could be reinterpreted through an analogy with Kurt Gödel's first incompleteness theorem. That is the reason we used Gödel's name to baptize the new type of physically embedded information, because it is this type of information that would account for things like intuition, the unique human trait that, according to Gödel, is required—above syntactic formalism—to decide on mathematical riddles.

Both the Santorini honeymoon and the phantom limb examples illustrate another key difference between Shannon and Gödelian information: while Shannon information deals mainly with the syntax of a message, Gödelian information accounts for our ability to confer meaning on external events and objects and to express semantics and even ambiguities in the messages we receive and transmit.

Different from Shannon information, which can be expressed independently of the medium in which it is transmitted—electrical cables, nerves, or radio waves—Gödelian information depends on its physical embedding in organic matter to exert its causal efficiency on the organism. Just think about the rings of our beloved tree: the ongoing deposits of wood, triggered by a continuous process of energy dissipation, which will eventually form a new tree ring every year and embed in the plant tissue Gödelian information about the occurrence of droughts, changes in sunspots, or periods of high precipitation. There is no way to dissociate this type of Gödelian information from the very organic scaffolding that defines the life history of this tree. Put differently, in our definition of Gödelian information, the medium in which it is embedded matters. And again, even though in trees the information contained in their rings may not be accessed directly by the same organism, in animals with brains that reading process can be performed very quickly and efficiently.

The causal efficiency of Gödelian information can be illustrated by a very familiar phenomenon: the placebo effect. Well known among health professionals, the placebo effect refers to the fact that a significant percentage of patients can exhibit substantial clinical improvements by taking an otherwise inert substance—like a tablet made of flour—that was labeled by a doctor as a "potential new treatment or cure" for the patient's ailment. In other words, once informed by a doctor they respect as an authority in the field that the pill they are about to ingest will certainly help them, many patients raise their expectations that the treatment will be truly effective. And indeed, in a large fraction of these patients some clinical improvement is documented. Interestingly, it seems that if the placebo treatment is administered using configurations that elicit expectations most people have of a very effective therapy, the effects are better. Thus, some studies suggest that if the placebo is delivered in large, "hot colors" (like red) capsules, effects will be maximized. This type of result points to the interpretation that cultural background about medicine may play an important role as well as be a motivating factor in the placebo effect.

In our terms, the placebo effect could be explained by the direct action on neuronal tissue triggered by the physician's message offering the new treatment to the patient. Although this message is initially transmitted by encapsulation of Shannon information in language, once it is received by the patient's brain it is confronted with her internal beliefs and expectations, and stored in the brain as Gödelian information. By reassuring the patient's original belief that a treatment or a cure for her ailment is available, the placebo message acts directly on neurons, triggering release of neurotransmitters, hormones, and leading to the production of neuronal electrical firing activity that can, for example, amplify the patient's immune system, just to mention one possible hypothesis under investigation to account for the actual placebo effect. According to our view, such a neuroimmunological linkage happens because of the causal efficiency effect that Gödelian information can have on neuronal tissue.

The placebo effect reinforces our proposal that while Shannon information is expressed with the rigid syntax provided by integer numbers, bits, and bytes, Gödelian information, because it is generated/stored by an integrated system (the brain), represents a rich analog range of cause-effect associations and semantics that amplifies the meaning and reach of a person's own language; the main way she communicates her own thoughts, emotions and feelings, expectations, and deeply entrenched beliefs.

Another important feature about Gödelian information is that its amount and complexity varies from organism to organism. That means that, in clear contrast to Shannon information, which rises with the increase in entropy in a system, Gödelian information increases in complexity with the reduction of entropy that results within far-from-equilibrium thermodynamic islands that we call living systems. That means that while Shannon information is all about measuring the degree of uncertainty and surprise in a transmitting channel, Gödelian information increases with higher levels of biological structure/function complexity, organism adaptability, stability, and survivability, reflected by an enhancement of the defenses against disintegration. The more complex an organism is, the more Gödelian information it has accumulated. Thus, according to our theory, by dissipating energy to embed Gödelian information, organisms try to maximize their existence by enhancing their search for sun-derived energy and eventually replicating themselves by passing their DNA to future generations.

This process reaches its pinnacle in human beings, since Gödelian information can be used to generate knowledge, culture, technologies, and recruit larger cooperating social groups that improve significantly our chances of adapting to changes in our surrounding environment.

Gödelian information may also explain why most of brain processing is carried out unconsciously. For example, a classic experiment carried out in the early 1980s by the American neuroscientist Benjamin Libet may help bring home the points made above. In Libet's experiment (figure 3.3), a person sits in front of a monitor showing a dot going through the round image of a wall

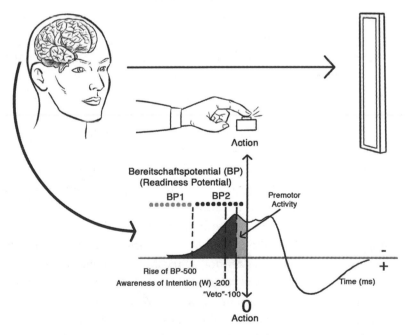

Figure 3.3. Benjamin Libet's classic experimental design. (Image credit to Custódio Rosa.)

clock. This individual is wearing a cap so that the experimenter can record his brain's electrical activity continuously using the classic electroencephalographic technique. The volunteer is then asked to perform a simple task: whenever he feels like it, he should press a button with the indicator finger. That seems easy enough. However, to make things more interesting, Libet asked his volunteers to use the dot circling the clock on the monitor to indicate the time in which they became conscious of the desire to flex their fingers. Using this simple apparatus, Libet could record three moments in time, represented in figure 3.3: when the person pressed the button; when the person decided to press the button, as reported by the subject calling out the time on the clock; and when the person's brain state, as recorded by the electroencephalogram, began to change. Figure 3.3 shows that while the conscious decision made by the person—and pinpointed by him—tends to happen about two hundred milliseconds prior to the actual finger press of the button, the increase in EEG activity took place about five hundred milliseconds prior to the button pressing.

There are many potential and somewhat conflicting interpretations of Libet's results. Indeed, most people tend to interpret this experiment as a categorical demonstration that a lot of what goes on in human brains is unconscious processing and that, because electroencephalogram modulation happens about three hundred milliseconds before the person is aware of her own decision to press the button, it shows that human beings do not have free will. The point here is not to dive into this controversial discussion. Yet Ronald and I have a completely different take on Libet's interesting findings. While everyone seems to focus on the fact that five hundred milliseconds before the subject presses the button, her brain is already busy at work, albeit in an unconscious mode of operation, Ronald and I ask the following questions: but what process(es) preceded and caused this unconscious EEG modulation to happen in the first place? Where does this signal originate from? We propose that prior to the five-hundred-millisecond period that separates the beginning of the EEG signal's rise and the finger movement, the subject's brain was busy accessing Gödelian information from the neocortex—and likely from subcortical structures, whose activity cannot be recorded from surface EEG. Once this Gödelian information was accessed—unconsciously—it was projected into streams of Shannon information, which could then be detected by the EEG measurement, five hundred milliseconds before the subject executed the finger movement to press the button. Once high-dimension Gödelian information was projected into low-dimensional Shannon information, an executable motor program was created and could now be transmitted by nerves—the neurobiological equivalent of Shannon's communication cables—from the primary motor cortex to the spinal cord, and from there to muscles, so that a movement was finally produced. According to our interpretation of this experiment, therefore, high-dimension Gödelian information was the true source that led to the Shannon information projection measured by the EEG signal detected five hundred milliseconds before the button press. By the way, in this view, the simple detection of EEG activity five hundred milliseconds prior to a movement does not imply that there is no free will. Free will may have manifested itself prior to the production of any measurable EEG activity, when Gödelian information was being accessed and read in preparation for a future movement.

In the past, Ronald and I have used yet another example to illustrate the disparity in what brain measures can tell us about the internal processing that goes on in any of our brains. Suppose a neuroscientist has designed an experiment to pinpoint what exactly happens when a volunteer sees on a computer screen a series of pictures containing upsetting images. To document the effect produced by these images on the volunteer's brain, this neuroscientist decides to measure both electrical brain activity, via EEG, and obtain high-resolution MRI images of the volunteer's brain while she watches the images on a monitor. Since the methods chosen by this experimenter sample the volunteer's brain from the outside, they will typically provide Shannon information only. As she gathers EEG and MRI signals, the neuroscientist also decides to simply ask the volunteer to express his feelings about the images in the format of free speech. Once both data sets are collected, the neuroscientist then seeks to correlate the quantitative measurements of brain activity—EEG and MRI data—with what the volunteer has expressed by language. What this neuroscientist will find is that her objective measurements of the pattern of brain activity do not necessarily correlate well with the feeling expressed through language. If we consider that even the language the volunteer uses to describe his feelings is just a low-dimension projection of the high-dimension Gödelian information stored in his brain, we begin to see how problematic it is to quantify the entirety of Gödelian information a brain like ours can store.

But this is not all. Because the brain is a dynamical complex integrated system, it can produce different emergent properties with immeasurably small changes in its initial conditions. Therefore, there is no hope for the dedicated and valiant neuroscientist of our story to measure all the necessary data in real time when dealing with a living brain. Even if we were able to obtain all the required measurements, we would not necessarily know how to translate them into the subject's feelings.

Because the human brain is capable of expressing Shannon and Gödelian information, as well as the impossibility of finding a perfect correlation between both, there is a unique challenge for the traditional scientific approach. This particular physical object we call a human brain occupies a very special position among the objects studied by natural sciences. In a brain, the external information (digital and formal) will never be able to fully account for the whole reality depicted by the internal information (analog and integrated). It is this internal information that includes the uniqueness that emerges from the brain's amalgamation of information and matter, arguably the most powerful computational endowment bestowed on us by evolution.

Overall, the differences between Shannon and Gödelian information can be described as follows: Shannon information is symbolic; that means that the recipient of a message containing Shannon information has to decode it to be able to extract some meaning from it. For this he obviously needs to know the code prior to receiving the message; if the code was not included in the message it would not be accessible. Without an external code, for instance, the very lines you are reading now would have no meaning for you. Meaning is essential for your brain to be able to do something with the message. Conversely, Gödelian information does not need any code to be processed; its meaning is recognized instantaneously by any human brain. That follows because the message's meaning is provided by the brain that generates or receives the message. As Chomsky says: "The most important thing in language is what is not said."

At this point you may be asking yourself, is the introduction of the Gödelian information required? My answer is categorically yes! As we saw, the introduction of the concept of Gödelian information allowed the derivation of a series of interesting new corollaries and hypotheses. First of all, it provided us with the basis to define organisms as a new class of computer. Traditionally, there are many types of man-made computing devices: mechanical computing devices, such as the abacus and the Charles Babbage differential engine; analog computers, such as the slide rule; digital computers, like the laptops and tablets that we are so fond of using these days; and, more recently, quantum computers. As we saw at the beginning of this chapter, Ronald and I propose that organisms can be considered as a different class of computing system altogether. We call them organic computers: devices in which computations are carried out by the very three-dimensional organic structure that defines them.

Our concept of organic computers can also be applied to the multiple levels of organization of living beings: from the very tiny nanomachines that operate by the synchronized collective work of multiple interlocked molecules (such as protein complexes—for example, the ATP-synthesized nanoturbine—or proteins and lipids—like the cell membrane) to groups of genes that need to work together to encode a particular physical trait or, on a slightly higher spatial scale, the highly complex energy microfactories (chloroplasts and mitochondria) that allow plants and animals to remain alive by producing energy to groups of cells that define a piece of organic tissue to vast networks of neurons that form an animal brain, all the way to networks of individual brains, or brainets, who synchronously interact to compute as part of an animal social group.

Although a distinction between hardware and software cannot be made in organic computers, these biological computing systems can use a mix of both Shannon and Gödelian information in their operation. Yet, as their organic complexity increases, so does the role of embedded Gödelian information which, because it is analog in nature—that is, it cannot be fully described or reduced by digital signals—cannot be properly uploaded, extracted, or simulated by a digital system. That does not mean, however, that organic computers cannot be programmed. The opposite is true. This very important topic is covered further in chapters 7 and 11.

As life evolved on Earth, simple organisms that were not capable of replicating themselves, because RNA and DNA were not available yet, were nothing but tiny vesicles contained by membranes, within which a few basic chemical reactions were carried out to sustain life for brief periods of time. At this stage, the cycle of sunlight and the conditions of the surrounding environment served as the programming influences of all living organisms on Earth. That view suggests that the Gödelian information (analog), deposited by energy dissipation into the first traces of organic matter to emerge on Earth, preceded the use of Shannon information (digital) by organisms, which became available only when mechanisms of self-replication, based on RNA or DNA, emerged. As such, before ribosomes could behave like a Turing machine to produce proteins from messenger RNA created from DNA strands, an analog membrane had to exist to allow tiny vesicles to form and sequester the stuff needed to sustain the earliest forms of life seen on our planet. In the case of the living, therefore, one needs to talk about "from BEing to BITing," meaning that first organisms needed to exist—organically speaking—and only after they accumulated some essential Gödelian information could they begin broadcasting bits of information in order to replicate themselves.

By the time "information molecules" like RNA and DNA began transmitting genetic information to hosts or organism descendants, they ascended to become the key "programmers" of the initial three-dimensional structure that defines an organism. Thus, when a virus infects a host cell, it is using its own RNA to reprogram the genetic machine of its victim to reproduce lots and lots of viral particles. By the same token, DNA carries the precise digital instructions to build any organism as a faithful three-dimensional replica of its progenitors. Using a modern analogy, RNA and DNA can be seen as carrying the programming instructions—in the format of Shannon information—that allow the three-dimensional printing of organic computers.

But further programming is necessary for complex living beings to function and survive. By promoting the increased accumulation of Gödelian information and the rise in biological complexity, evolution eventually gave rise to nervous systems that can store information in memory and learn from interactions with the outside world. At one point in this evolutionary riddle, our primate nervous system emerged. And since then, for each human being who has ever lived, after the brain's initial three-dimensional structure had been laid down into organic matter by instructions contained in our genome, the movements of our own body, our social interactions, language, human culture, and eventually technology, assumed the role of programming the most elaborate and sophisticated of all organic computers: the True Creator of Everything.

4 • Fueling the Brain with Dynamics

Biological Solenoids and Functional Principles

By one hundred thousand years ago, each human nervous system already boasted about 86 billion organic processing units, or neurons, which altogether could establish between 100 trillion to 1 quadrillion direct contacts, or synapses, among themselves. From within this incredible neuronal atelier the True Creator of Everything began sculpting the human universe as we know it today.

The cortex, whose evolution we just traced, represents about 82 percent of the overall human brain mass. Surprisingly, that mass contains only 19 percent, some 16 billion, of the brain's neurons. For comparison, the human cerebellum, a key collection of gray matter that regulates motor control, packs about 69 billion neurons into only 10 percent of the brain mass, making it a very dense neuronal cluster. But the cerebellum, as far as we can say, did not conceive of the sonnets and plays of Shakespeare, nor the spaceships used by us to explore outer space (although it helped us build them). This is why, from now on, we will focus mainly on the neocortex to describe how the True Creator of Everything accomplishes its most elaborate feats.

The complex mesh of white matter plays a crucial role in optimizing the functioning of the cortex. Some of the dense packs of nerve fibers (figure 4.1) that form the white matter are organized in loops that reciprocally connect pools of gray matter. I call these loops biological solenoids, after the coils of wire used in electromagnets. The largest of these biological coils is the corpus callosum.

Formed by a thick slab of roughly 200 million nerves, spread throughout the longitudinal axis of the brain, the corpus callosum enables the two cortical hemispheres to exchange information and coordinate their activities. There

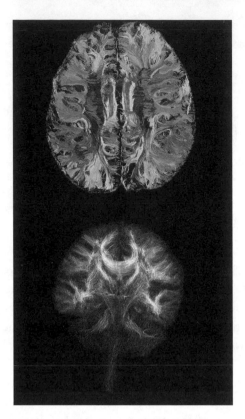

Figure 4.1. Typical examples of cortical
white matter loops as seen through diffusion tension imaging. (Courtesy of
Allen Song.)

is significant variation in the structure of the corpus callosum from the front
to the back of the cortex, including the density and diameter of the neuronal
axons that conduct electrical impulses—called action potentials—as well as
the level of axonal myelinization. Particular types of supporting brain cells
produce sheets of myelin that wrap around nerve fibers. The great advantage
of adding myelin sheets to a nerve fiber is that the resulting myelinated axons
transmit action potentials much more quickly. As a consequence, myelinated
nerves need less energy to conduct action potentials. For example, while an
unmyelinated C nerve fiber, with a diameter of 0.2–1.5 micrometers, conducts
action potentials at roughly 1 meter per second, the same electrical impulse
moves at about 120 meters per second, or more than 400 kilometers per hour,
in a large myelinated fiber. Thus, across their length, the time for transmit-

ting information between the cortical hemispheres can vary dramatically, depending on where in the cortex neural signals originate. Overall, this variation in conduction velocity is characterized statistically by a broad bell-shaped distribution. According to this distribution, the interhemispheric exchange of information between motor and sensory areas, for example, happens very quickly, as the corpus callosum connects them with thick myelinated axons. Conversely, interactions between the so-called association areas in the frontal and parietal lobes are much slower.

How the 200 million fibers of the corpus callosum actually coordinate the two hemispheres so precisely remains unknown. That the corpus callosum does synchronize them is clear, however, because if it is cut, the two hemispheres actually work independently. Extensive study of such split-brain patients, as they are called, began when such a surgery was adopted many decades ago to treat serious seizures by preventing them from spreading from one hemisphere to the other. The American neuroscientist Roger Sperry shared the Nobel Prize in medicine in 1981 for his pioneering work on split-brain patients and the corpus callosum.

In most people, some key brain functions such as language are lateralized in the cortex; that is, those functions are predominately produced by one hemisphere (in the case of language, the left one in right-handed people). Because of this lateralization, split-brain patients cannot always verbally describe what they see. For example, when presented with images that are restricted to the left visual field or when asked to hold with their left hand an object they cannot see, they simply cannot name or describe the image or object. That is not because they do not know the answer to these questions. They do. The problem is that the left-side stimulus is processed by the right side of the brain. Because the corpus callosum is cut, the right hemisphere cannot communicate with the language areas on the left side of the brain in these patients. Indeed, split-brain patients can use their left hand to choose, from a set, an object that is identical to the one they held minutes before; they are consciously aware of what they have seen or touched. They simply cannot speak of it.

Many other sizable white matter loops and bundles connect different cortical regions within each cerebral hemisphere. One of these systems, which provide an important connection between the frontal, parietal, and temporal lobes, is formed by three major tightly bundled nerve highways. The first one is the so-called extreme capsule that connects key regions in the temporal lobe—like those located in the superior temporal sulcus (STS) and the inferior temporal cortex—to the inferior prefrontal cortex. The second one, which connects the STS and the region of the parietal cortex, is formed by the bundles of the so-called inferior and middle longitudinal fasciculus. Finally, there is the superior longitudinal fasciculus, which mediates the communication between the parietal and frontal lobes. All together, these three pathways are involved in the circuits that mediate key functions such as language, tool making, and motor imitation.

Another major brain highway, the cortico-thalamic-cortical loop, mediates a reciprocal link between the cortex and thalamus, a fundamental subcortical structure that receives most of the sensory stream data generated by peripheral nerves. This multimodal sensory pathway, therefore, is essential for continuous comparison between the brain's own point of view and the sample of raw information coming from the external world. This loop also contributes decisively to synchronizing the electrical activity of the cortex and the thalamus.

Another important characteristic of human white matter is how it matures. Compared to those of our chimpanzee cousins, our human brains are relatively immature at birth, requiring two decades to reach mature size. Furthermore, although we are born with pretty much all the neurons we will ever have, it takes about three to four decades for white matter to reach its full functional maturity. Particularly in the prefrontal area of the frontal lobe, connections between neurons—both the synapses that transmit potentials from neurons and the dendrites that receive those transmissions—reach a mature level only by the third decade of life. Taken together, this means that most postnatal brain enlargement is due to white matter growth and refinement. This lengthy maturation process—and the possibility that it may be disturbed—may explain why humans are vulnerable to mental disorders such as schizophrenia and autism during their early postnatal and teenage years. Delayed maturation of the white matter also goes a long way to explain the changes in behavior and mental functions that we all go through in the early decades of our lives. So, next time you have a "friendly" argument with your revolutionary teenager, take a deep breath and, in earnest, blame it all on the delay of white matter maturation!

One of the most remarkable discoveries of brain research in the past half century was made by a group of neuroscientists led by Jon Kaas, from Vanderbilt University, and Michael Merzenich, from the University of California, San Francisco, who in the early 1980s demonstrated categorically that the elaborate neuronal circuits that define the brains of mammals and primates are in continuous dynamic flux throughout one's life. Our brains change themselves, both anatomically and physiologically, in response to everyone and everything we interact with, as we learn new skills, and even when significant modifications take place around and inside our bodies. Neuroscientists call this property brain plasticity, and it is of paramount relevance to unveiling the deep mysteries of the True Creator of Everything.

At the synaptic level, neuronal plastic changes happen in multiple ways. For example, the number and distribution of synapses on neurons can change significantly as a result of learning a new task, or as part of the recovery process triggered by damage to the body periphery or to the brain itself. Even in adult animals, individual neurons can sprout new synapses, enabling greater connectivity with some or all of their target neurons. Going in the opposite direction, neurons can also prune synapses and hence reduce their connectivity with target neurons. The individual strength of each synapse's influence on target neurons can also vary significantly as a result of what our brains are exposed to. Essentially any stimulus can cause changes in the delicate microstructure and function of the hundreds of trillions of synaptic connections by which the tens of billions of neurons of the cortex communicate.

After studying brain plasticity for more than a decade, during the summer of 2005, I proposed to Eric Thomson, a senior neuroscientist working in my lab at Duke University, a very unorthodox idea aimed at investigating how far we could stretch this phenomenon. To do that, we designed an experiment aimed at testing whether, by pushing plasticity to the extreme, adult rats could acquire a complete new sense in addition to the traditional ones (touch, vision, audition, taste, olfaction, vestibular) with which they are born. We decided that we should attempt to induce the rat brain to learn how to "touch" otherwise invisible infrared light. This required building a device that would transduce infrared light generated in the outside world into a stream of electrical pulses, the language the brain uses to transfer messages, that can then be delivered into the animal's primary somatosensory cortex, a major region involved in the generation of the sense of touch in mammals. By delivering this new electrical message to the primary somatosensory cortex, we wanted to determine whether our "cyborg rats" could learn to process infrared light as part of an expanded tactile perceptual repertoire.

To test this idea, Eric built devices consisting of one to four infrared sensors that could be easily placed on a rat's skull (figure 4.2). Each sensor could detect infrared light in a spatial sector of about 90 degrees, meaning a device with four sensors would endow a rat with a 360-degree infrared view of its surrounding environment. Our target in the somatosensory cortex was a subsection called the barrel cortex, which processes incoming tactile signals

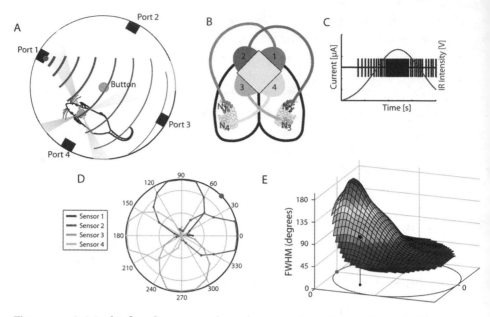

Figure 4.2. Original infrared neuroprosthetic device configuration used in Eric Thomson's experiments in my laboratory. *A:* Schematic of the behavioral chamber used for the infrared (IR) light discrimination task. Four ports are arranged symmetrically around the inner surface of a large (24-inch) cylinder. Each port contains a nose poke, an IR light, and a visible light. *B:* Topographical organization of the four implants into the rat primary somatosensory cortex (S1) that convey electrical signals from four IR detectors. The IR sensors were placed 90 degrees apart, with each sensor coupled to a different stimulating electrode pair in the S1. *C:* Stimulation frequency depended on IR intensity in each sensor. The intensity of each IR light was translated into different stimulation frequencies, in real time, in its corresponding stimulation channel. *D:* Polar plot showing the response of each IR sensor as a function of angle in the chamber when the sensor array is at a fixed position in the chamber relative to a single activated IR source. The point on the circle perimeter (*top right*) indicates the relative location of the IR source. *E:* Full-width at half-maximum (FWHM) of the response profiles as a function of position in the chamber. The black point represents the position of the active IR source, while the FWHM is the mean FWHM of all four sensors at the given position (see *D*). As you move further way, or obliquely, from the source, the response profiles become narrower. The black point indicates the position of the data represented in *D*. (Reproduced with permission. Originally published in K. Hartmann, E. E. Thomson, R. Yun, P. Mullen, J. Canarick, A. Huh, and M. A. Nicolelis, "Embedding a Novel Representation of Infrared Light in the Adult Rat Somatosensory Cortex through a Sensory Neuroprosthesis," *Journal of Neuroscience* 36, no. 8 [February 2016]: 2406–24.)

generated by stimulation of the facial whiskers of rats. As fingertips are for primates, the facial whiskers are the rat's most sensitive tactile organs and, as a consequence, a large area of the rat somatosensory cortex is allocated to processing tactile signals generated by these facial hairs.

We began our first experiments by training the rats to track a beam of visible light that led them to a reward. Once they were trained in that basic task, we attached Eric's infrared sensors to see whether they could detect and track a beam of infrared light by touch alone to find their reward. To run these experiments Eric placed four infrared sources at the 0-, 90-, 180-, and 270-degree locations on the inner wall of a round chamber where our cyborg rats were placed during the experiments. The placement of the emitters allowed us to randomly vary the source of the infrared beams during the trials, so that we could be certain the rats wouldn't learn how to find their rewards by one of their other, normal senses. At first we implanted only one infrared sensor in the rats. It took the animals about four weeks to learn to successfully "touch" and track the infrared light to the reward on more than 90 percent of trials.

Our cyborg rats exhibited a very interesting behavior during the initial trials: first they moved their heads laterally as if they were scanning the world around them for a signal; when the infrared beam appeared, rats invariably groomed their faces with their forepaws before starting to track the infrared beam toward the emitting source located in a given port. Although the first observation suggested that the rats developed a search strategy of their own to detect the first signs of the infrared beam, the second indicated that they felt the infrared light as if their whiskers had been touched by something in the outside world. But nothing had touched their whiskers. Instead, the brains of our rats learned to treat the incoming infrared beam as some sort of tactile whisker stimulation!

While all these results were very encouraging, the biggest surprise came a bit later, when Eric began to analyze the recordings of the electrical activity of individual neurons located in the somatosensory cortex of our infrared-tracking rats. A large percentage of those neurons, which originally fired only when the animals' whiskers touched something, had now acquired the ability to respond to the presence of infrared light in the environment (figure 4.3).

Our next experiments used four infrared sensors to allow a panoramic infrared view of the chamber. In those trials, the rats required only three days, instead of four weeks, to master the same task. Control experiments revealed that even when the spatial relationship between the infrared sensors' output and the different subregions of the somatosensory cortex were scrambled, rats

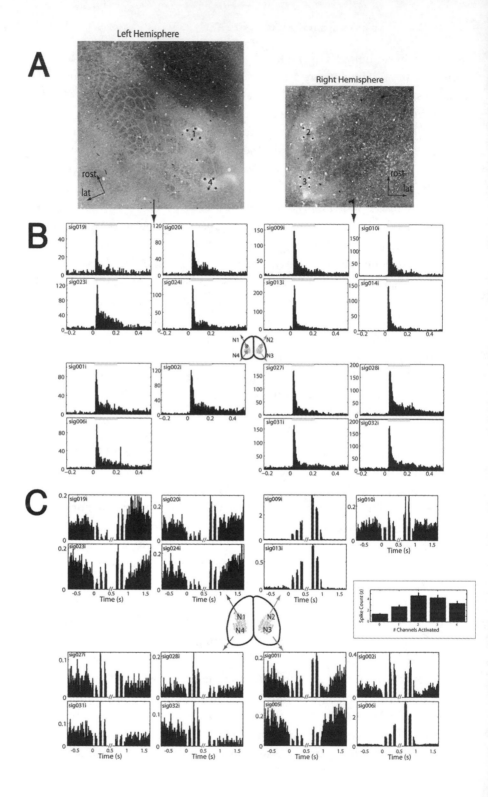

were able to quickly relearn how to track infrared beams to the reward more than 90 percent of the time.

Overall, these two studies thoroughly confirmed that we could endow rats with a new sense. Remarkably, this did not come at the cost of the animal's old perceptual repertoire: by the summer of 2016, Eric had already demonstrated that none of these infrared rats had lost the ability to use their long facial whiskers to execute the routine tactile discrimination tasks they are so famous for performing. In other words, a piece of the cortex once devoted exclusively to processing one crucial type of signal—tactile information, in this case—had morphed into a multimodal brain region—even though no rat, in the long evolutionary history of that sturdy species, had ever experienced that type of signal before. Basically, through the employment of our cortical sensory neuroprosthesis, the brains of our augmented rats were able to create a new infrared-based image of their surrounding world on top of a previously existing tactile representation.

Like the discoveries we made in the Walk Again Project, our experiments with rats and infrared light represent a very tangible outcome of a long series

Figure 4.3. Individual neurons in the rat's somatosensory cortex (S1, A) respond to both mechanical stimulation of the animal's facial whiskers (top shelf, B) and IR light in rats implanted with an IR neuroprostheic device (bottom shelf, C) that delivers electrical stimulation to the primary somatosensory cortex (S1). For the top shelf, A: Flattened cortical sections through both S1 hemispheres in one animal show the location of electrodes. The asterisks mark the electrode implant locations. B: Very robust sensory-evoked responses for 15 S1 neurons in the same animal, following mechanical whisker deflections, are illustrated by clear peaks of neuronal electrical activity plotted in peri-stimulus time histograms (PSTHs). Such tactile neuronal-evoked responses were obtained after the animal was trained in the IR discrimination task. PSTHs bin width equals 1 ms. C: PSTHs depict the electrical response of S1 neurons to IR stimulation sequences. Arrows indicate location of the neurons in the S1 cortex. The graph to the right shows the spike count z-score as a function of the number of stimulating channels activated. This is a typical profile, with maximum response occurring when two channels are co-activated. (Modified with permission. Originally published in K. Hartmann, E. E. Thomson, R. Yun, P. Mullen, J. Canarick, A. Huh, and M. A. Nicolelis, "Embedding a Novel Representation of Infrared Light in the Adult Rat Somatosensory Cortex through a Sensory Neuroprosthesis," *Journal of Neuroscience* 36, no. 8 [February 2016]: 2406–24.)

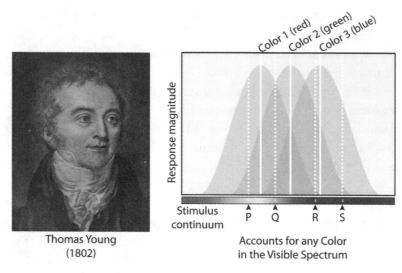

Figure 4.4. Schematic representation of Thomas Young's classic trichromatic color theory. (Reproduced with permission. Originally published in M. A. L. Nicolelis, "Brain-Machine Interfaces to Restore Motor Function and Probe Neural Circuits," *Nature Reviews Neuroscience* 4, no. 5 [May 2003]: 417–22. Young's portrait ©National Portrait Gallery, London.)

of scientific discoveries aimed at identifying and characterizing the key functional principles that define the workings of the human brain.

This fascination with brain circuits dates back to the foundation of modern neuroscience. The key founder of this pursuit was the nineteenth-century British genius Thomas Young, a polymath who, among other accomplishments, performed the now-classic double-slit experiment that demonstrated the wave nature of light. Young made some forays into neuroscience even before the field received that particular name. One such effort was the proposition of his trichromatic hypothesis for explaining color vision: Young postulated that the human retina could encode any color using only three types of color receptors, each responsive to partially overlapping ranges of wavelengths of light. According to Young's theory, this would happen because the response profile to color of each of these three retinal color receptors should follow bell-shaped curves with different maximum response peaks—meaning that they responded maximally to a given color—and partially overlapping response ranges (figure 4.4). That latter property indicated that each receptor would also respond in smaller magnitudes to multiple other colors. As time showed, Young was right on all accounts, despite the fact that all his predictions were made without his ever touching a real retina to examine its histological composition.

Young's model of neuronal function was the first example of a neuronal population-based or distributionist model of the nervous system. This model basically proposes that any brain function requires the collaboration of large numbers of neurons, distributed across multiple brain areas, to be enacted. The alternative interpretation, the one in which individual brain areas account for a particular neurological function, is known as the localizationist model. A complete reconstruction of the two-hundred-year battle between the proponents of the distributionist versus the localizationist models can be found in my previous book, *Beyond Boundaries*. For the purposes of this book, it suffices to say that it would take two centuries to decide which of these two models better describes how the human brain operates its wonders.

More definitive support for the distributionist model was gathered only in the past thirty years or so when neuroscientists acquired the technological means to investigate the detailed neurophysiological properties of brain circuits in freely behaving animals or in human subjects. Indeed, thanks to the introduction of new neurophysiological methods and, in the last two decades, a variety of brain-imaging techniques, more and more the focus of modern neuroscience has moved from the single neuron to populations of interconnected neurons that define the vast neural circuits that carry out the true business of the brain. In that regard, one could at last declare in mid-2018 that Young's view of the human brain had finally triumphed.

Among the new technologies employed to investigate the properties of animal brains, the method known as chronic, multisite, multielectrode recordings (CMMR) has provided the most comprehensive neurophysiological data in favor of the notion that distributed populations of neurons define the true functional unit of the mammalian brain, including ours. I have quite a bit of experience with the technique: during my five years of postdoctoral training, spent in the lab of John K. Chapin, one of the greatest American neurophysiologists of the past fifty years, my main job was to develop and implement one of the first versions of this new method in behaving rats. Thanks to all this work, and the efforts of a couple of other generations of neuroscientists working in my lab and many others around the world, today this neurophysiological method allows hundreds of hair-like flexible metal filaments, known as microelectrodes, to be implanted in the brains of rats or nonhuman primates. These microelectrodes enable us to simultaneously record the electrical action potentials produced by up to a couple of thousand individual neurons within a particular neural circuit, like the motor system, which is responsible for generating the higher motor plan needed for producing limb movements. Because of the characteristics of the material used to produce these micro-electrodes, the multielectrode neuronal recordings we perform in my lab can continue for many months (in rats) or even several years (in monkeys). This essential technical feature allows us not just to track the electrical brain activity of our animals as they learn a new task, but also to document how brain plasticity manifests itself during this learning period.

This technique proved critical to my work on brain-machine interfaces (figure 4.5), which I pioneered together with John Chapin and his lab some twenty years ago. In this paradigm, the recordings of the collective electrical activity of a population of neurons, located in one or many interconnected cortical areas, are used as the source of motor information needed to control the movements of artificial devices, such as robotic arms or legs, or even entire virtual bodies. Using a real-time computation interface, the recorded

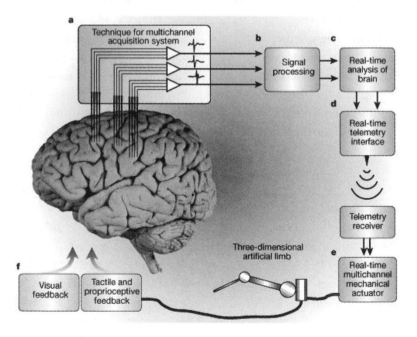

Figure 4.5. Classic representation of a typical brain-machine interface. (Reproduced with permission. Originally published in M. A. Lebedev and M. A. Nicolelis, "Brain-Machine Interfaces: From Basic Science to Neuro-prostheses and Neurorehabilitation," *Physiological Reviews* 97, no. 2 [April 2017]: 767–837.)

brain signals are fed into a series of mathematical models, translated into computational algorithms especially designed to extract these motor commands from brain electrical activity and transform them into digital control signals that artificial devices can understand. The development of this approach served as the experimental seed that led directly, ten years later, to the Walk Again Project.

Two decades of brain-machine interface research has generated a huge amount of experimental data related to how brain circuits operate in freely behaving animals, like rats and monkeys, and even in humans. Altogether, these findings support a rather dynamic view of the cortex unlike anything held by the majority of neuroscientists just a couple of decades ago.

Building on the analysis of simultaneous neuronal recordings performed in my laboratory over a quarter century at Duke University, I began to enumerate a series of neurophysiological rules, which I call the principles of neural-ensemble physiology, to describe the dynamic roots of the human brain.

At the top of this list is the distributed principle, which holds that all functions and behaviors generated by complex animal brains like ours depend on the coordinated work of vast ensembles of neurons distributed across multiple regions of the central nervous system. In our experimental setup, the distributed principle was clearly demonstrated when monkeys were trained to employ a brain-machine interface to control the movements of a robotic arm using only their electrical brain activity, without any overt movement of their own bodies. In these experiments, animals could succeed only when the combined electrical activity of a population of cortical neurons was fed into the interface. Any attempt to use a single or a small sample of neurons as the source of the motor control signals to the interface failed to produce the correct robot arm movements. Moreover, we noticed that neurons distributed across multiple areas of the frontal and even parietal lobe, in both cerebral hemispheres, could contribute significantly to the population needed to execute this motor task via a brain-machine interface.

Further quantification of these results led to a second principle, the neural-mass principle. This principle describes the fact that the contribution of any population of cortical neurons to encoding a behavioral parameter, such as one of the motor outputs generated by our brain-machine interfaces to produce robotic arm movements, grows as a function of the logarithm of the number of neurons added to the population. Because different cortical areas exhibited distinct levels of specialization, the logarithm relationship varied from region to region (figure 4.6). In support of the distributed principle, this

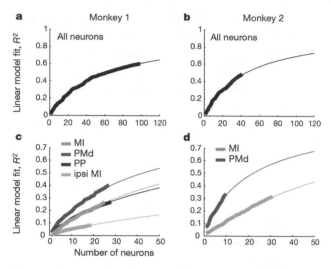

Figure 4.6. Examples of neuronal dropping curves (NDCs) relating accuracy of prediction of arm movement through the employment of a linear decoder. Decoding accuracy was measured as coefficient of determination, $R2$. NDCs represent $R2$ as a function of neuronal ensemble size. They were constructed by calculating $R2$ for the entire neuronal population, then removing one neuron from the population and calculating $R2$ again, and so on until only one neuron was left. MI = primary motor cortex, PMd = premotor dorsal cortex, PP = posterior parietal cortex, ipsi MI = ipsilateral primary motor cortex. (Originally published in J. C. Wessberg, C. R. Stambaugh, J. D. Kralik, P. D. Beck, M. Laubach, J. K. Chapin, J. Kim, et al. "Real-Time Prediction of Hand Trajectory by Ensembles of Cortical Neurons in Primates," *Nature* 408, no. 6810 [November 2000]: 361–65.)

finding indicates that all these cortical areas can contribute some meaningful information to the final goal: moving the robot arm just by thinking.

The multitasking principle holds that the electrical activity generated by a single neuron can contribute to the operation of multiple neural ensembles simultaneously; that is, individual neurons can participate simultaneously in multiple circuits involved in the encoding and computation of several brain functions or behavioral parameters at once. For instance, in the brain-machine interface experiment described above, the same cortical neurons could contribute to the generation of two distinct motor parameters at the same time—

that is, the calculation of the direction of arm movement and the production of the exact amount of hand-gripping force.

The neural-degeneracy principle posits that a given behavioral outcome, such as moving your arm to reach for a glass of water, can be produced at different moments in time by distinct combinations of cortical neurons. This happens both within single cortical areas and across multiple ones, since a motor behavior may require the coordinated activity of several cortical regions, not to mention a set of subcortical structures, like the basal ganglia, the thalamus, and the cerebellum. In other words, multiple combinations of cortical neurons working together, within and between cortical areas, can yield the same behavioral outcome at different moments in time; there is no fixed neuronal activity pattern responsible for controlling the lifting of your right arm, or any other action you might undertake. In fact, some preliminary evidence obtained in my lab suggests that the same combination of neurons is never repeated to produce the same movement.

A few years ago, I came up with a model to describe how the brain recruits and combines large numbers of cortical neurons, distributed across vast territories of the cortex, to generate a particular body movement. For any given action, there is a huge initial pool of cortical neurons—hundreds of millions of them—that could potentially participate in the action. From them, only a few thousand to a few million will actually take part in the computation of all the parameters needed to generate a movement. The recruitment of this reduced pool of neurons does not happen at the same instant; the process stretches throughout the hundreds of milliseconds needed to plan, define, and broadcast the voluntary cortical motor program out to the subcortical structures that will take care of its implementation. I think of this as the brain creating a "temporary organic computer" inside the cortex before any overt movement of the body can be produced by the subject. From moment to moment, however, the neuronal composition of this intracortical organic computer varies significantly because, from moment to moment, some or all the neurons that take part in computing the program of an earlier motor task may not be available again to participate in a new one; some may be in their resting refractory period, where they cannot fire electrical sparks for a few milliseconds, others may be inhibited by other neurons, or some may have died since their last contribution to this analog cortical computer was made.

Such ad hoc combinations of neurons add another dimension to the dynamic robustness that characterizes the distributed mode of operation of our cortex. Indeed, it seems clear to me that the tremendous gain in flexibility exemplified by this process isn't just inherent to why evolution favored population/distributed coding in the brain, but also to why it selected such distributed action across multiple levels of organization of biological systems: from proteins and genes to cells and tissue, all the way to the level of social interactions between individuals of a given species. In the specific case of the neocortex, distributed neuronal coding affords us the luxury of being able to continuously generate movements or perceive stimuli even after a great deal of the cortical tissue involved in such tasks is destroyed by disease or trauma. In other words, a distributed neural encoding scheme offers great protection against catastrophic failure. Indeed, when I was in medical school I had the opportunity to see several patients who, despite having suffered localized losses of cortical gray matter due to a minor stroke, showed none of the clinical motor symptoms we normally associate with this horrible condition. It so happens that the patients who display the typical stroke symptoms usually suffer damage not only to large portions of the gray matter of the motor cortex but also to the underlying white matter. That means that, when the connectivity that links the vast cortical circuits involved in motor planning and execution is compromised, things get really bad. Yet if the stroke is circumscribed to a small and localized region of cortical gray matter, provided that it does not wipe out the entire primary motor cortex, patients may still be able to move their limbs somewhat normally.

Next in my list comes the context principle, which holds that at any point in time, the global internal state of the brain determines how it is going to respond to some incoming sensory stimulus. In a sense, the context principle is complementary to the degeneracy principle, as it describes why and how, during different internal brain states (that is, when animals are fully awake, versus sleeping or under the effects of anesthesia), the same neurons can respond to an incoming sensory stimulus—let's say, a touch on its whiskers, in the case of rats—in a completely distinct way.

Whereas this may sound obvious to some, it took a lot of work to rigorously demonstrate the context principle from a neurophysiological point of view. And this was a major result because, if we put it in slightly different terms, the context principle basically postulates that the brain relies on "its own point of view" to make any decision regarding any new event occurring in the outside world. In my definition, the "brain's point of view" is determined by a series of contributing and interacting factors that include the accumulated evolutionary and individual perceptual history of the subject, which summarizes the brain's multiple previous encounters with similar and dissimilar stimuli;

the particular internal dynamic brain state at the moment of the encounter with a novel stimulus; the internal expectations set by the brain just prior to the encounter; the emotional and hedonic value associated with the potential incoming stimulus; and the exploratory motor program, manifested in terms of coordinated eye, hand, head, and body movements aimed at sampling a given stimulus.

Over the years, my laboratory has documented the manifestation of the brain's internal model of reality in a series of animal studies. For instance, in our rat experiments, we demonstrated the occurrence of "anticipatory" neuronal electrical activity across most cortical and subcortical structures that define the somatosensory system of rats when these animals are performing an active tactile discrimination task. This anticipatory neuronal activity is represented by the occurrence of large increases or decreases in the firing rate of single neurons prior to the moment rats begin to engage their facial whiskers to touch an object (figure 4.7). Embedded in this widespread anticipatory neuronal activity one can identify signals related to planning the whisker and body movements needed to perform the task as well as brain-generated expectations of what the animal is about to encounter as it uses its whiskers to explore the outside world. This latter component includes expectations about both the tactile attributes of objects and the amount of reward the animal should receive for successfully completing this tactile discrimination task. For me, this anticipatory activity depicts the rat brain's own point of view setting up a broad initial hypothesis for what it expects to encounter in the near future. Supporting this view, further studies in monkeys in our lab have recently showed that if one alters the amount of reward delivered at the end of a task trial, generating a clear mismatch with what was initially anticipated by the animal's brain at the beginning of the same trial, individual cortical neurons tend to change significantly their firing rate in response to this deviation from their brain's expectation. Such a "neuronal surprise" has been documented in many other brain regions under similar experimental conditions. According to the view of many neuroscientists, these post-reward changes in neuronal firing rate depict the deviation from what the brain had originally expected to happen and what really took place in terms of reward at the end of the task trial. Once such a mismatch happens, however, the brain uses the new information to reconfigure its internal point of view so that its expectations for the next trial can be upgraded.

What I am trying to say, therefore, is that by comparing experiences accumulated throughout the subject's lifetime with information gathered on a

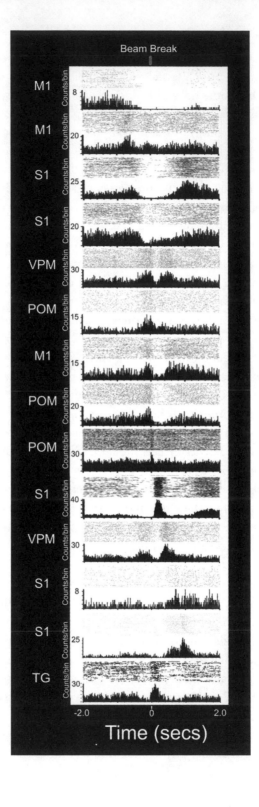

Beam Break

M1
M1
S1
S1
VPM
POM
M1
POM
POM
S1
VPM
S1
S1
TG

Time (secs)

-2.0   0   2.0

moment-to-moment basis, the brain continuously reshapes and updates its internal point of view in order to refine its neuron-based model of the surrounding world's statistics. In the case of humans, this also involves a continuous updating of the subject's own sense of self.

A clear demonstration of the context principle can be seen in figure 4.8, which illustrates how the same individual neuron in the rat somatosensory cortex responds differently when an equivalent mechanical stimulus is delivered to the animal's facial whiskers while it is either anesthetized, fully awake but immobile, or moving and using its whiskers to actively sample the same object. This striking difference in how the same cortical neuron in the rat

Figure 4.7. Individual neurons located at different processing levels of the rat's somatosensory pathway exhibit extensive anticipatory firing modulation (either increases or decreases in firing) prior to the time the animal's facial whiskers contact a pair of lateral bars. Peri-stimulus time histograms (PSTHs) are used to display different periods of increased or decreased neuronal firing activity across cortical and subcortical structures spanning the whole length of a task trial. Time 0 corresponds to the time when the rat breaks the beam just in front of the discrimination bars. The top four neurons, recorded in the primary motor (M1) and somatosensory cortex (S1), show a period of increased (anticipatory) firing activity before the trial started. As soon as the door opened, three of these neurons reduced their firing significantly. The onset of this decreased activity matched the beginning of firing increases observed in other neurons, for instance, in the primary somatosensory (S1) neurons (10th row). This suggests an initial role for M1 at the preparatory stages of a trial, followed by a second class of cells both in M1 and S1 related to early anticipatory activity as the door opens (approximately 0.5 s). As the animal moved from the door to the discrimination bars, anticipatory cells in the ventral posterior medial (VPM) and posterior medial (POM) nuclei of the thalamus (a subcortical structure), and M1 (5th, 6th, 7th, and 8th rows) exhibited a sharp increase in activity that ended as the whiskers contacted the bars (time 0). As this group of anticipatory cells decreased its activity, a different subset of neurons in POM, S1, and VPM (9th through 11th rows) presented an increase in firing activity. This period coincided with the rat's facial whiskers sampling the discrimination bars. Also, as the animal's whiskers touched the center nose poke and the rat chose one of the reward ports, firing increases were observed in the S1 (12th and 13th rows). The 14th row shows how a trigeminal ganglion (TG) neuron, the cell that innervates the whisker follicle, responds vigorously to the mechanical displacement of a single whisker. (Reproduced with permission. Originally published in M. Pais-Vieira, M. A. Lebedev, and M. A. Nicolelis, "Simultaneous Top-Down Modulation of the Primary Somatosensory Cortex and Thalamic Nuclei during Active Tactile Discrimination," *Journal of Neuroscience* 33, no. 9 [February 2013]: 4076–93.)

primary somatosensory cortex responds to a similar tactile stimulus takes place because the expression of the brain's own point of view is dramatically different in each of the three experimental conditions: going from nonexistent when animals are fully anesthetized to different levels of manifestation when the same rats are awake and immobile to a maximum expression capacity when these animals are free to move around and explore an object at their will.

Overall, the demonstration of the context principle accounts for some of the fundamental differences between the model of brain function that I introduce in this book and some more classic theories. For example, the pyramid graph of figure 4.9 compares the classic Hubel-Wiesel model of vision, originally

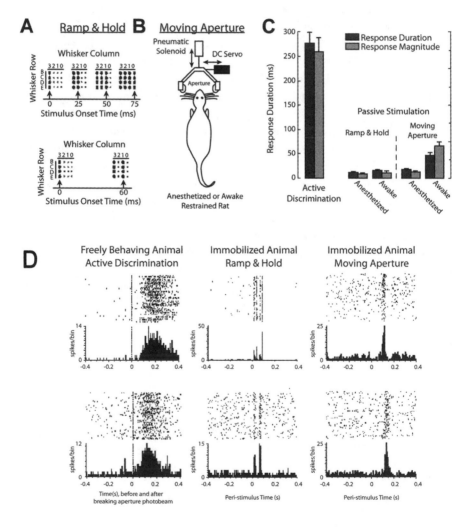

derived from data obtained in deeply anesthetized animals, with my relativistic brain theory, which was derived entirely on neurophysiological data collected from fully awake and freely mobile animals.

Coming back to our experiments with augmented rats, I can now say that a radical updating in their brain's own point of view took place to allow our cyborg rats to learn how to interpret infrared signals delivered to their somatosensory cortex. But once their brain's own point of view was updated, they assumed that "touching" infrared light belonged to their natural repertoire of perceptual skills. Essentially, that suggests that once the brain's own point of view is updated to include a new series of world statistics, what was once considered unexpected and unusual—like touching otherwise invisible infrared light—becomes just part of a new version of a brain-generated reality.

Underlying all of the stunning malleability of neuronal circuits is the phenomenon of brain plasticity, the exquisite property that not only enables us to

Figure 4.8. A: The upper schematic shows the pattern of multi-whisker ramp-and-hold passive stimuli delivered to anesthetized rats. Large black dots represent stimu lation of a particular whisker. Upward arrows show stimulation onsets. The lower schematic shows the stimulation pattern of the awake restrained rats. B: Schematic of the moving-aperture stimulus. The aperture is accelerated across the facial whiskers (with variable onsets and velocities) by the pneumatic solenoid and also simultaneously deflected laterally in varying amounts in order to accurately replicate the range of whisker-deflection dynamics that occurred during active discrimination. C: Mean (SEM) excitatory response duration (left Y axis) and magnitude (right Y axis) evoked during active discrimination and by the different passive stimuli delivered to anesthetized or awake restrained rats. D: Left panel, a representative neuronal response in S1 cortex showing long-duration tonic activation while rat performs an active discrimination. The upper portion of the panel is a raster plot where each line represents a consecutive trial in a recording session, and each dot is a unit spike; the lower portion of each panel shows a PSTH depicting the summed neuronal activity for all trials in 5-ms bins. The 0 time point represents the moment when rats disrupted the aperture photobeam. Middle panel, a neuronal response evoked by passive ramp-and-hold stimulation of 16 whiskers in lightly anesthetized rats. The 0 time point represents stimulus onset. Right panel, neuronal response evoked by moving-aperture stimulation of an awake restrained rat (the 0 time point represents the onset of aperture movement). (Reproduced with permission. Originally published in D. J. Krupa, M. C. Wiest, M. G. Shuler, M. Laubach, and M. A. Nicolelis, "Layer-Specific Somatosensory Cortical Activation during Active Tactile Discrimination," Science 304, no. 5679 [June 2004]: 1989–92.)

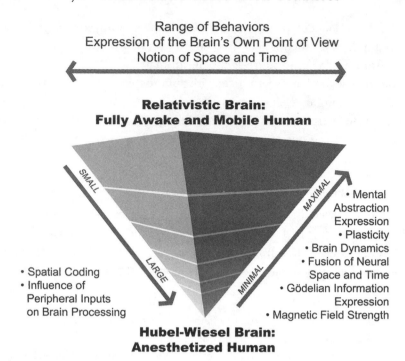

Figure 4.9. Pyramid graph compares the properties of the classic David Hubel and Torsten Wiesel model of vision, originally derived from data obtained in deeply anesthetized animals, with the main tenets of the relativistic brain theory (RBT). (Image credit to Custódio Rosa.)

learn and adapt but also creates a profound and unassailable chasm between the brain and other computing systems. It is thanks to plasticity that animal brains continuously adapt their micromorphology and function in response to new experiences. According to the plasticity principle, the internal brain representation of the world, and even our own sense of self, remains in continuous flux throughout our lives. It is because of this principle that we maintain our ability to learn until we die. Plasticity, for example, explains why, in blind patients, neurons in the visual cortex can become responsive to touch.

During early development, brain plasticity can accomplish truly astonishing feats. For instance, babies suffering from an autoimmune inflammation of the brain, a condition known as Rasmussen's syndrome, can undergo terminal damage of an entire cerebral hemisphere. As a result, they can suffer from epileptic seizures that do not respond to medication. Sometimes the only treatment possible in these cases is the complete removal of the affected cerebral hemisphere. This might seem like a treatment that would result in profound neurological deficits in the affected individual. Indeed, that is what the first physicians attempting such treatments expected. Nevertheless, most of these children—if the surgery is done early enough in postnatal life—grow up to lead almost totally normal lives. As a matter of fact, any external observer who comes in contact with these patients when they are adults may never realize that they are literally missing an entire cerebral cortex—such is the power of the brain to adapt to trauma. Sometimes such a discovery is made only when an emergency head scan is conducted in an unconscious subject—perhaps brought to the hospital because of a traffic accident—and, to the astonishment of the radiologist, a gigantic empty space is found inside the patient's skull.

Plasticity can also be observed to occur in the white matter bundles that connect cortical areas. For example, in a study carried out by Hecht and colleagues, brain scans were obtained before, during, and after a group of volunteers underwent a long (two years) and intensive period of hands-on training to become proficient makers of Paleolithic-like stone tools. In an amazing discovery, the authors were able to demonstrate that such tool-making training induced significant metabolic and structural changes in the superior longitudinal fasciculus and its vicinity, represented by changes in nerve density, nerve caliber, and level of myelination of axons.

One of the more surprising results of our multielectrode recording experiments in freely behaving rodents and monkeys was the discovery of the conservation of energy principle. As animals learn to perform a variety of different tasks, there is a continuous variation in an individual neuron's firing rate. Nevertheless, across large cortical circuits the global electrical activity tends to remain constant. To be more technical about it, the total number of action potentials produced by a pseudo-random sample containing hundreds of neurons that belong to a given circuit—let's say the somatosensory system— tend to hover tightly around a mean. This finding has now been validated by recordings obtained from multiple cortical areas in several animal species, including mice, rats, and monkeys. Indeed, just a couple of years ago, Allen Song, a professor of neuroradiology at Duke University—one of the leading brain-imaging experts in the world and one of my best friends—showed me that when examining the magnetic resonance imaging of human brains, one can identify not only areas where oxygen consumption and neuronal firing increase above baseline but also regions where oxygen consumption is proportionally reduced, suggesting that the overall level of energy consumption by the brain is kept constant. These human findings further corroborate the principle of energy conservation observed in our neurophysiological experiments in animals.

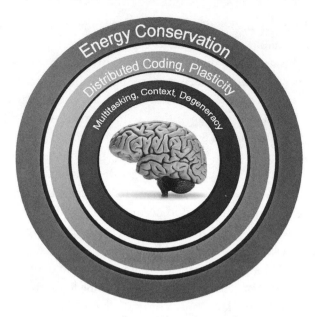

Figure 4.10. Schematic summary representing the
hierarchy of the different principles of neural ensemble
physiology. The outside ring (energy conservation principle) represents the most general principle. Subsequent
inner rings rank the other principles from most general
to more specific. (Image credit to Custódio Rosa.)

A major implication of this principle is that, since the brain has a fixed energy budget, neural circuits have to maintain a firing rate cap. Thus, if some
cortical neurons increase their instantaneous firing rate to signal a particular
sensory stimulus or to participate in the generation of a movement or other behavior, other neighboring cells will have to reduce their firing rate proportionally, so that the overall activity of the entire neural ensemble remains constant.

To summarize the discussion on neural ensemble principles, figure 4.10
represents a potential hierarchy between these principles, moving from the
more general (outer circle) to the more specific principles (subsequent inner
circles).

Although I have derived a few other principles from my almost thirty
years of multielectrode experiments, what I have just reviewed is sufficient
to portray the kind of dilemma facing neuroscientists who seek to find some

synthetic theory on how complex animal brains operate. Certainly, none of the classical theories of mainstream neuroscience could explain the findings that have emerged from the multielectrode recording experiments of the past three decades. For starters, most of these theories do not take into account any notion of brain dynamics: from the millisecond scale, in which neural circuits operate, to the temporal scale, in which brain plasticity occurs, to the seconds and minutes needed to produce behaviors, brain dynamics have been utterly ignored for almost a full century of brain research. Thus, the various manifestations of neuronal timing were never part of the classical central dogma of neuroscience, which remained dominated by static concepts such as cytoarchitectonic diagrams, cortical maps, and the never-ending cataloging of particular neuronal tuning properties. Furthermore, competing theories of brain function do not take into account the other principles I derived while recording the activities of large populations of cortical neurons.

For the past decade, I have been attempting to formulate a theory of brain function that would explain all the principles and experimental data summarized above. A key feature of this new theory is that it must account for why there are no fixed spatial borders that constrain the operation of the cortex as a functional whole. My response to this demand was to imagine the cortex as a continuum entity along which neurological functions and behaviors can be generated by recruiting widely distributed populations of neurons as part of an ensemble. The operation of such neuronal ensembles would be bounded by a series of constraints, among which are the evolutionary history of the species, the layout of the brain determined by genetic and postnatal development, the state of sensory periphery, the state of internal brain dynamics, other body constraints, the task context, the energy available to the brain, and the maximum speed of neuronal firing.

Another major challenge in building a new theory of brain function is the identification of a robust physiological mechanism that could account for how vast territories of cortical tissue can precisely synchronize their interactions and form a functional continuum in order to accomplish all major tasks a brain performs routinely. This led me to consider the brain's white matter loops, those biological solenoids that allow various regions of the brain to communicate with each other, as the place to look for this potential synchronization mechanism. A solenoid is a loop that acts as an electromagnet when a current is passed through it. It seemed to me our brains are full of them. And so I asked, what might electromagnetic fields, generated by action potentials running through bundles of white matter, contribute to the functioning of our brain?

5 • The Relativistic Brain Theory

It All Comes Down to One Picotesla of Magnetic Power

The discovery that the human brain relies on dynamic interactions between large, widely distributed populations of neurons, organized in elaborate circuits, promises to unveil a series of answers to fundamental and wide-open questions in modern brain research. For instance: what neurophysiological mechanisms were responsible during evolution for the fusion of our many unique mental capacities (language, theory of mind, tool making, general and social intelligence, a sense of self) so that they could merge into a single cohesive functional mind? How does the brain synchronize the action of its different anatomical components, functionally binding the entire neocortex so that the multiple sensory signals, actions, abstractions, and thoughts we experience are welded together as a continuum? How can we update and maintain our memories for a lifetime?

Finding the final answers to these questions will likely take much more time than I have left in my own lifetime, but they are the ones that drive my work as a systems neuroscientist. Indeed, I see no endeavor worthier than trying to pursue a few breakthroughs that might help us achieve a more definitive understanding of these fundamental problems and their solutions, assuming—and this is a big assumption—that the human mind can fully comprehend itself.

Based on the neurophysiological principles discussed in chapter 4, if I had to offer a few sentences to describe my best possible guess of what might solve those problems, it would go like this. The brain works by recursively mixing analog and digital neuronal signals. This dynamic process allows the fusion of the neuronal tissue into an operating continuum that is engaged in a bidirectional process of Shannon and Gödelian information conversion (see figure 3.2). By dissipating energy to physically embed Gödelian information into neuronal tissue (making information have a direct cause efficiency action on its anatomical structure), the brain is capable of using incoming new signals describing the surrounding world to continuously update its internal model of reality. Ultimately, it is this process of checking and updating the brain's own point of view that guides the operation of our central nervous system from moment to moment.

At this point, this "gut feeling" statement may not make sense to many readers. But there is no need to despair. In this and the next few chapters I will dissect this paragraph and explain, as clearly as possible, what I have in mind, both literally and metaphorically.

My solution to the central problem of neuroscience is what I call the relativistic brain theory. I proposed the initial statements of the relativistic brain theory in *Beyond Boundaries*, and for the past eight years I have joined forces with my great friend Ronald Cicurel to further elucidate its tenets; in 2015 we coauthored an academic monograph on the subject, *The Relativistic Brain: How It Works and Why It Cannot Be Reproduced by a Turing Machine*. I chose the term *relativistic* inspired by its historical use to suggest the inexistence of an absolute frame of reference for natural phenomena. Albeit in different fields, Aristotle and Galileo, among others, have also defended a "relativistic" view of human constructs (that is, ethics and morals) and natural phenomena (falling objects). The German philosopher Emmanuel Kant introduced what could be considered a relativistic view of perception by proposing that we cannot directly understand what is out there in the universe, only create mental representations of such a reality by relying on our senses and reasoning. Sharing this notion, the distinguished Austrian physicist Ernst Mach believed that all movement could only be described as being relative to the rest of the universe. Mach also applied his relativistic views to discuss human perception. In his 1886 book, *The Analysis of Sensations*, he echoed Kant: "The objects that we perceive consist merely of bundles of sense-data linked together in regular ways. There exists no further object independent of our sensations—no Things-in-Itself. . . . We thus know only appearances, never a Thing-in-Itself— just the world of our own sensations. Therefore, we can never know whether there exists a Thing-in-Itself. Consequently, it makes no sense to talk about such notions."

Interestingly, Mach's view of perception was clearly aligned with the new way of looking at the world proposed by a revolutionary group of painters who created the impressionism movement in France in the late nineteenth century.

In their frontal rebuke of the realism school, which believed in the faithful detailed transcription of the outside reality of the world in photographic precision, impressionists strongly believed that their main job was to represent their internal, subjective personal view of the world. As precisely put by the Brazilian art critic Mário Pedrosa, the impressionists proposed to "liquefy the solids and corrode the angles, transforming everything, from the façade of cathedrals to the structure of bridges, into the same colorful and itching paste spread, without any plane hierarchy, all over the canvas surface."

My kind of people, those impressionists!

All in all, Mach's views resonate very well with the choice of the word *relativistic* to name a brand-new brain theory. And even though one could argue that Albert Einstein can be credited with finally bringing a human observer into a relativistic framework to describe the entire fabric of the cosmos, neither he nor his antecessors and followers have attempted to go one step further and try to pinpoint the intrinsic relativistic mechanisms of the observer's brain. Hopefully, the introduction of a relativistic theory of brain function can now open the doors for that to happen in earnest.

Following what I believe is a neurophysiological version of Machian thinking, the central axiom of the relativistic brain theory states that the general mammalian brain mode of operation is based on the continuous comparison of an internal model of the world (and the subject's body) with the incessant multidimensional flow of sensory information that reaches our central nervous system at each instant of our lives. From this comparison the human brain chisels for each of us a sense of self and a brain-centered description of the universe that surrounds us. Therefore, to achieve any task—from the act of calculating an arm movement to the mapping of the most complex chain of causal relationships needed to build a spaceship—the human brain continuously builds mental abstraction and analogies, searching for the best fit between its internal neuronal-based simulation—its view of the world and the job it needs to execute. Anything that has ever materialized inside the human universe, from a spoken word to the creation of new tools to the composition of a symphony to the planning and execution of a horrible genocide, had to first happen, in the shape of a mental abstraction or analogy, inside someone's head. Thus, before I can begin producing a complex movement of my hand, thousands or even millions of cortical neurons have to come together transiently to form the organic computing entity (thousands of subcortical neurons will also become recruited into this functional unit, but for the sake of clarity let's ignore them for now and concentrate on the cortex). This entity, a functionally integrated neuronal network, is responsible for calculating the motor program that leads to this action. I call this neuronal-based motor program an internal mental analogy of the movement that will be executed by the body a few hundred milliseconds into the future. As such, following the principles of neural ensemble physiology, this neurobiological entity represents a true analog computer that simulates a body movement using a particular distributed pattern of neuronal activity. According to the degeneracy principle, however, every time a movement has to be executed, a different combination of neurons will do this mental work ahead of the real action.

A major challenge raised by this view is how the brain would be capable of forming these ad hoc analog computers so quickly, and how different organic entities are able to reliably produce precise movements, whether generated by a violinist, a ballerina, a baseball pitcher, or a surgeon.

A second major question is how to reconcile the local and global modes of operation of the brain. On one level, the brain utilizes electrical pulses, known as action potentials, to exchange messages from neuron to neuron. The digital nature of such communication is defined by both the all-or-none binary way in which action potentials are created and by the precise timing of their production by each individual neuron belonging to a neural circuit. Sequences of such action potentials are transmitted by the axons of neurons; when they reach a synapse—the terminal contact established by the axon with another neuron—these electrical messages trigger the release of a neurotransmitter in the synaptic cleft. The transmission and processing of these digital signals can be described with Claude Shannon's theory of information—that is, we can measure the information in the signals with bits and bytes, as we would describe the information transmitted in phone lines or the symbols represented by the computer on your desk.

But the brain relies on neuronal analog signals, too, as only they would be capable of fully underlying the type of information processing our central nervous system needs to carry out to generate human-like behaviors. As discussed in chapter 3, in addition to Shannon-information, I propose that animal—and particularly human—brains utilize analog Gödelian information to produce the functions and behaviors that distinguish them from digital machines. Simply put, only an analog signal can represent a perfect analogy of the physical parameters we encounter in nature, such as electric voltage or current, temperature, pressure, or magnetic fields. Like those physical entities, signals generated by neurons also must vary continuously in time to allow the brain to perform its job properly. As such, a digital version of these neuronal signals would represent only discrete samples of an otherwise continuous signal taken at some predefined time interval. And, although the precise time in which a neuron produces an electrical pulse can be represented digitally, all the electrical signals generated by these brain cells, such as their membrane, synaptic and action potentials themselves, are all analog waves in which electric voltage varies in time. Furthermore, the global electrical activity of the brain, which results from the mixture of synaptic and action potentials produced by billions of neurons, is also an analog signal. Taken together, I propose that animal and human brains operate through the combination of a hybrid digital-analog computation engine.

After some years of reflection, it became clear to me that the maximum velocity at which nerves conduct action potentials—approximately 120 meters per second—was insufficient to explain the speed at which the brain performs some of its most fundamental functions, like integrating many cognitive skills into a cohesive mind. As a result, I began searching for an analog signal that could propagate across the entire brain at a speed close to the fastest thing we know in the universe—no, not a Philadelphia Eagles' wide receiver, but something even faster, like the speed of light!

One of the most fundamental architectural features of the human brain is the presence of tightly packed bundles and loops of nerves, formed by tens of millions of axons, which are responsible for transmitting fast sequences of action potentials from one brain area to another (chapters 2 and 4). As Michael Faraday discovered in the early nineteenth century, electrical currents can induce magnetic fields. Likewise, a changing magnetic field will spontaneously induce a current in a conductor. With this in mind, I began reasoning that all those loops of white matter in our brains are not just conducting electricity, they are wrapping the brain in a multitude of time-varying neuronal electromagnetic fields. That is why I like to refer to the white matter connecting cortical and subcortical structures as biological solenoids.

Cortical electrical fields have been measured since the mid-1920s through a technique known as electroencephalography. In addition, brain magnetic fields have also been measured for several decades now through another method known as magnetoencephalography. Those latter measurements, however, have been mainly confined to the cortex due to the current lack of sensitive methods that can reach deeper into the brain.

The relativistic brain theory proposes that very complex spatiotemporal neuronal electromagnetic field patterns can emerge as electrical potentials flow through the many biological solenoids that are found all over our brains.

It is important to mention that these biological solenoids include not only very large nerve loops but also a myriad of other white matter rings of different sizes, including microscopic ones formed by the dendrites and axons of small networks of neurons. Based on this pervasive anatomical arrangement, the relativistic brain theory predicts the existence of widespread subcortical electromagnetic fields in addition to the well-documented cortical ones.

In my view, the recursive interaction between those two classes of brain signals, the digitally generated action potentials and the analog electromagnetic fields that result from them moving through nerves, is at the heart of our brain's unique computation abilities (figure 5.1). In this context, I propose that neuronal electromagnetic fields enable the emergent neural properties that we believe are essential for the manifestation of the higher mental and cognitive skills of the human brain. This would happen because these electromagnetic fields would provide the physiological glue needed to fuse all the neocortex into a single organic computational entity capable of combining all our mental capacities as well as enabling very rapid coordination between cortical and subcortical regions of the brain. The end result of this process would enable the brain to compute as a whole. This would happen because the far-from-equilibrium combination of a multitude of analog brain electromagnetic fields could conspire to create what I call the neuronal space-time continuum. In this context, neuronal space and time could become fused, just as Albert Einstein's general theory of relativity did for the entire universe.

Overall, in my view, this electromagnetic binding enables the brain to coordinate and precisely synchronize the activities of its disparate areas, whether those regions are separated by distance or time. As in Einstein's theory, where space and time themselves are folded by the presence of mass, changing the space-time distance between objects, I argue that this neuronal space-time continuum can also, in a neurophysiological sense, "fold" itself. As a result, this folding would bring together parts of the brain that are—when simply measured in inches—apparently quite distant into a single neurophysiological/computational entity. I believe that this phenomenon exists, in more rudimentary ways, in all higher mammals. In humans, however, I suggest that the resulting neuronal continuum—or mental space, as I like to call it—is the analog neuronal substrate from which all higher human brain functions emerge.

Several factors would shape the dynamics of the mental space: the spatial distribution and composition of neuronal pools in the brain; the structural features of the nerve pathways and loops of the white matter that connect

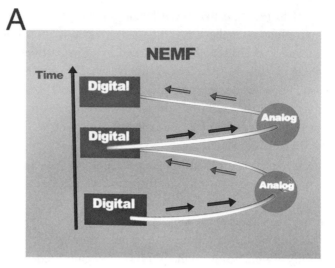

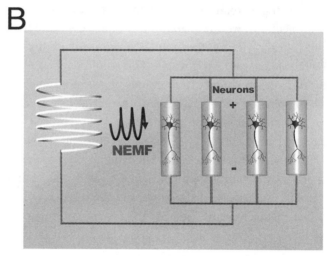

Figure 5.1. Two schematic illustrations describing the recurrent analog-digital interaction that takes place in the cortex, according to the relativistic brain theory, mediated by neuronal electromagnetic fields (NEMF). *A:* Neurons generate trains of electrical action potentials, the main type of digital-like signal produced by the brain, which can then generate electromagnetic fields, an analog signal, as they are transmitted by bundles of nerves. *B:* Such electromagnetic signals can then influence the generation of new action potentials by neighboring neurons. (Image credit to Custódio Rosa.)

these neuronal clusters; the energy available to the brain; the different types of neurotransmitters available to nervous tissue; and our memories, which are a key component in defining the brain's own point of view. Indeed, changes to one, some, or many of these individual components—such as spatial configuration, density of axons, and myelin levels of some white matter loops—in addition to brain volume and numbers of neurons, could have accounted for the significant changes in brain capabilities that took place over the 6-million-year evolution of hominids.

To investigate some of the ideas raised by the relativistic brain theory, one of my Duke University graduate students, Vivek Subramanian, built a simulation of a recurrent analog-digital computational system in which individual neurons fire digital-like action potentials, which can then produce electromagnetic fields that, by induction, influence the next firing cycle of the same original neurons. By letting this system run for a few cycles, Vivek observed that once a very small set of neurons fire a single action potential, the entire neural network—dispersed over space—tends to quickly evolve to a state of high synchronization, meaning that most of its neurons tend to fire together, creating a perfect rhythmic oscillation. Such tight synchronization of individual neurons is also reflected in the electromagnetic fields generated by the combined activity of the same neuronal ensemble. Although not a definitive demonstration by any means, this simple simulation can be used as a proof of the principle that recurrent analog-digital neuronal interactions could account for the type of large-scale synchronizing mechanism needed for the binding of many cortical and subcortical structures into a single computational entity. Moreover, this research offers the possibility of creating brain-based analog-digital computing applications that could in the future be more efficient than the current digital-only machine-learning algorithms used by artificial intelligence to try to mimic human behavior. I believe that this could happen because recursive analog-digital computing architectures may be able to solve problems that today are considered beyond the reach of contemporary digital computers.

Following these initial results, Vivek, Gary Lehew (another member of my lab), and I decided to build a physical version of this computer simulation. We accomplished this goal by directly routing the electrical outputs produced by a digital simulation of a large neuronal network to a three-dimensional printout of a diffusion tension image of a subset of white matter coils of the human brain, such as the ones illustrated in figure 5.2. Each coil in this physical model generates an electromagnetic field as an electrical charge runs through it. In return, the electromagnetic fields generated by these biological coils

A

B

Figure 5.2. *A:* The analog component of a brain-inspired digital-analog computer based on *B,* a 3-D printed representation of the organization of cortical white matter bundles related to motor control as originally imaged using diffusion tension imaging. (Image credit to Custódio Rosa.)

induced the firing of the digital neurons in the system. This physical rendition of such a "neuromagnetic reactor" defines a hybrid analog-digital computer; experimenting with it will enable us to observe and measure in great detail the dynamic operations we think may be going on inside our own brains.

Interestingly, as I write this description of our new hybrid analog-digital, brain-based computer, a group at the U.S. National Institute of Standards Technology in Boulder, Colorado, has just reported on its own experiences in using magnetic fields to add a new dimension of information encoding to build a "neuromorphic" device, or a machine that attempts to mimic more closely the operation of the human brain. This effort, combined with our own, shows that neuronal electromagnetic fields may become a hot area of research in neuromorphic computing in the near future.

One major question confronting such an analog-digital model of brain operation is whether the magnetic fields that surround us, such as the one produced by the Earth, are capable of influencing the activities of our brains. This is a pertinent question because researchers have discovered that various organisms have the capacity to detect the Earth's magnetic field: bacteria, such as *Magnetococcus marinus*; insects; nematodes; mollusks; eels; birds; and even mammals, including wood mice, the Zambian mole-rat, the big brown bat, and the red fox. The fox exhibits a unique hunting behavior: it tracks small rodents as they move in underground tunnels until a point at which it uses a high-jump to dive, head first, into the ground to catch its meal. These high jumps are performed along a northeasterly direction.

The widespread presence of animal magnetoreception strongly suggests that the Earth's magnetic field played some significant selective role in the evolutionary process, although to date this is a topic that has not received the attention it deserves.

The widespread reliance on magnetoreception by animals also means that any drastic variation of the Earth's magnetic field, such as the many geomagnetic reversals that have happened to our planet in the past, may cause havoc among these species, impacting dramatically their ability to forage and navigate. An interesting corollary of this notion is the hypothesis that some of the transient minor cognitive impairments experienced by astronauts bound for the moon during the Apollo program may have resulted from some neurological effect produced when they left the essential embrace of the Earth's magnetic field, which has been around them since their conception. That, however, remains to be demonstrated.

By the same token, assuming that the human brain relies on tiny neuronal electromagnetic fields to operate normally, you might readily expect human-made magnetic fields, such as those generated by an MRI machine, to have some important effect on our mental activities. After all, these devices generate fields some trillion times more powerful than those found in our brains.

One reason neither the Earth's nor most of the MRI-generated magnetic fields affect our brains is that both are static and hence cannot induce any neurons to fire electrical pulses as a result of our exposure to them. Moreover, those MRI gradient magnetic fields that do oscillate are set to do so at much higher frequencies than the low-frequency (0–100 hertz) electrical signals encountered in the brain. In other words, the human brain is basically blind to most of the magnetic fields that either exist in nature or are created artificially. Nevertheless, when exposed to the fields of a magnetic resonance imaging machine, some patients report mild neurological effects, such as dizziness or a metallic taste in their mouths. If humans are subjected to magnetic fields much higher than those from regular MRI machines, these effects can be exacerbated and others can be manifested.

Other evidence for the role of neuromagnetic fields in the functioning of the brain arose from the introduction of a new technology, transcranial magnetic stimulation. When conductive metal coils of a particular shape are applied to the scalps of subjects and electrical currents are passed through them, the resulting low-frequency magnetic fields can both induce cortical neurons to fire and inhibit their firing. As such, there is a long and growing list of neurophysiological and behavioral effects induced by the application of transcranial magnetic stimulation to different regions of the human cortex.

In addition to circuit-level synchronization, there is another potential effect of neuronal magnetic fields that has been largely ignored so far. Figure 5.3 illustrates how the brain can be considered as a multilayered structure that works by tightly integrating multiple levels of information processing, ranging from the atomic/quantum level to the molecular, genetic, chemical, subcellular, cellular, and circuit levels. To work properly, the brain has to ensure that information flows in perfect synchrony across these levels, which are linked by multiple feed-forward and feedback loops. Each of these levels defines an open system whose reciprocal interactions are likely to be very nonlinear or even noncomputable, meaning that they cannot be mediated simply by algorithmic and/or digital processes. Instead, the job of integrating all these levels

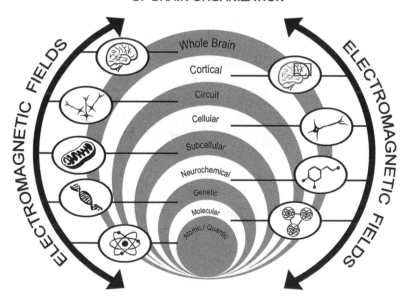

Figure 5.3. The multiple levels of brain organization that can be influenced directly and simultaneously by neuronal electromagnetic fields. (Image credit to Custódio Rosa.)

of information processing into a single operating unit could be achieved only by an analog signal that can elicit effects at all these levels of resolution simultaneously. Electromagnetic fields fulfill this strict prerequisite. Accordingly, neuronal electromagnetic signals would ensure that the brain operates as an integrated computing system by mediating the operation and exchange of information between all its processing levels, from quantum to circuit levels.

In general terms, the relativistic brain theory tries to account for a variety of findings that are beyond the reach of traditional theories in neuroscience, such as the classic feed-forward model of vision proposed by David Hubel and Torsten Wiesel. For instance, by introducing the concept of the brain's own point of view, the relativistic brain theory provides a physiological explanation for the findings that led to the formulation of the context principle. The theory proposes that under different animal behavioral states (anesthetized, awake and fully mobile, awake but immobile), the internal dynamic state of the brain is different. As such, the manifestation of the "brain's own point of view" varies dramatically from an anesthetized animal, where it is basically nil, to a subject fully engaged in sampling its immediate whereabouts, where the brain's own point of view is fully expressed. Since the brain's response to the same sensory stimulus depends on the comparison of the incoming sensory volley with the brain's internal model of the world, neuronal sensory-evoked responses should vary dramatically from anesthesia to fully awake/mobile conditions. That is precisely what has now been observed in a variety of animal experiments involving the tactile, gustatory, auditory, visual, and olfactory systems. The same would be true for human beings subjected to different emotional states. For example, it is well known that soldiers involved in intense combat on a battlefield may be able to temporarily avoid feeling pain that in a regular setting would be considered excruciating and unbearable.

Indeed, the sensation of pain is an example that illustrates well the view that complex mental experiences can be generated by the interaction of neuronal electromagnetic fields that define the mental space. Although neurons related to different aspects of nociception (that is, pain-related information processing) have been identified, how a complex integrated sensation of pain (involving a variety of factors, including a range of emotions) emerges from a distributed neural circuitry formed by multiple cortical and subcortical structures remains elusive. For example, it is not possible to elicit the complete range of sensations and emotions associated with pain experiences by electrically stimulating any particular cortical region that has been identified as being implicated in the genesis of pain.

According to the relativistic brain theory, the difficulty in pinpointing a precise source for pain sensation results from the fact that pain, or any other complex mental or cognitive function, emerges as a result of widely distributed interactions of neural tissue and the electromagnetic fields generated by them. In this relativistic terminology, the sensation of pain results from the seamless combination of multiple factors (location, intensity, stored memories of previous nociceptive stimuli, and emotional state). Thus, by assuming that pain emerges in the analog component of the brain as a result of neural digital signals and mnemonic traces that combine to generate particular electromagnetic fields, we can identify a mechanism through which a subject's emotional, contextual, and historical factors could play such an important role in modulating incoming nociceptive signals from the body periphery and define why the same peripheral nociceptive signals don't always generate the same subjective experience of pain.

Other clinical findings also support the existence of an analog component of brain processing. For example, an interesting set of phenomena, known collectively as alterations of the body schema, is consistent with the relativistic brain theory and the potential physiological role played by neuronal electromagnetic fields. The most well known of these phenomena is the phantom limb sensation, already discussed in chapter 3. This phenomenon refers to the ubiquitous finding that patients who suffer the loss of a limb tend to continue to experience its presence. Most amputees not only feel the presence of the missing limb but also report the presence of excruciating pain in a limb that does not exist any longer.

During the Walk Again Project, I came in contact again with the phantom limb phenomenon. That happened because all paraplegic patients who enrolled in our training protocol experienced phantom sensations in their lower body as soon as they began practicing how to use a brain-machine interface to control the leg movements of an avatar soccer player. For this first phase of training, patients were immersed in a virtual reality environment that allowed them to use their EEG activity to control the walking of an avatar soccer player, while they received synchronized visual and tactile feedback describing this virtual stroll on a soccer field. Visual feedback was delivered via a virtual reality goggle, while tactile information describing the moment the avatar's feet contacted the ground was presented by stimulating the skin surface of the patients' forearms. As they interacted with this brain-machine interface and virtual reality apparatus, all subjects experienced a clear sensation of having legs again. In fact, they reported that they felt their legs moving and touching the ground, even though their legs had remained paralyzed and only the avatar player moved. This was a major surprise for us, given that the tactile feedback described was delivered to the patients' forearms. Somehow, by seeing the avatar player walking on a virtual soccer field while experiencing coherent tactile feedback on their forearms that described the precise contact of the avatar feet with the ground, the brain of our paraplegic patients basically synthesized a vivid phantom sensation. In some cases, this sensation moved our patients to tears because of the emotion of feeling they were walking again on their own.

Going in the totally opposite direction, patients suffering from a high-order cognitive deficit known as hemispatial neglect ignore and fail to act on space located on the opposite side of a lesion of the parietal lobe. Hemispatial neglect occurs most often in patients who have suffered large lesions of the right cerebral hemisphere. Following a major stroke or traumatic lesion of right parietal areas, patients do not recognize the left side of their bodies nor the external space around it. As a result, neglect victims are easily identified since they tend to leave the left side of their bodies undressed and uncared for. Moreover, when asked to make a left turn and enter a door while walking in a long corridor, these patients usually walk a bit further, turn right, and then, by the time they reach the designated door, turn right again to comply with the instruction. If asked to draw a clock they can see on a wall in front of them, neglect patients can draw a closed circle, but then they proceed to clump all numbers representing hours on the right half of the drawing.

Another fascinating example, the rubber hand illusion, in which normal subjects report that a mannequin's hand feels like their own biological hand, also lends support to the relativistic brain theory. This illusion is produced by first occluding one of the subject's hands from her view and then placing a mannequin arm and hand in front of the subject. Next, both the subject's occluded hand and the mannequin's hand are touched synchronously for a period of three to five minutes by the experimenter. When the experimenter stops touching the occluded subject's hand but continues to touch only the mannequin's hand, most subjects experience the mannequin's hand as their own.

The phantom limb sensation, hemispatial neglect, and the rubber hand illusion suggest that the brain contains an a priori internal and continuous body image that can be reshaped very quickly as a function of the subject's experience. This internal body representation would account for all the peculiar sensory and affective ways in which we experience having a body of our own. The Canadian neuroscientist Ronald Melzack named this body image the *neuromatrix* and proposed that some of its foundation was defined by inherited genetic factors. Yet Melzack did not elaborate on what could be the potential neurophysiological mechanism that maintains this internal neuronal-based image of the body from the time we are born until we die.

Since obviously no peripheral tactile or proprioceptive inputs are generated by either an amputated limb or stimulation of a mannequin's rubber hand, the classic explanation proposed by Hubel and Wiesel for the genesis of perception cannot account for these phenomena at all. That happens because their classic theory presupposes that in order to perceive any fine somatosensory sensation, pain, or movement emanating from one's limb, corresponding tactile, nociceptive, or proprioceptive signals have to be generated in the limb itself and then transmitted by peripheral nerves and sensory pathways to our brains. Once there, key sensory features are first extracted from these inputs and later somewhat bound into a whole perceptual description of the limb. Hubel and Wiesel's theory also does not account for what binding mechanism would be employed to achieve this task of creating a whole, multidimensional perception of an object or our own bodies. Because this prerequisite is absent in phantom limb, hemispatial neglect, and the rubber hand illusion, another explanation needs to be put forward to account for these illusions. Furthermore, nothing in the Hubel and Wiesel model accounts for the amalgamation of the multiple sensory and affective sensations we normally use to describe our sense of self.

In my view, the many phenomena associated with the existence of a body schema (and the sense of self) in our brains can be described only as a brain-derived expectation—an analog mental abstraction—of the subject's own body configuration, which despite having its primordial roots in our genetic inheritance needs to be actively updated and maintained throughout our lives. According to this view, the brain internally generates an expectation of what the subject's body should contain, based on the combination of stored memories initially laid down by our genetic endowment—that is, a body having two arms and two legs—and perceptual experiences accumulated throughout our lives. During every moment of our lives the brain is continuously testing the accuracy of this internal body image—contained in the brain's own point of view—by analyzing incoming sensory signals that are transmitted continuously from our body to the central nervous system. As long as this body image is confirmed by peripheral signals, everything is fine, and we experience our bodies as a whole. But after there is a dramatic change in the flow of peripheral sensory information (for example, when a limb is amputated or occluded from vision), a mismatch occurs between the neuron-based body image held by the brain and the real physical configuration of the body under those conditions. As a result of this mismatch, amputees experience a vivid sensation emanating from a limb that does not exist anymore or, in the case of the rubber hand illusion, the mannequin hand feels like their own. Lesions of a component of the cortical circuit responsible for generating this body expectation, as in the case of hemispatial neglect, will alter profoundly what we believe are the physical borders of our own body.

In the case of the rubber hand illusion, the initial conditioning phase likely induces the subject to experience a subsequent isolated touch of the rubber hand as if it were delivered to the skin of his own biological hand. That may happen because during the conditioning phase, the subject could see the brush touching the rubber hand and feel the tactile stimulation of his hand, which was occluded from his view. That creates a visual-tactile association that can now be triggered every time the rubber hand is touched in isolation. We found support for this hypothesis by showing that individual neurons in the primary somatosensory cortex of monkeys trained in an equivalent task become responsive to visual inputs after the conditioning phase included synchronous stimulation of a virtual arm and the monkey's own limb. Before this conditioning these cells did not respond to visual inputs, only tactile signals coming from their arms.

Overall, the relativistic theory proposes to explain these phenomena by postulating that the sense of self and the body image arise from a widely distributed electromagnetic field generated by the many cortical and subcortical structures involved in the definition of the brain's body schema.

Preliminary and encouraging support for the hypothesis that neuronal electromagnetic fields may be involved in the definition of highly complex cognitive functions, such as the definition of the body schema and the generation of pain sensation, comes from a growing literature on the application of low-frequency (typically 1 hertz) transcranial magnetic stimulation (TMS) to the cortex of subjects suffering from phantom limb sensation/pain, hemispatial neglect, and chronic neuropathic pain. To my delight, TMS has also been applied to the cortex of subjects experiencing the rubber hand illusion. Briefly, this literature indicates that such simulation applied to different cortical areas can reduce phantom limb pain in a significant number of subjects. Transcranial magnetic stimulation applied to the left parietal cortex has also been reported as producing a clinical improvement of left hemispatial neglect. Moreover, stimulation applied to a region at the border of the occipital and temporal lobes has been shown to clearly exacerbate the rubber hand illusion when compared to sham stimulation. Finally, transcranial magnetic stimulation has also been implicated with the improvement of neuropathic pain.

Interestingly, there is growing evidence that TMS can act at multiple levels in the brain: genetic, molecular, synaptic, and cellular. And although most researchers believe that TMS effects are mainly mediated by induction of electrical current on neurons, the possibility that TMS also exerts a direct magnetic effect on neuronal tissue cannot be ruled out. Such an effect would follow the notion that induced magnetic fields can act on physical, chemical, and biological systems. Indeed, a 2015 review article on TMS by Alexander Chervyakov and colleagues published in *Frontiers in Human Neuroscience* raised the interesting idea that low-frequency electromagnetic waves produced by this technique could affect brain tissue at quantum, genetic, and molecular levels simultaneously. According to this proposal, since large molecules and even cell organelles are known to be deformed or oriented by magnetic fields, TMS could modulate or even alter multiple neuronal functions mediated by them. This would be particularly crucial in the case of protein complexes, which are known to be involved in essential brain functions such as plasticity, learning, and memory acquisition, storage, and maintenance. This latter possibility is very relevant and plausible, given that the effects produced by TMS can last for as long as six months after the end of the treatment. That basically means that TMS application can trigger long-term plastic changes on neuronal circuits, a highly relevant finding for our present discussion.

Although the potential existence of direct magnetic effects of TMS in the brain gives credence to my view that our internal body image is shaped by analog processing, the discovery that TMS can induce neuronal plasticity supports the hypothesis that neuronal electromagnetic fields could also have a causal efficiency effect on neuronal tissue. That would happen because such electromagnetic fields play a key role in the process of physically embedding Gödelian information into neuronal circuits. If confirmed by further experimentation, this concept would also raise the notion that the neurophysiological processes through which our memories are laid down involve some sort of electromagnetic etching of neuronal tissue. Indeed, I envision that this process could happen through the widespread influence that electromagnetic fields would have in synchronously modulating the three-dimensional structure—and hence the function—of a large number of intracellular neuronal and synaptic proteins all over the cortex. By acting simultaneously throughout the cortical mantle, such a mechanism could account for up and down modulations in the number of synapses and their individual synaptic strength. Moreover, this mechanism could explain the well-known nonlocality nature of our memories which, instead of being stored in a single restricted location, are typically distributed throughout vast regions of the neocortex.

Conversely, electromagnetic fields could also participate in the readout of these memories and their translation into widely distributed spatiotemporal patterns of neuronal electrical activity. Thus, through the process of induction, electromagnetic waves carrying high-dimension Gödelian information would project into low-dimension Shannon information (see figure 3.2), defined by streams of neuronal electrical pulses that can be readily translated into body movements, language, and other forms of communication that rely mainly on digital signals.

I believe that the notion that long-term memories are stored in a distributed fashion across the cortical tissue is much more easily explained by an analog-digital model than by a purely digital one. Indeed, without the existence of the analog brain component, it would be very difficult to explain how cortical circuits, characterized by complex micro-connectivity that is continuously changing, could recall the type of precise information needed for memories to emerge, virtually instantaneously, throughout one's life.

The potential role of neuronal electromagnetic fields in embedding Gödelian information into neuronal tissue is also consistent with the pervasive notion that one of the main functions of the sleep cycle is to help consolidate memories acquired during the previous period of wakefulness. Overall, one can identify a variety of highly synchronous neuronal oscillations in the electroencephalogram (EEG) as subjects move through the different phases of the sleep cycle. As we fall deeper into sleep, a high-amplitude, slow-frequency cortical oscillation (0.5–4.0 hertz) known as delta waves pass to dominate the EEG. Such slow-frequency oscillations are widely considered a fundamental component of a mechanism to downscale and remove undesired metabolites. During the night, slow-wave sleep episodes are followed by brief periods of rapid eye movement (or REM) sleep in which cortical activity is dominated by fast gamma neuronal oscillations (30–60 hertz) that resemble those observed when we are fully awake. It is during REM sleep episodes that we can experience dreams. REM sleep has been associated with memory consolidation and motor learning. According to the relativistic theory, during the sleep cycle, neuronal electromagnetic fields may not only provide the glue needed to establish the different states of widespread brain synchrony, they may also offer the driving force needed to lay down memories by contributing to the process of consolidating or eliminating synapses made during the course of the day. In this view, dreams would emerge as one of the by-products of the operation of the analog-digital computation engine that is responsible for the fine sculpturing of neuronal microcircuits every night of our lives, as a way to maintain and refine our mnemonic records.

All in all, the relativistic brain theory proposes a new biological mechanism—recursive analog-digital computing—for the generation of highly complex and likely noncomputable human cognitive skills such as intuition, insight, creativity, and problem-solving generalization. The so far insurmountable difficulty scientists working in artificial intelligence have experienced for the past half a century in trying to emulate any of these basic human cognitive functions in digital platforms offers a testimony as to why I refer to them as noncomputable entities. As discussed in chapter 6, I propose that these and many other unique human mental attributes can neither be reduced to an algorithm formulation nor simulated or mimicked in any digital system. As such, the establishment of a recursive analog-digital computational strategy, combined with the capability of physically embedding Gödelian information, which exerts causal efficiency on neuronal tissue and can be readily projected into Shannon outputs, may be part of the neurophysiological mechanism behind the emergence of such mental capabilities in our brains.

Altogether, the existence of an analog domain endows the animal brain with yet another level of plastic adaptation capability. Indeed, if electromagnetic fields can fuse the cortex into a neuronal continuum, in principle any part of the cortex could be recruited for mediating, at least partially, a particularly demanding task. For example, when humans go blind, either temporarily or permanently, their visual cortex is quickly recruited—in a matter of a few seconds or minutes—to process tactile information, particularly when they begin learning to read Braille's embossed characters by rubbing their fingertips on top of them. If this were purely a matter of new connections forming between previously disjointed neurons, it would be difficult to explain the fast rate at which the visual cortex is put to a new purpose. Indeed, this could not be achieved at all if our central nervous system relied simply on a digital mode of operation and transmission of Shannon information through streams of action potentials conducted by our nerves. By adding the analog mechanism I am proposing here, represented by the action at a distance—at light speed of neuronal electromagnetic fields—the human brain may have acquired a powerful extra level of flexibility and redundancy to perform such tricks, just in the nick of time.

According to the relativistic brain theory, the perceptual experiences taking place during wakefulness require the full engagement of the brain's biological solenoids at a high synchronous frequency in order to generate the complex combination of electromagnetic fields within our brains that ultimately account for the richness and unpredictability of our conscious experiences. The immaturity of such neuronal electromagnetic fields in early postnatal life in humans may also explain why a clear sense of self does not develop in babies until a few months after birth; that would be the time required for enough of the brain's white matter to mature to a level capable of generating electromagnetic fields sufficiently strong to bind the brain into a neuronal continuum, from which a sense of self materializes and Descartes's central motto of our kind, *Cogito ergo sum*, can emerge.

Major disruptions in the normal operation of the neuronal space-time continuum may also explain why a large variety of brain disorders can be observed in humans. Just as the normal functioning of the brain relies on proper levels of brain synchronization, most, if not all, brain disorders may result from pathological hyper- or hyposynchronization of different spatial components of the neuronal space-time continuum. This is not to say that there are not some genetic, metabolic, or cellular factors responsible for triggering these pathological neurophysiological states, but that the main signs and symptoms of any brain disorder may result from improper levels of neuronal synchronization between some regions of neuronal continuum that define the central nervous system. For example, further work carried out in my laboratory and others during the past decade has revealed that Parkinson's disease involves the emergence of chronic mild epileptic-like neuronal activity, characterized by pathologically high synchronous neuronal firing in the beta range (12–30 hertz). These abnormal neuronal oscillations have been observed throughout the motor circuit formed by the frontal cortex—where the motor and premotor cortical areas are located—and the basal ganglia and thalamus.

As a result of this discovery, in 2009 my lab published a paper in *Science* (see Fuentes et al.) showing that if one delivers high-frequency electrical stimulation through a microchip chronically implanted on the surface of the spinal cord, one can significantly reduce the type of Parkinson-like movement freezing observed in rodents (mice and rats). In these experiments, genetic or pharmacological manipulations were used to induce severe depletions of dopamine that lead to the clinical manifestation of Parkinson's disease.

These experiments showed that before the electrical stimulation was delivered, animals could not move at all as a result of the body freezing that occurred simultaneously with the presence of a widespread beta frequency hypersynchronization in their motor system. However, as soon as the microchip was turned on, high-frequency electrical signals were transmitted through the spinal cord to the entire brain and the beta epileptic-like activity was disrupted. At once, the animals began to move as if they were perfectly normal. One of the most important findings of this study was that such electrical stimulation of the spinal cord did not have to be continuous to be effective. Indeed, just about an hour a day of treatment was enough to keep mice and rats moving for a few days or even a whole week.

Five years later we reproduced these findings in a primate model of Parkinson's disease. And since 2009, the effects of this potential new therapy have been investigated in close to fifty patients suffering from advanced Parkinson's disease that produced severe body freezing. With the exception of two cases in which the lack of therapeutic effect likely resulted from technical problems in properly adapting our method to human subjects, all other Parkinson's patients tested experienced significant improvements in locomotion and even in other cardinal symptoms of Parkinson's disease. This example illustrates well how useful a reinterpretation of the pathophysiology of Parkinson's disease, as well as other brain disorders, proposed by the relativistic brain theory may be the first step toward the development of future therapies for otherwise untreatable neurological and psychiatric disorders.

Since given that the spinal cord has never been implicated in the genesis of Parkinson's disease, our results were received with great surprise since, until then, all proposed nonpharmacological therapies for this disease involved electrical stimulation of motor structures, such as the basal ganglia, that are more closely involved with its genesis. Yet, if these initial clinical results are confirmed in larger, randomized clinical trials, spinal cord electrical stimulation could become a very important alternative to the current dominant surgical treatment for Parkinson's disease, known as deep brain stimulation. I say that because not only does spinal cord stimulation require a much easier, shorter in duration, and less risky surgical procedure, it has no major side effects associated with it. That means that any neurosurgeon could perform such implants without highly specialized training. Moreover, if necessary, such spinal cord implants can be easily removed. Finally, such spinal cord implants cost much less than deep brain stimulation, a factor that cannot be ignored these days.

Following this line of reasoning, over the past decade, my former graduate and postdoctoral student Kafui Dzirasa and I have shown that abnormal levels of neuronal synchronization may be present in animal models of several neurological and psychiatric disorders. Again, these observations were obtained in a series of experiments using transgenic mice and rat models of brain diseases. In any animal model of a brain pathology we chose to study—mania, depression, and obsessive-compulsive disorder—we invariably identified the presence of pathological levels of neuronal synchrony in different brain areas or even in entire brain circuits. These animal studies yielded strong support for the hypothesis proposed by the relativistic brain theory that a large number of brain diseases are nothing but expressions of disturbances in neuronal timing, also known in clinical neurology as focal, partial chronic epilepsy. Indeed, one upshot of this proposition is that it removes the strict classic borders that medicine typically draws between neurological and psychiatric disorders. Essentially, from the relativistic brain theory's point of view, these are all diseases of neuronal timing, so they should be lumped together simply as different types of brain pathologies.

In more technical terms, the relativistic brain theory proposes that the particular clinical signs and symptoms that characterize each brain disorder result from an improper (pathological) folding of the neuronal continuum that defines the mental space. By improper folding I mean the recruitment of a particular brain circuit—a spatial subcomponent of the entire mental space—into abnormal levels of synchronization. Neuronal hypersynchrony like the one seen in Parkinson's would result from an excessive folding of the mental space, whereas hyposynchrony would emerge due to its insufficient folding. Therefore, the introduction of the concept of a continuous mental space becomes useful in a very practical sense because it may allow us to import into clinical neuroscience the same type of mathematics—that is, non-Euclidean, Riemannian geometry—used by Einstein in his general theory of relativity. Combined with the principles of neural ensemble physiology described in the previous chapter, this effort my even allow us to create a particular algebra to describe the folding of the cortex in normal and pathological circumstances.

Thinking in these terms also makes it easier to explain why, in most cases, it is difficult to establish a clear differential diagnosis, particularly when we talk about the many types of psychiatric diseases that have been reported. Just as the normal functions of the brain involve interactions between cortical and subcortical structures, so too should the pathological functioning of the brain. Thus, the signs and symptoms that a given patient exhibits can be dispersed broadly among different types of psychiatric diseases. Thinking relativistically about the brain helps us see why we shouldn't expect to observe exactly the same set of clinical signs and symptoms in two different individuals. Instead, when compared to each other, the clinical symptoms of individual patients show a lot of variability, resulting in a broad range of emergent behavioral phenotypes. That would explain why it is so difficult to find typical "textbook" cases of the traditional classes of psychiatric diseases.

Further support for the notion that a significant number of neurological and psychiatric disorders derive from abnormal levels of neuronal synchronization comes from the well-known observation that several anticonvulsant drugs have proved effective in treating some of their clinical manifestations (for example, bipolar disease), even though there hasn't been a clear explanation for their usefulness besides empirical practice. The relativistic brain theory sheds some light on this phenomenon by suggesting that these drugs may be acting by reducing the underlying partial epileptic activity, produced by the pathological folding of the mental space, that may generate the patient's main symptoms.

Until now, I have restricted most of my description to the pathological changes in the levels of electrical synchronization in different brain circuits and how they can underlie some of the symptoms and signs experienced by patients suffering from one or another brain disorder. Within the framework of the relativistic brain theory, one needs to ascertain that these pathological levels of synchronous neuronal firing would also interfere with the generation of optimal neuronal electromagnetic fields. If they do interfere, then the main tenets of the relativistic brain theory could account for the profound disruptions in mood and sleep cycle, altered sense of reality, personality disturbances, hallucinations, delirium, and paranoid thinking that are well-known components of psychiatric disorders.

The potential clinical role played by the abnormal generation of electromagnetic fields can be illustrated by citing another example of a very prevalent brain disorder: autism. During the last decade, many brain-imaging studies have shown a considerable level of functional disconnection between multiple cortical areas in the brains of autistic children. This happens because of a developmental disruption in the establishment of long-range connections that link cortical areas that are far apart. According to the relativistic theory, therefore, the main symptoms of autism could emerge as a direct consequence of the disruption in the formation of the white matter coils, like the superior longitudinal fasciculus, that generate the electromagnetic fields responsible for the fusing of the cortical neuronal continuum. Such inability would lead to an improper lower level of cortical neuronal synchrony, or hyposynchrony (resulting from insufficient folding of the mental space). This is consistent with the theory that such a functional cortical disconnection accounts for the occurrence of the communication, cognitive, and social deficits that autistic children experience. I should mention, however, that autistic children also show a much higher than normal incidence of epileptic activity, which could take place locally, within individual cortical areas, perhaps as a result of the overall reduction in cortico-cortical connectivity.

Support for this latter view of autism was obtained in my lab in the last couple of years through the work of Bobae An, a postdoctoral fellow from South Korea. In these experiments, Bobae first observed that during courtship male mice tend to sing, like songbirds, complex ultrasound melodies to the female with whom they want to mate. By recording simultaneously from the brains of the female and male, Bobae observed the emergence of a complex pattern of synchronization between the two animals. Interestingly, this interbrain synchronization produces a wave that spreads from the back to the front of the animals' brains. Next, Bobae repeated these experiments using genetically modified male mice that exhibit social deficits, while interacting with regular females, that resemble those seen in autism. Bobae showed that these genetically modified male mice do not sing as much as normal mice, which may explain why they do not establish physical contact with the females. Interestingly, when Bobae recorded the brains of the socially dysfunctional male and the female simultaneously, she observed that there was no wave of interbrain synchronous activity flowing from the back to the front. This type of hyposynchronization could be exactly what happens in autistic children when they interact with their siblings or other people.

But if epilepsy is so prevalent and may be associated with most of the disorders of the central nervous system, why hasn't it been diagnosed through laboratory tests more often? It turns out that the universal method employed to diagnose epileptic activity, scalp electroencephalography, is known to be very good at detecting pathological levels of neuronal synchronous activity only at the most superficial region of the human brain, the cortex. If someone suffers from chronic, mild partial epilepsy that is restricted to deep subcortical areas of the brain, EEG will not be able to detect any electrical anomaly, at least in the early stages of the disease. These seizures would simply happen and be beyond the range of detection by the current electrophysiological technology applied to human subjects.

The same problem, however, would not apply to experimental animals. For example, in my lab we routinely implant tens to hundreds of hair-like micro-wire electrodes deep into the brains of mice, rats, or monkeys to measure patterns of neuronal activity that cannot be studied with EEG recordings. Using this approach, one can investigate whether mild partial seizures can be confined to subcortical territories and generate the kind of behavioral outputs observed in patients. That is how Kafui and I have discovered a variety of distinct mild epilepsy-like activity, occurring in different neuronal circuits, associated with different rodent models of brain disorders.

My theory that neuronal seizures define a common neurophysiological pathway for the clinical manifestation of most brain diseases is further supported by a series of clinical studies that showed that epileptic activity is often observed in patients suffering from Alzheimer's disease, one of the most prevalent brain disorders of our times. In a 2017 review of this literature, Keith Vossel and colleagues indicated that the presence of seizures may lead to an acceleration of the process of cognitive decline. Further support for such an association comes from the observation that use of low doses of anticonvulsant medication in Alzheimer's patients who exhibit EEG alteration compatible with epilepsy can be beneficial. If confirmed, altogether these findings could become a real game changer in the way we approach the development of future therapies for Alzheimer's. I say this because I am convinced that in the future neuroimplants, like the one we designed for Parkinson's disease, or advanced noninvasive techniques, such as TMS, instead of drugs, may become the therapies of choice to treat a larger number of brain disorders, including those today classified as being of the psychiatric variety.

Such a future can be anticipated by many recent encouraging developments and findings in the emergent field of neuromodulation. For example, currently there is a growing consensus that repetitive sessions of TMS applied to the dorsolateral prefrontal cortex are effective in improving the symptoms of chronic depression. Although not yet as effective as electro-convulsion therapy—still the most efficient method to treat severe cases of depression—TMS has been shown to be beneficial in a series of randomized trials. The main inconvenience of this novel approach, however, is that patients have to come often to a hospital or clinic to receive a TMS session under medical supervision. Because of this important limitation, I believe that our method for electrical stimulation of the spinal cord may quickly become an alternative for these patients too. Indeed, one preliminary study has shown already that it may actually alleviate the symptoms of depressed patients. Since this therapy is delivered without the need of medical supervision through a chronically implanted neurochip in the spinal cord, patients could receive their treatment through a continuous or intermittent (one hour per day, for example) stimulation protocol at home, without the need to come periodically to a hospital. By the same token, if the role of cortical seizures in Alzheimer's disease is confirmed, it is theoretically conceivable that spinal cord electrical stimulation could also be used to try to improve some of the cognitive deficits of these patients or even to delay the disease's progression.

Interestingly, there is also the possibility that in the future TMS applied to the spinal cord may be able to reproduce the findings we have obtained with our chronic implants. Indeed, I can envision a future scenario in which patients suffering from Parkinson's, depression, Alzheimer's, and many other brain disorders may be able to receive their daily therapy at home by sitting on a therapeutic chair whose backrest contains a portable TMS system. According to this view, while the patient sits comfortably and reads a book for an hour or so, the chair-embedded TMS system would deliver, in a noninvasive way, the required electromagnetic stimulation of the spinal cord needed for treating the patient's brain disorder. If such a future of home-based brain therapy one day materializes, one will witness huge gains in clinical management and in the quality of life for millions of patients suffering from a variety of brain disorders, not to mention an enormous reduction in costs in our health care system.

Ironically, if magnetic-based brain therapy ever reaches the kind of widespread acceptance I foresee, it will be fulfilling a widespread belief, born centuries ago, that lodestones—naturally magnetic rocks—carry in them some sort of magic therapeutic power. Such a view was summarized by Bartholomew the Englishman, who wrote in the thirteenth century: "This kind of stone [the magnet] restores husbands to wives and increases the elegance and charm of speech. Moreover, along with honey, it cures dropsy, spleen, fox mange, and burns. . . . When placed on the head of a chaste woman [the magnet] causes its poisons to surround her immediately [but] if she is an adulteress, she will remove herself from bed for fear of an apparition."

If Bartholomew could see how far we have come in using electromagnetism for medical ends, how astonished would he feel?

Before moving on, I want to recognize those who have explored neuronal electromagnetism before me as a potential substrate for a general theory of brain function. Several researchers in the past sixty years have raised the hypothesis that small neuronal electromagnetic fields may play a crucial role in human brain function. One of the first attempts to develop a field theory of the brain originated from adepts of the gestalt movement, who believed that one should investigate the brain holistically, not as a mosaic of individual parts, if one wanted to understand the neurophysiological mechanisms involved in higher cognitive functions. Based on this philosophical approach, in the early 1950s two distinguished gestalt psychologists, Aron Gurwitsch and Wolfgang Köhler, pioneered the idea that electric fields generated by large populations of neurons could contain the secrets to understanding human perception. Gurwitsch and Köhler's thesis was adamantly rejected by some contemporary American neuroscientists (such as Karl Lashley and the Nobel laureate Robert Sperry) who designed and carried out some animal experiments in the late 1950s aimed at disproving Köhler's claim. Although most modern psychology textbooks state that these latter experiments succeeded in refuting Köhler's thesis, as I look again at Sperry's and Lashley's own results sixty years later, I cannot see how this could be the case. Interestingly, neither could Köhler himself back in the 1950s. The reason for skepticism from Köhler and me is that neither Sperry's nor Lashley's experiments ruled out at all whether electromagnetic fields play any role in brain function. For example, in his experiments Lashley spread multiple gold strips over most of the surface of one monkey's brain. In another animal, he inserted a dozen golden pins into a restricted portion of the visual cortices in both hemispheres. Lashley postulated that these manipulations should shorten the electric fields postulated by Köhler and hence disrupt the monkey's ability to perform visual tasks. Lashley then proceeded to test the two monkeys, in a single session, on their ability to perform a very simple visual task they had learned prior to the gold strips and pin implants. Since the two monkeys performed as well as they had before, Lashley concluded that he had falsified Köhler's theory. Curiously, Lashley never bothered to use more difficult visual tasks or even to record any brain activity during these experiments. Sperry, although he was less confrontational about the interpretation of his own experiments, reported that cortical implants of tantalum pins in the brains of cats did not disturb these animals' visual perception.

Knowing what we know about the brain today, both such rough cortical manipulations tell us next to nothing about the relevance (or not) of electromagnetic fields in brain function. Simply put, contrary to Lashley's and Sperry's beliefs, the limited gold or tantalum implants used by these researchers would have no significant effect on brain electromagnetic fields. Therefore, one cannot conclude anything from their experiments. Curiously, for the past seventy years, any proposition of a new field theory of brain function is instantaneously dismissed by most of the neuroscientific community using these crude, flawed, and inconclusive experiments. Yet the idea continues to survive, albeit in the underground alleys of modern neuroscience.

Later on, in an attempt to explain the nonlocality aspect of memory, the American neuroscientist Karl Pribram, a former collaborator of Karl Lashley, proposed that the brain could work like a laser-produced hologram. In his model, local electric waves of cortical neuronal activity, generated primarily at the level of dendrites, would interfere with one another in order to store information in a series of regional holograms. According to Pribram, the cortex would contain, therefore, not one but many of these regional holograms, a patch arrangement known as holonomy. Based on this, Pribram's theory became known as the holonomic theory of brain function. In proposing this theory, Pribram was also heavily influenced by the work of the American physicist David Bohm.

It is also important to highlight that back in 1942, Angelique Arvanitaki showed that when a giant squid's axons were placed in close proximity to a medium with reduced conductivity, one axon could be depolarized by the activity generated in a neighboring nerve fiber. This became known as the ephaptic neuronal interaction. Recent studies have shown that similar interactions can be induced or modulated by applying electromagnetic fields to neuronal tissue.

In the 1990s, a distinguished American neurophysiologist at New York University, Erwin Roy John, rekindled interest in the role played by neuronal electromagnetism by suggesting that neuronal electromagnetic fields could push individual neurons already near the firing threshold to produce action potentials. E. R. John already believed that populations, not single neurons, are the functional entities that compute in an animal brain and, ultimately, generate a conscious being. Thus, for the brain to produce the type of perfect synchronization of a huge number of widely distributed neurons, a task needed to underlie all its major neurological functions, the only plausible solution would be to take advantage of its weak—but sufficient for the mission at hand—electromagnetic fields. Using those fields, perfect neuronal synchronization across the entire cortex could be attained very quickly. Many years ago, E. R. John sent me one of his latest reviews on this theme, which I rediscovered while researching this book. There I found very similar ideas to the ones presented here as part of my own theory.

About fifteen years ago, the molecular geneticist Johnjoe McFadden of the University of Surrey introduced what he called the conscious electromagnetic information theory, which proposes that consciousness and other high-order brain functions are determined by neuronal electromagnetic activity. McFadden published a series of articles detailing his theory and the innumerous findings from many other labs that may support it. Yet, as happened with Gurwitsch and Köhler, and then with E. R. John's ideas, the vast majority of the neuroscience community continues to dismiss any potential role for electromagnetic fields in brain function.

Electromagnetism is one of the four fundamental forces of nature. As such, electromagnetic fields are pervasively found everywhere in the cosmos, ranging in magnitude from the humongous gigateslas produced by a magnetar—a massive neutron star—to the microtesla field that envelopes the Earth, working as a protective shield, without which life on our planet would not be possible. At the edge of the heliosphere, the huge magnetic bubble that defines the range of the solar magnetic field, which extends beyond Pluto's orbit, the magnitude of the sun's magnetic field reaches a minimum value of one hundred picotesla. If you divide this solar minimum by a hundred you obtain a value close to the magnitude of the human brain magnetic field: one picotesla. It is no wonder, therefore, that very few neuroscientists ever bothered to consider such a tiny signal as potentially capable of playing any fundamental role in generating most, if not all, of our most cherished brain functions. Obviously, I do not believe that such a quick dismissal has been fully validated experimentally. Rather the opposite: the hypothesis that neuronal electromagnetism is essential for brain function remains as open as it was in the early 1950s. Therefore, I cannot stop wondering how absolutely stunned we will all feel if, one day in the near future, categorical experimental proof is obtained to demonstrate that all that it took to build the entirety of the human universe was a miserable one picotesla of magnetic power.

6 • Why the True Creator Is Not a Turing Machine

In the summer of 2016, a single-sentence tweet posted by the prestigious American magazine *Scientific American* startled me out of my midmorning torpor. The understated message read:

Artificial synapses could let supercomputers mimic the human brain.

A Korean materials scientist, Dr. Tae-Woo Lee, had told the magazine that, now that scientists could manufacture tiny transistors capable of mimicking neuronal synapses, the long-held dream of building brain-like machines was just around the corner. Exuding enthusiasm, Dr. Lee said that this development "could lead to better robots, self-driving cars, data mining, medical diagnosis, stock-trading analysis and other smart human-interactive systems and machines in the future." The article reported that thanks to the estimated 1 quadrillion connections that link roughly 100 billion neurons (the real number is closer to 86 billion), the human brain can execute about 10 quadrillion operations per second. In comparison, the fastest supercomputer in the world at the time, the Chinese Tianhe-2, could reach 55 quadrillion operations per second at peak performance. Of course, the Tianhe-2 needs 1 million times the energy of a human brain to operate. Understandably, the reason Dr. Lee was so enthusiastic was because his latest implementation of an artificial synapse needs only 1.23 femtojoules to produce a single synaptic transmission event—about one-eighth of what a human synapse requires. Therefore, Dr. Lee thought that by packing about 144 of these artificial synapses on four-inch wafers and connecting them with wires 200–300 nanometers in diameter, he and his colleagues would take a major step toward reproducing the operation of a real human brain. For that, he said, they only needed to wait for some three-dimensional printing advances to stack their wafers into tridimensional structures and, almost from nowhere, an artificial brain capable of surpassing our very own gray matter's computational skills would emerge.

It wasn't the first time the world had read a forecast of the imminent demise of our True Creator; similar claims have been regularly made since the beginning of the industrial revolution. Admittedly, none of these previous attempts came with a 1.23 femtojoules per synapse benchmark. However, for more than three centuries, whatever the most advanced technology of the day was—steam machines, mechanical devices, electronic gizmos and, since 1936, elaborate digital machines, including supercomputers made of thousands of interconnected microprocessors—prognosticators have claimed that specific skills of the human brain would soon be reproduced by man-made tools.

Inevitably, all these adventures failed miserably.

Nevertheless, since the dawning of the information age, there has been an inexorable buildup of the notion that digital computers will eventually supplant the human brain at its own game. Sometimes, judging by the fervor with which this claim is made, one gets the impression that its proponents believe their prediction is almost a divine prophecy, and that nothing will stop us from fulfilling it in the near future. And yet, despite one too many predictions made by futurologists and artificial intelligence practitioners and enthusiasts, no concrete evidence has been offered for what would be the most disruptive technological development in the history of humankind.

In place of such a categorical demonstration, what often is offered, particularly in the last decade or so, is a rather naïve argument, like the one that opens this chapter, that holds that to reproduce the complex mental capabilities of our brains one needs only to properly connect hundreds of billions of neuron-like and energy-efficient transistors and then press the power button.

I beg to disagree.

The notion that the intimate works of the human brain can be reduced to an algorithm and reproduced by digital logic has to be considered simply as yet another postmodern myth, an urban legend of sorts, or an example of the age of the post-truth, a time when a false or fake statement, because it is repeated so many times and disseminated so widely among the public, becomes accepted as true. The notion that complexity like that exhibited by our brains can be re-created simply by wiring up an extraordinarily large number of efficient electronic elements is not only very far removed from reality but, when examined in depth, has no credible chance of succeeding: not now, not ever.

Very few of those who believe in this view have stopped to think that human brains are the true creators of both digital hardware and software, not the other way around. This blind belief in the possibility that man-made technology can turn itself against its creator and surpass him basically states that a system of any sort—let's say, our human brain—can create something more complex than itself! However, what the proponents of this thesis cannot offer, other than simply relentless promoting of their almost religious belief, is a credible explanation of how this excess complexity would arise. My view is that such a proposition is clearly false since it violates many well-accepted logical theorems, including Kurt Gödel's two incompleteness theorems and a more recent formulation, known as the complexity theorem, proposed by the Argentinean American mathematician Gregory Chaitin. According to Chaitin, a formal system—like a computer program—cannot generate a subsystem—another program—that is more complex than itself. In a more formal version, described by John Casti and Werner Depauli in *Gödel: A Life of Logic, the Mind, and Mathematics,* Chaitin's complexity theorem could be formulated as: There exist numbers having complexity so great that no computer program can generate them.

Together, Gödel's work and Chaitin's, which are intimately related to each other, provide a clear logic barrier to the hypothesis that, if a human brain was a computer-like device, expressing complexity X, it could not actually generate something—like a super-intelligent artificial device—that exhibited a larger than X level of complexity.

Because the digital computer is the benchmark of this comparison, it is only fair that we begin this discussion by returning to the historical origin of this incredible machine. Every digital computer that exists today represents one of the huge variety of possible concrete implementations of an abstract computing device, originally proposed by the British mathematician and logician Alan Turing in 1936. Named in his honor the universal Turing machine (UTM), this mental construct still defines the operations of every digital machine from a laptop to the most powerful supercomputer on the planet. A universal Turing machine operates by using an internal table of instructions, programmed by the user, to sequentially read and manipulate a list of symbols contained on a tape fed to the machine. As it reads the symbols from the tape, one by one in a sequence, a Turing machine employs this internal table of instructions—or software—to execute a variety of logical operations and then write down its results.

Sounds simple, doesn't it? Yet, for better or worse, most of the technological breakthroughs of the past eighty years, including the emergence of the most disruptive mass communication tool of our species' history, the internet, can be considered as spin-offs of an abstract mental toy crafted in the depths of a genial mathematician's mind.

The original idea that all natural phenomena can be simulated on a digital computer largely borrows its credibility from a peculiar erroneous interpretation of the so-called Church-Turing conjecture, originally proposed by Turing and the American mathematician Alonzo Church. In essence, the conjecture says that if one can propose a series of well-defined steps to solve a given mathematical equation or problem, a procedure known as an algorithm, a digital computer can reproduce this operation and compute the solution to the same equation. This equation can then be classified as a computable function.

And here all the confusion starts.

Originally, the Church-Turing hypothesis was intended to focus exclusively on issues related to formal mathematical modeling. However, many authors since have interpreted the Church-Turing as if it would set a computation limit for all natural phenomena. Basically, these authors concluded that no physical computing device could exceed the capacity of a Turing machine. This may sound innocuous, but by simply ignoring that Turing's computability relates to questions arising only in formal mathematics, one risks producing a lot of trouble and misunderstandings. Indeed, when we focus on the debate of whether a human or other animal brain is simply a Turing machine, we soon find out that Turing's computation theory makes a series of assumptions that rule out its immediate applicability to complex biological systems like brains. For example, in a Turing machine the representation of information is formal—that is, abstract and syntactic, such as 1+1—rather than physical and semantic, as is the case for most biological systems. In brains like ours, a peculiar type of information, Gödelian information, is physically embedded in the neural tissue from which the central nervous system is made (see chapter 3). Semantics refers to the fact that even a simple phrase such as "You really robbed me!" can acquire many distinct meanings depending on context: it can be a joke among friends or a serious accusation. Humans can easily distinguish between these meanings, but a Turing machine, reliant on bits, would have serious problems dealing with this sentence.

Nevertheless, many computer scientists and neuroscientists have borrowed the Church-Turing assumption as their main theoretical justification to propose that any animal brain, including our own, can be reduced to an algorithm

and simulated on a digital computer. These scientists argue that the successful approach of using simulations for the study of mechanical systems can
be seamlessly extended to the study of biological systems whose complexity
is far superior to any man-made device. This philosophical position is known
as computationalism, a term attributed to Hilary Putnam, who proposed it
in "Brains and Behavior" in 1963, and has been defended by many other philosophers, such as Jerry Fodor. Critics of computationalism regard this thesis
as a purely mystical view. Because so many people now think that brains are
like digital computers, the use of my organic computer definition to talk about
animal brains becomes particularly relevant for our discussion.

Taken to its extreme limit, computationalism not only predicts that the
entire spectrum of human experiences can be reproduced and initiated by
a digital simulation, it also implies that in the near future, because of exponentially growing computer power, machines could supplant the totality of
human mental capabilities. This latter notion, put forward by Ray Kurzweil
and others, has become known as the singularity hypothesis. In *In the Age
of Spiritual Machines: When Computers Exceed Human Intelligence*, Kurzweil
states a radical version of Church-Turing: "If a problem is not solvable by a
Turing machine, it is also not solvable by a human mind." The origins of this
kind of thinking, however, go back to the 1940s and 1950s, when several former colleagues of Claude Shannon at MIT, people such as Norbert Wiener and
Warren McCulloch as well many other distinguished scientists—John von
Neumann among them—began to take a broad look at the many disruptive
ideas that were popping out around them in order to forge a completely new
paradigm to define human intelligence and how information is processed by
the human brain. This movement was called cybernetics, and for the next
decade or so it provided the intellectual basis and rationale for what is known
today as the field of artificial intelligence.

As my colleague at Duke University N. Katherine Hayles discusses in her
extraordinary book, *How We Became Posthuman*, this group met at a series
of conferences, known as the Macy conferences on cybernetics, to forge a
completely new field. They mixed Claude Shannon's theory of information,
Warren McCulloch's model of individual neurons as information-processing
units, John von Neumann's new architecture for digital computers based on
binary logic and digital circuits, and Norbert Wiener's way of conceptualiz-
ing machines and human beings as members of the same class of autonomous, self-directed devices. According to Hayles: "The result of this breathtak-
ing enterprise was nothing less than a new way of looking at human beings.

Henceforth, humans were to be seen primarily as information-processing entities who are *essentially* similar to intelligent machines."

Humans suddenly also seemed to be made of bits, albeit a lot of them, and as such, their minds, their life history, their unique perceptual experiences and memories, their choices and tastes, their loves and hates, up to the very organic matter that made them, could be—and in due time would be—reproduced in machines. Future digital machines, cybernetics believed, would be able to upload, assimilate, replicate, reproduce at will, and above all, simulate and mimic all that is human. Such intelligent machines were not available at the time the Macy conferences took place (and, of course, they still are not), but as with today's prophets of artificial intelligence, some members of the cybernetics movement seemed to believe that this was just a matter of time and, primarily, a matter of proper technology development. Using a similar framework, many research programs, including the one known as strong artificial intelligence, which has come significantly short of fulfilling previous optimistic predictions, have emerged in the pursuit of creating brain-like machines or, at the very least, simulating the physiological behavior of entire animal brains using supercomputers, such as the IBM Brain Project and the European Union's Human Brain Project. In 1968 Marvin Minsky, head of the MIT artificial intelligence lab, had announced: "Within a generation we will have intelligent computers like HAL in the film *2001*." Certainly, his prediction did not materialize and Minsky has recently declared that brain simulation programs have very little chance to succeed.

Interestingly enough, as Hayles reveals in her book, Claude Shannon himself was not very keen on extrapolating his rather focused definition of information to other fields in which communication took place. As history proved, Shannon was absolutely right in issuing words of caution. After all, his definition of information was devoid of any account for meaning, context, semantics, or, for that matter, medium peculiarity. Furthermore, by relying exclusively on binary logic and a rigid digital syntax, which facilitated tremendously the implementation of algorithms into digital machines, Shannon also distanced its creation from the semantic-rich and context-dependent nature of human thinking and brain functioning.

In general, neuroscientists believe that higher neurological functions in both animals and humans derive from complex emergent properties of the brain, even though the origin and nature of these properties remain debatable. Emergent properties are usually considered as global system attributes, which do not result from the description of the system's individual components. Such emergent properties occur everywhere in nature where elements interact and coalesce to form an entity, such as a flock of birds, a school of fish, or a stock market. These entities are usually referred to as complex systems. Hence, the investigation of complex systems has become the focus of a large spectrum of disciplines: from the natural sciences, such as chemistry and biology, to the social sciences, including economics and sociology.

Animal brains are archetypical examples of complex systems. Complex brain behavior, however, extends across the brain's different organizational levels—that is, from its molecular, cellular, and circuitry scaffolding all the way to the entire nervous system as a whole. Thus, to be really precise in our modeling of a particular animal's brain, we should also include in the definition of its complexity the exchanges our central nervous system makes with external entities, such as the surrounding environment and other subjects' brains, since these also interact and continuously modify the particular brain under investigation.

As we saw in chapter 4, brains also exhibit plasticity; information acts in a causally efficient way in the human brain by reconfiguring its structure and function, creating a perpetual recursive integration between information and the blob of organic matter that defines our central nervous system. That is the reason neuroscientists usually refer to systems like the human brain as complex self-adaptive systems. Importantly, the very characteristics that define a complex self-adaptive system are the ones that undermine our capacity to accurately predict or simulate its dynamic behavior. For example, at the beginning of the twentieth century, the genius French mathematician Henri Poincaré showed that the emergent behaviors of a system composed of even a few interconnected elements—let alone tens of billions of hyperconnected neurons—cannot be formally predicted through the analysis of its individual members. In a complex system like the brain, individual elements dynamically interact with one another in order to generate new behaviors of the system as a whole. In return, such emergent behaviors directly influence the system's various elements. As such, the elaborate brains of animals, including our own, have to be viewed as *integrated* systems, a particular continuum that processes information as a whole and for which one can distinguish neither software from hardware, nor memory from processing.

In one of the most fascinating passages of her book, Hayles reveals that Donald MacKay, a British scientist, strongly defended this view, one in which the reception of information causes changes in the receiver's mindset, at the Macy conferences. Because of this causal efficiency effect, according to

Mackay, no comprehensive theory of information could preclude the inclusion of meaning. For that, MacKay postulated that the receiver's mental states had to be measured and the impact of information quantified, a deed that, as Hayles acknowledges, we can barely dream of accomplishing even today.

Instead, the way animal brains generate, represent, memorize, and handle information (see chapter 3) is significantly different from the way computer scientists normally conceptualize how various material realizations of the universal Turing machine, such as digital computers, handle computing through the employment of algorithmic programs (software) dissociated from the machine's hardware. In this new context, when one examines the brain's operations by using both a mathematical and a computational point of view, emergent behaviors cannot be fully reproduced via classical, syntactically abstracted software procedures running on fixed hardware. In other words, the rich dynamic semantics that characterize brain functions cannot be reduced to the limited algorithmic syntax employed by digital computers. That happens because emergent properties that simultaneously encompass different levels of the brain's physical organization, involving the precise coordination of billions of top-down and bottom-up interacting events, are not effectively computable in the context proposed by the Church-Turing conjecture. Instead, they can only be temporarily approximated by a digital simulation. And this is a crucial point because if one accepts that brains behave like *integrated* and self-adaptive complex systems, these digital approximations will immediately diverge from the natural behavior of a given brain. The end result of this divergence is that no matter how powerful a particular digital implementation of a Turing machine is—not even if it is the 55 quadrillion operations per second Tianhe-2 supercomputer—its internal logic will not allow the typical strategy used by modelers to reproduce the full complex dynamic richness that endows living brains, including our own, with their ultimate functions and capabilities.

In the monograph Ronald Cicurel and I wrote, we made several further arguments against the idea that the brain could be reduced to the actions of a Turing machine. We clustered the arguments in favor of refuting such a hypothesis into three main categories: evolutionary, mathematical, and computational.

Our evolutionary argument highlights a fundamental difference between an organism and a mechanism such as a digital computer. This point is often ignored, despite the fact that it is a pivotal issue in this debate. Mechanisms are engineered and intelligently built according to a preexisting plan or blueprint. That is why a mechanism can be encoded by an algorithm, simulated on a machine and, consequently, be reverse engineered.

Organisms, on the other hand, emerge as the result of a huge number of evolutionary steps happening at multiple levels of organization (from molecules to the entire organism), which do not obey any previously established plan or intelligent blueprint. Rather, these steps take place through a series of random events. Organisms, therefore, are closely related to their environment because they are continuously shaped by changes in the statistics of the external world. Given the ever-changing nature of the surrounding environment, this task can be achieved only by continuously using the data that organisms collect about themselves and the world to reshape and optimize the very organic matter substrate that defines such a living form and from which the information produced by it emerges. Without this perpetual expression of information causal efficiency, the organism would progressively disaggregate and die. As we saw in chapter 3, death occurs when an organism can no longer maintain full operation of its homeostatic mechanisms, leading the entire system to decay to thermodynamic equilibrium.

This is obviously true for the brain. Thus, the idea of substrate-independent, or disembodied, information cannot apply when considering the information flow within organisms. While in a typical Turing machine, information flow is provided by software or the input tape, which are independent of the hardware that defines the physical structure of a digital machine, in the case of organisms, and especially in the brain, information is truly embedded in the organic matter and information flow is handled at a large series of different organizational levels. In addition, information produced by an organism continuously modifies the very material substrate (neurons, dendrites, spines, or proteins) that has generated it. This unique process binds both organic matter and information in an irreducible single entity. Thus, Gödelian information in organisms is substrate-dependent, a conclusion that confirms the integrated nature of the brain and overtly exposes the unsurpassable difficulties of applying the software/hardware dichotomy to an animal's central nervous system. In fact, these differences clearly indicate why the brain has to be considered a completely distinct type of computing system: an organic computer.

John Searles exemplifies this by saying that one can simulate the chemical reaction that transforms carbon dioxide into sugar, but as the information is not integrated, this simulation will not result in the natural process of photosynthesis. In support of this view, Prigogine insists that dissipative systems, like animal brains, survive far from thermodynamic equilibrium. As such, these systems are characterized by instability and time irreversibility in information processing. Overall, that makes organisms resistant to standard deterministic causal explanations. Instead, they can be described only statistically, in probabilistic terms, as a process whose temporal evolution is not reversible at all scales. Conversely, C. H. Bennett has shown that a Turing machine can be made logically reversible at every step, simply by saving its intermediary results. This is generally called the irreversibility argument, which has been previously put forward by Selmer Bringsjord and Michael Zenzen.

Examining one aspect of this temporal irreversibility, the American paleontologist and evolutionary biologist Stephen J. Gould proposed a thought experiment that nicely illustrates the dilemma faced by those who believe that "reverse engineering" of complex biological organisms is possible through a digital deterministic platform. Gould named this the "tape of life experiment" and indicated that, if a theoretical tape containing the record of all evolutionary events that led to the emergence of the human species could be rewound and then let go again, the chances that playing this tape would generate the same sequence of events that culminated with the appearance of the human race would be equal to zero. In other words, since the tape of life would follow a path made of a huge sequence of random events that had never happened before in the history of Earth, there is no hope that the precise combination that originally gave rise to humankind, millions of years ago, could be reproduced. This argument also validates the claim I made in the beginning of the book about the chances that Mr. Spock's brain would in all likelihood differ significantly from ours. And hence, his own cosmological view of the universe.

Essentially, the logic behind the tape of life experiment strongly suggests that it is impossible to employ deterministic and reversible models to reproduce a process that emerges as a sequence of random events. Accordingly, any model running on a Turing machine (a deterministic entity) that intends to track the evolutionary path of our species would diverge very quickly from the real process from which our species emerged. This basically means that there is no way to reverse engineer something that was never engineered in the first place. Thus, paradoxical as it may sound, the proponents of the reverse engineering view, which is considered by some to be at the very edge of modern biology, may not realize that by assuming this theoretical position they are frontally challenging the most enduring framework ever conceived in their own field: Darwin's theory of evolution by natural selection. Instead, acceptance of the reverse engineering thesis would directly favor the notion that some sort of intelligent design scheme was involved in the process that led to the emergence of humans and their brains.

If the evolutionary argument against building digital replicas of the human brain has been utterly ignored until recently, to some degree the logical basis for the mathematical and computational arguments described below relies on the work of Turing himself and another genius, the Austrian mathematician and logician Kurt Gödel, in the 1930s. Gödel himself sustained the idea that his famous incompleteness theorems provided a precise and explicit indication that the human mind exceeds the limitations of Turing machines and that algorithmic procedures could not describe the entirety of the human brain's capacities. As Gödel noted, "My theorems only show that the mechanization of mathematics, i.e. the elimination of mind and abstract entities, is impossible if one wishes to establish a clear foundation. I have not shown that there are non-decidable questions for the human mind, but only that there are no machines that can decide all questions of number theory."

In his famous Gibbs lecture, Gödel also asserted the belief that his incompleteness theorems imply that the human mind far exceeds the power of a Turing machine: indeed, the limits of a formal system do not affect the human brain, since the central nervous system can generate and establish truths that cannot be proven to be true by a coherent formal system, that is, an algorithm running on a Turing machine. Roger Penrose's description of the first incompleteness theorem makes this clear: "If you believe that a given formal system is non-contradictory, you must also believe that there are true proposals within the system that cannot be proved to be true by the formal system."

Roger Penrose has maintained that the Gödelian arguments offer a clear indication of some kind of limitation for digital computers that is not imposed on the human mind. In support of Penrose's position, Selmer Bringsjord and Konstantine Arkoudas have given very convincing arguments to sustain the Gödelian thesis by showing that it is possible that the human mind works as what they call a hypercomputer, because the human brain can exhibit capacities—like recognizing or believing that some statement is truthful—that cannot be simulated by an algorithm running on a Turing machine.

The straightforward conclusion of all these statements is clear: the full repertoire of human mental activities cannot be reduced to digital systems running algorithms. They are noncomputable entities. Accordingly, the central premise of the singularity hypothesis can be totally falsified simply because no digital machine will ever solve what became known appropriately as the Gödel argument.

We don't need to rely only on logic to make the case. In *The Relativistic Brain*, Ronald and I list mathematical and computational objections to counter the thesis that our brains will be soon surpassed by digital machines. What follows is a summary of our argument.

Building a digital simulation relies on many preconceptions and assumptions, such as the type of information representation involved. Moreover, various obstacles have to be overcome. These assumptions may at the end completely invalidate the model. For example, let us consider any physical system S whose evolution we want to simulate. The first approximation is to consider S as an isolated system. At once we hit a wall since in real life, biological systems cannot be isolated from their surroundings without losing many of their functionalities. For instance, if S is a living system, at any given moment its structure is totally dependent on its exchange of matter and information with its environment. S is an integrated system. Considering S as an isolated system, therefore, can completely bias the simulation, especially when a living system such as the brain is considered. That constraint would, for instance, invalidate any attempt to build a realistic model of a living adult mouse brain based on data collected from experimental preparations, such as brain slices obtained from juvenile mice. Such an experimental preparation dramatically reduces the true complexity of the original system and destroys its interactions with the surrounding environment. Translating results obtained from such a reduced model to the real behavior of a living brain is simply meaningless, even when the model yields some trivial emergent behavior, like neuronal oscillations.

This is only the first of a series of vital problems in applying the classical reductionist approach to understand a complex system such as the human brain. As you reduce the system to smaller and smaller modules, you basically destroy the intimate core of the operational structure that allows the system to generate its unique level of complexity. And without being able to express its inherent complexity, whatever is left is useless to explain how the entire system actually works.

The next step in a computer simulation involves selecting the data measured directly from S, knowing that we are neglecting a wide variety of other data and computations at different observation levels of S. By option or necessity, we usually consider all these other data to be irrelevant for a particular simulation. But with an integrated system such as the brain, one can never be certain that some further observation levels—say, a quantum description of the system—are truly irrelevant. Thus, we are certain to use a very incomplete sample of S to run our simulation.

Once observations or measurements are made about the behavior of a given natural phenomenon related to S, one then tries to select a mathematical formulation that can fit the selected data. As a rule, this mathematical formulation is defined by a set of time-dependent differential equations. Differential equations were developed primarily for applications in physics and they do not necessarily apply well to biological systems. Furthermore, it is important to emphasize that in most cases this mathematical formulation is already an approximation that does not render the natural system completely at all its many organizational levels. Besides, most physical processes can at best be approximated only by a mathematical function. If this comes as a surprise to you, you are not alone. Most people who believe that computer simulations can reproduce any natural phenomenon in the universe are, to my astonishment, not aware of this simple fact.

Next, we need to try to reduce the chosen mathematical formulation to an algorithm that can run on a digital machine. Altogether, that means that a computer simulation is an attempt to simulate the mathematical formulation of a set of observations made of a natural phenomenon, not the whole natural phenomenon itself. Because the evolution of a biological system is not governed by the binary logic used in a digital computer, the outcome obtained by a computer simulation may, in many circumstances, evolve very differently than the natural phenomenon itself. This is particularly true when one considers complex adaptive systems where emergent properties are essential for the proper operation of the whole system. Thus our algorithmic approximation may diverge quickly from the real behavior of the natural system, yielding only nonsensical results from its very beginning.

For example, most models that claim to have created artificial life employ combinations of various algorithmic techniques, from object-oriented to process-driven programming to iterative grammars, in order to try to mimic human behavior. According to the evolutionary computer scientist Peter J. Bentley, this is a flawed strategy because "there is no coherent method to correlate these programmer tricks with biological entities. As such, this approach results in opaque and largely unsustainable models that rely on subjective metaphors and wishful thinking to provide relevance to biology."

These issues are not limited to biology. The mathematician Michael Berry gives a simple example to illustrate the difficulties related to simulating any physical system, even one as apparently simple as a pool game. Calculating what happens during the first impact of the billiard ball is relatively simple. But estimating the second impact gets more complicated because one has to be more precise in estimating the initial states in order to get an acceptable estimation of the ball's trajectory. Things only get worse from there on. For instance, to compute the ninth impact with great precision, you will need to take into account the gravitational pull of somebody standing near the table. If you think this is bad, wait until you confront what is needed to compute the fifty-sixth impact—for that, you will have to take into account every single particle in the universe.

Another interesting way of illustrating the limitations of predicting the behavior of complex systems, particularly biological ones, is given by the now widely used approach known as big data. For the past few years, we have been bombarded with the idea that if one could build extremely large databases containing huge amounts of data about a particular domain, one could, by using machine-learning algorithms, predict the future behavior of the same system with a great level of accuracy. There is a huge amount of literature on the subject, so I will not have space to cover it completely here. But I do want to point to two apparent failures of the big data approach: in election prediction and the management of baseball teams.

During the 2016 U.S. presidential elections, tens of millions of dollars were thrown into creating big data–based systems that were supposed to predict the winner of the election even before the votes were cast, let alone counted. By the time millions of people had voted and the polls were closed on the U.S. East Coast, multiple traditional media outlets, including the *New York Times*, *CNN*, and the three major U.S. television networks, began disclosing the predictions of their big data systems which, almost unanimously, pointed to a landslide win by Hillary Clinton, the Democratic Party's candidate. As we all know by now, Donald Trump won the election in one of the most unexpected upsets in the history of U.S. presidential elections. The media's waffling on Trump's imminent victory was even more flagrant and humiliating than the famous headline the *Chicago Tribune* printed on its front page the morning of November 3, 1948—"Dewey Defeats Truman"—which erroneously proclaimed that Thomas Dewey had beaten the incumbent president, Harry Truman, when in reality the opposite had happened.

But how could these powerful media organizations and all the money they invested in big data have made predictions that were as bad as or even worse than the one made in 1948? Although the details are not known at the time I write this chapter, what happened illustrates very well the core problem of the big data approach: all predictions made by these systems assume that a future event will reproduce the statistics of past events used to build the big data database and the correlations derived from them. Predictions made by these systems can be accurate only as long as future events do not behave differently from the ones that preceded them. However, in highly volatile complex dynamic systems, big data predictions can easily become useless since the relevant variables are either different from past events or are interacting in a completely distinct way. As we know from experience, human social groups perfectly fulfill the definition of a highly volatile complex system, so there is little reason to expect that future elections will turn out as past ones did.

In the United States, the big data approach became very popular when the movie *Moneyball*, starring Brad Pitt, became a box office hit in 2011. Based on the 2003 book *Moneyball: The Art of Winning an Unfair Game,* by Michael Lewis, the story chronicled the unorthodox approach used by Billy Beane, general manager of the major league but small-pocketed Oakland Athletics baseball team to build a competitive team. Beane became convinced that for the "A's," a small-market team, to compete with the juggernauts of the major leagues—the Yankees, the Red Sox, and my own favorite, the Phillies—he had to defy the industry's standard methods of identifying the most talented players so he could get the biggest "bang" for the fewest possible bucks. To achieve that, Billy Beane became a believer in sabermetrics, a big data–like approach pioneered by the baseball writer and statistician George William James, which relies on empirical analysis of baseball statistics to predict who will be the best players to form a winning baseball team. Going against the advice of his seasoned scouting team, Beane relied on the main conclusions of sabermetrics, which argued that on-base and slugging percentages were better predictors for selecting players who could form a high-scoring team.

It turns out that, under the tutelage of Beane and his faithful embrace of sabermetrics, the Oakland A's reached the playoffs in consecutive years (2002 and 2003). Soon after that, other teams began to emulate Beane's strategy. One can only imagine how much money—likely hundreds of millions of dollars—has been poured into this strategy since the U.S. major league teams decided that the twenty-first century's version of the game is all about statistics.

Curiously, what is often neglected in this discussion is that, like any other sport, baseball is not only about offensive power. Above all, pitching is key to victory and, as a 2017 article in the *Guardian* pointed out, that season the A's had great pitchers on the lineup almost every day. Defense matters too, as do tactics and player smarts and chemistry. Moreover, there are many other human factors in addition to playing performance that determine whether a team formed by talented players will "gel" into a championship team. I mention this because very little emphasis has been given to actually measuring whether there is a causal dependency between the parameters highlighted by sabermetrics and winning a major league championship, which I naïvely assume is the central objective of any team (although some owners may care only about making money).

It turns out that the Oakland Athletics did well against fierce opposition, but the team did not win anything. Neither did other teams that dropped a lot of cash to adopt this new approach. Granted, the New York Mets, whose management adopted Beane's approach a decade later, did win the 2015 World Series. Yet there is no real scientific proof that this or any other World Series was won because of the sabermetrics method. Once again, it seems that the "inevitability" was created more as an abstraction, like a zeitgeist, in the minds of those who adopted the big data approach than as a result of tangible evidence that proved a real phenomenon.

If elections and baseball are complicated enough processes to simulate and predict, the difficulties grow much worse when we deal with the dynamics of an 86-billion-neuron brain. In fact, it is obvious that when one considers a simulation of an entire animal brain, which requires exquisite coherence from billions of neurons and multiple levels of organization to exert its functions, the possibility of one's simulation diverging is overwhelmingly high.

Mathematics also poses a problem to simulating brains. The first issue that needs to be dealt with is computability. Computability refers to whether it is possible to translate a mathematical formulation into an effective algorithm that can be run on a digital machine. Computability is related to the possibility of generating an alpha-numerical construct and not to any physical property of the system. And here we hit a major wall: because most mathematical formulations of natural phenomena cannot be reduced to an algorithm, they are defined as noncomputable functions. For instance, there is no general procedure that allows a systematic debugging of a digital computer: there is no algorithmic expression of a function F that can detect in advance any possible future bug that may hamper the work of a computer. Whatever one does, the machine will always exhibit unexpected faulty behaviors that could not be predicted when the computer and the software were manufactured. This function F is, therefore, classified as noncomputable. As such, it does not pass the Church-Turing proposition that defines what kind of functions cannot be simulated by a Turing machine.

It is also well known that there is no such a thing as universal antivirus software. The reason for this is because the function F, whose output is all programs that do not contain a virus, is also noncomputable. The same type of reasoning also justifies why there is neither a universal encryption system on a digital machine nor algorithmic procedures to tell whether dynamical systems are chaotic or not.

Living brains are the same: they can generate behaviors that can be fully described only by noncomputable functions. Because a Turing machine cannot handle those functions, there is no possibility of a precise simulation of one on a digital computer.

The examples above represent just a minute sample of the pervasiveness of noncomputability in mathematical representations of natural phenomena. These examples are all consequences or variants of the famous halting problem, one version of which is known as David Hilbert's tenth problem. The halting problem asks whether there is a general algorithm that will enable us to predict whether a computer program will either halt at some point or run forever. Alan Turing demonstrated that there is no such algorithm. Accordingly, Hilbert's halting problem has since become the primordial model of noncomputable functions.

The halting problem means that there is no way of deciding in advance which functions are computable and which are not. This is also why the Church-Turing hypothesis remains a hypothesis: it could never be proved or disproved by any Turing machine. Actually, nearly all functions cannot be computed by a Turing machine, including the majority of functions that should be used to describe the natural world and, in Ronald's and my own view, those generated by highly evolved animal brains.

Being already aware of the limitations of his computing machine, in his PhD thesis published in 1939 Alan Turing himself attempted to overcome them by conceiving what he called an Oracle machine. The whole point of the Oracle machine was to introduce a real-world tool for reacting to what "*could not* be done mechanically*" by the Turing machine. After the Oracle responded, the Turing machine could resume the computation. Turing showed that some Oracle machines are more powerful than Turing machines. He concluded, "We shall not go any further into the nature of this Oracle apart from saying that it cannot be a machine."

Turing's statement is simply stunning. Even at the very onset of the digital information age, one of its founding fathers had already realized that computers were limited. Perhaps more shocking is the realization that, at that same moment, Turing had already convinced himself that the computation power of the human brain surpassed by far that of his own computation creation. As he put it, "The class of problems capable of solution by the machine can be defined fairly specifically. They are those problems, which can be solved by human clerical labour, working to fixed rules, and without understanding"—and, it turns out, an unlimited paper supply. In reaching that conclusion, Turing inadvertently launched the field of hypercomputing.

I should emphasize, however, that Turing himself never suggested that something like the Oracle could be built; he repeatedly insisted that intuition (a noncomputable human property) is present in every part of a mathematician's thinking. In saying that, he basically corroborated Gödel's own conclusion expressed in his theorems. For Gödel, when a mathematical proof is formalized, intuition has an explicit manifestation in those steps where the mathematician sees the truth of a formerly unprovable statement. Turing, however, did not offer any suggestion as to what, in his opinion, the brain was physically doing in a moment of such intuition.

Many decades after the introduction of the Oracle machine, Gregory Chaitin, working with his Brazilian counterparts Newton Carneiro Affonso da Costa and Francisco Antônio Dória, proposed a related idea: that "analog devices, not digital ones, can decide some undecidable arithmetic sentences." That would happen because analog computational engines physically compute, meaning they compute by simply obeying the laws of physics rather than by running a pre-given algorithm within a formal system. Put differently, in analog computers there is no separation between hardware and software because the computer's hardware configuration is in charge of performing all the computing and can modify itself. This is precisely what we have defined above as an integrated system.

According to Chaitin, da Costa, and Dória, analog devices could serve as the basis of hypercomputers, or "real-world devices that settle questions which cannot be solved by a Turing machine." These authors further suggest that the possibility of effectively building a prototype of such a hypercomputer, by coupling a Turing machine with an analog device, is only a matter of developing the proper technology. That means that the entire question can be reduced to an engineering problem. Now you know why my lab is actively testing this hypothesis by building a recursive analog-digital computing device, the neuromagnetic reactor, inspired by the main tenets of my relativistic brain theory: to test some of these ideas.

In this theoretical context, it is not surprising at all that integrated systems such as the brain do overcome the computational limitations of the Turing machine. Indeed, the very existence of animal brains can be used to disprove the "physical version" of the Church-Turing hypothesis. Inspected from this point of view, the human brain qualifies as a type of hypercomputer. By the same token, by linking brains to machines via brain-machine interfaces, one is also creating another type of hypercomputer, or brainets, when multiple brains are interconnected (see chapter 7).

There are other mathematical issues that influence the computability of biological systems. For example, early in the twentieth century Henri Poincaré demonstrated that complex dynamical systems—entities in which individual elements are themselves complex interacting elements—cannot be described with integrable functions, that is, derivative functions that can be integrated, allowing us to figure out relations between the quantities themselves. Such dynamical systems are characterized in terms of the sum of the kinetic energy of their particles, to which one has to add the potential energy resulting from the particles' (elements') interactions. In fact, this second term is responsible for the loss of linearity and integrability of these functions. Poincaré not only demonstrated the nonintegrability of these functions, he provided an explanation for it: the resonances (interactions) between the degrees of freedom (number of particles).

That means that the richness of the dynamic behaviors of complex systems cannot be captured by solvable sets of simple differential equations because their interactions will, in most cases, lead to the appearance of infinite terms. Infinite terms account for the daily professional nightmares of mathematicians since they cause a lot of trouble in their attempts to analytically solve equations.

As we saw before, animal brains are formed by intrinsically complex, self-adaptable (plastic) individual neurons whose elaborate connectivity and functional integration with billions of other cells add many other levels of complexity to the entirety of a nervous system. Furthermore, the behavior of each neuron at the various observational levels of a given neural circuit cannot be understood except with reference to the global pattern of brain activity. As such, even the most rudimentary animal brain fulfills Poincaré's criteria to be considered a complex dynamical system with resonances between different organization levels or composing biological elements (neurons, glia, and so on). In this context, one can say that it is highly unlikely that an integrable mathematical description can be found to describe the brain's operation overall.

Moreover, if one assumes that vital computations in the brain—indeed, the ones that are responsible for its emergent properties—are taking place, even partially, in the analog domain, as the relativistic brain theory proposes, it follows that a digitalization process would be capable of neither approximating the brain's physiological behavior at a precise moment in time nor predicting how it would evolve in the immediate future.

Poincaré also showed that dynamical complex systems can be very sensitive to initial conditions and are subject to instabilities and unpredictable behaviors, a phenomenon known today as chaos. In other words, to make a prediction with a digital machine, concerning the behavior of Poincaré's time-varying analog system, one would need to know precisely the initial state of the system and have an integrable computable function that can compute a prediction of its future state. Neither of these conditions can be met when we talk about brains.

Put in other words, the critical and unsolvable problem that any modeler aiming at reproducing any animal brain's behavior in a digital simulation faces is that, given the inherently dynamic nature of nervous systems, it is impossible to estimate precisely the initial conditions of billions of neurons at various organizational levels; every time a measurement is taken, the initial conditions change. Moreover, most of the equations selected to describe the brain's dynamic behavior would be nonintegrable functions.

In light of those constraints, typical simulations on a Turing machine, even if that machine is a modern supercomputer with thousands of microprocessors, are not likely to reveal any relevant physiological attributes of real brains. Essentially, such simulations will likely diverge from the dynamic behavior of real brains as soon as they start, rendering their results absolutely useless for the goal of learning something new about how brains operate.

Simulating the brain on digital machines also involves dealing with numerous nontractable problems. Tractability in a digital computation relates to the number of computer cycles required to conclude a given calculation as well as other physical limitations, such as available memory or energy resources. Thus, even if an algorithmic representation of a mathematical function that describes a natural phenomenon can be found, the computing time required to run a simulation with this algorithm may not be viable in practical terms, that is, it may require more than the life of the entire universe to yield a solution. These kinds of problems are known as nontractable. Since a universal Turing machine is capable of solving any problem that another Turing machine could solve, the simple increase in computing power or speed does not transform a nontractable problem into a tractable one. It can only make a better approximation in a given time.

Let's examine an example of a nontractable problem. Protein structures embedded in the neuron's membrane, known as ion channels, are fundamental for the transmission of information between brain cells. To enact their effects, proteins must assume a particular optimal tridimensional configuration. The final three-dimensional shape of proteins, achieved by a process known as protein folding, is a critical element for the proper function of neurons. It includes expanding, bending, twisting, and flexing of the amino acid chain that forms the protein's primary structure. Each individual neuron has the potential to express some twenty thousand different protein coding genes as well as tens of thousands of noncoding RNA. As such, proteins are part of the integrated system that generates information in brains. Let us then consider a simple protein formed by a linear sequence of about one hundred amino acids and suppose that each amino acid can assume only three different conformations. According to the minimum energy model normally used to attempt to estimate the three-dimensional structure of proteins, we would have to examine 3^{100} or 10^{47} possible states to reach a final result. Since the solution landscape of our protein folding model grows exponentially with the number of amino acids and with the number of considered conformations, this becomes a nontractable problem; if the protein should find its native state by random search, visiting one state each picosecond, the overall search could take longer than the current age of the universe.

Protein folding is an optimization problem, that is, it involves searching for an optimal solution in a landscape of possible solutions. This is usually expressed as the minimum or maximum of a mathematical function. Most optimization problems happen to fall in the category of intractable problems, usually named NP hard problems. NP problems are those problems for which solutions can be checked in polynomial time by a deterministic Turing machine. All the problems that complex brains are good at solving fall in this category. In simulations, these problems are generally dealt with by using approximation algorithms that could give near-optimal solutions. In the case of a brain simulation, however, an approximation solution would have to be found simultaneously at different organizational levels (for example, molecular, pharmacological, cellular, circuit, atomic quantum level), making the question even more complicated since optimizing a complex adaptive system often implies suboptimizing its own subsystems. For instance, by limiting the organizational levels that one considers in simulating an integrated system, as is traditionally done during a coarse simulation of a brain, one would likely miss crucial phenomena lying at lower levels of the integrated system, which can be critical for the whole system optimization.

This example illustrates well what Turing intended by a "real-world Oracle": in real life, a protein, an integrated biological system, solves the problem in milliseconds, whereas the algorithmic computer translation can take more time than the whole life of the universe to reach the same solution. The difference here is that the protein "hardware" computes the optimal solution and "finds" its three-dimensional configuration by simply following the laws of physics in the analog domain, while a Turing machine would have to run an algorithm created to solve the same problem on a digital device. Real-world organisms, being integrated systems, can handle their complexity in an analog way, a process that cannot be properly captured by a formal system, ergo, neither by algorithms.

Usually, tractable algorithms are designed as approximations in order to allow some estimation of future states of a natural system, given some initial conditions. This is, for instance, how meteorologists attempt to model the weather and make predictions whose probabilities of realization are known to rapidly decrease with time. In the case of simulations of the brain, the tractability problem becomes even more critical because of the huge number of interconnected neurons interacting in a precise time sequence. For instance, given that a digital computer has a clock, which runs on a step-by-step function, the problem of updating, in a precise timely order, billions or even trillions of parameters that define the current state of the brain becomes totally nontractable. Yet again, any further attempt at predicting the next state of a brain from arbitrarily chosen initial conditions will produce a poor approximation. Consequently, no meaningful predictions of the emerging properties can be obtained in the long run, even in time scales as short as a few milliseconds.

Again, if one accepts the notion that there is some fundamental aspect of brain function mediated by analog fields, like the neuronal electromagnetic field of the relativistic brain theory, a digital machine would neither be able to simulate these functions nor be capable of updating all the huge parameter space (billions or trillions of operations) in precise synchrony during the same clock cycle. In other words, a digital simulation would not generate any realistic brain emergent property.

At this point, it is important to observe that if one wishes to simulate a whole brain (that is, a dissipative, highly connected system interacting with the animal's body and the external environment), any processing speed that does not exactly match in real time should be automatically disqualified. A brain simulation running at a speed—even at a supercomputer speed—that is lower than the "real" environment to which it is connected and with which it will be in constant interaction will not produce anything similar to what a naturally evolved brain can produce or, for that matter, feel. For instance, a real animal brain must detect, in a fraction of a second, whether it is about to be attacked by a predator. If one's "simulated brain" reacts at a much slower speed, the simulation will not be of any practical use to understand how brains deal with the natural phenomenon of predator-prey interaction. These observations apply to a broad spectrum of brains in the phylogenetic scale, ranging from the most rudimentary brain of an invertebrate animal such as the nematode *Caenorhabditis elegans*, which contains only 302 neurons, all the way to the human brain, which is formed by up to 86 billion neurons.

All the objections raised in this chapter are well known and basically recognized as sensible and difficult to ignore, even among the practitioners of the field of artificial intelligence. Nevertheless, they insist on selling the utopia that digital machines will not only become able to simulate human-like intelligence but, eventually, exceed all of us at our own game, the game of thinking, behaving, and living as humans.

During my public talks, I usually use a hypothetical dialogue between a neuroscientist (N) and an artificial intelligence researcher (AIR) to illustrate the chasm that exists today between people who, like me, believe that our use of state-of-the-art technology for the betterment of humankind and the alleviation of people's suffering is welcome, and those who are working toward fulfilling Kurzweil's dystopian views of the future. The dialogue goes like this:

N: Tell me, how will you program the concept of beauty in a Turing machine?

AIR: Define beauty for me, and I can program it.

N: That is the central problem. I cannot define it. Neither can you, nor any human who has ever lived and experienced it.

AIR: Well, if you cannot define it precisely, I cannot program it. In fact, if you cannot define something precisely, it simply does not matter. It does not exist. And, as a computer scientist, I do not care about it at all.

N: Do you mean it does not exist? Or you do not care about it? Likely there are as many definitions of beauty as there have been living human brains in the history of our species. Each one of us, because of the different conditions in which our lives unfolded, has a peculiar definition of beauty. We cannot describe it precisely, but we know when we find it, when we see it, when we touch it or hear it. Is your mother or your daughter beautiful?

AIR: Yes, they are.

N: Well, can you define why?

AIR: No, I cannot. But I cannot program this personal and subjective experience of mine in my computer. Therefore, it does not exist or mean anything from a scientific point of view. I am a materialist. I cannot define precisely, in a quantitative or procedural way, what my experience of beauty is. It simply does not exist in my materialistic, scientific world.

N: Are you trying to tell me that just because you cannot quantify the sensation of encountering a beautiful face—the face of your mother or daughter—this sensation is meaningless?

AIR: Pretty much. Yeah. You got it right.

Terrifying as it may sound, many in our modern times have already decided that anything a Turing machine can't do isn't important, neither for science nor for humankind. I fear, therefore, that my hypothetical artificial intelligence researcher is not by any means alone in his prejudice. Worse, I am growing very afraid that by becoming so cozy and reliant on the way digital machines operate, our highly adaptive primate brains are at risk of emulating the way these machines work. That is why I believe there is a concrete possibility that if this trend continues, the True Creator of Everything may progressively decay and morph into some sort of biological digital machine, condemning our entire species to become moderately smart zombies.

7 • Brainets

Coupling Brains to Generate Social Behaviors

Nobody really knew what to expect from the long day ahead, neither the neuroscientists nor the subjects who congregated in our lab that morning in Durham, North Carolina. But even if the researchers could not anticipate the outcome of the experiment they had planned weeks before, they knew that the day promised to be unlike any before it. For starters, our scientific team would have to simultaneously handle three subjects, each one placed in a separate, soundproof room in our lab, unable to communicate with—indeed, unaware of the presence of—the other two participants. Nevertheless, to succeed in this new experiment these three pioneers would have to find a way to intimately collaborate among themselves, in a way that neither they nor anyone else had done before.

Ever!

To make things really interesting, we never gave the participants any instruction or hint about what kind of social interaction we were requiring of them. The only thing they knew during the hour they spent in the lab was that the job at hand was to move a virtual arm, which looked pretty similar to their own biological arms, projected on a computer screen placed in front of their faces. Once they realized they could move the virtual arm, their job was to make the virtual hand reach the center of a sphere that at the beginning of each trial appeared in different randomly chosen locations of the screen. Every time they completed this task correctly they would get a tasty reward, which each participant enjoyed very much.

Sounds pretty simple, right? It so happens, however, that the experiment was a bit more complicated. First, although each subject could see only a two-dimensional representation of the virtual arm, in reality that tool had to be moved in a three-dimensional virtual space if there was any hope of reaching the target. Second, they could not produce any overt movement of their own bodies to guide the virtual arm—using a joystick, for example—toward the target. Indeed, there was no joystick or any other mechanical or electronic actuator available for them to use—at least not with their own limbs. Instead, they could achieve this goal only if they learned a very different strategy: they had, quite literally, to use their collective brains' electrical activity to get this job done.

Such a feat was possible only because, for the past few weeks, these subjects had learned to interact with a new type of brain-machine interface created in our lab just for this experiment. However, the experiment we were about to run that afternoon introduced a substantial innovation to this now-classic paradigm. The original brain-machine interface enabled a single subject to learn, through a variety of feedback signals, how to control the movements of a single artificial device using his electrical brain activity alone. For several years, each of our three subjects had interacted with different brain-machine interfaces built in our lab. Indeed, each of them could be considered a world-renowned expert in operating such devices; after all, they had participated in enough studies on the topic to fill an extensive list of scientific publications in the field. But that day they were going to try to operate for the first time a shared brain-machine interface, their three distinct brains connected to a computer so that they could collectively move the virtual arm.

Years prior to that day I had named this shared brain-machine interface a brainet. I created this concept as a theoretical framework, thinking that it would take many years to actually try it out in a real experiment. Unpredictably, as is often the case with real experimental science, it turned out that a series of unexpected events allowed us to bring this idea into fruition in our lab around 2013. The first version of the brainet was tested in experiments carried out by one of my brightest postdoctoral fellows, the Portuguese neuroscientist Miguel Pais-Vieira. In a series of groundbreaking studies, Miguel was able to show that pairs of rats whose brains were directly connected could exchange very simple binary electrical messages (figure 7.1). In these experiments, one rat, called the encoder, performed a behavior, such as pressing one of two possible levers in order to get a food reward. Meanwhile, a second rat, called the decoder, received, directly in its somatosensory or motor cortex, a brief electrical message generated by the encoder's brain that described what this latter rat had just done. This electrical message informed the decoder what it had to do—that is, mimic the encoder's action—in order to receive a reward too. In

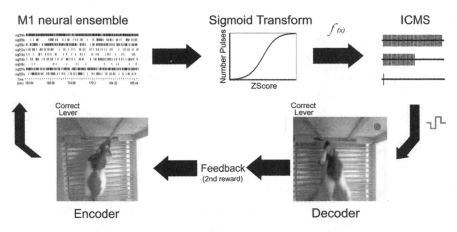

Figure 7.1. Experimental apparatus scheme of a brain-to-brain interface for transferring cortical motor signals. Arrows represent the flow of information from the encoder to the decoder rat. (Originally published in M. Pais-Vieira, M. Lebedev, C. Kunicki, J. Wang, and M. A. Nicolelis, "A Brain-to-Brain Interface for Real-Time Sharing of Sensorimotor Information," *Scientific Reports* 3 [2013]: 1319).

about 70 percent of trials, that is exactly what happened. The decoder decided which lever to press based on an electrical instruction that originated in the motor cortex of another animal (that is, the encoder rat)!

To add some spice to this first demo of brain-to-brain communication, Miguel Pais-Vieira performed some experiments in which the encoder rat was located in a lab of the Neuroscience Institute I created in the city of Natal, Brazil, in 2005, while the decoder remained in our lab at Duke in the United States. Using a regular internet connection, the communication between this rodent dyad worked as if the two subjects were next to each other.

By 2014, I had decided to test yet another brainet configuration, this time a shared brain-machine interface for motor control. To lead this work, I recruited Arjun Ramakrishnan, a brilliant young neuroscientist from Bangalore, India, who had joined my laboratory around 2012. Once we agreed on the general goal of this new experiment, Arjun and I got down to the business of detailing the key features that would be part of this first shared-brain-machine interface for motor control built in a laboratory. For starters, different from Miguel Pais-Vieira's approach, which called for a direct brain-to-brain link, we decided to employ a computer to mix up the raw electrical activity generated simultaneously by the three individual brains. In this particular arrangement, which we called a B3-brainet, each of the three subjects would be able to use her brain electrical activity to control only two of the three dimensions required to move an avatar arm properly in a virtual environment. For example, subject 1 would be in charge of generating a brain-derived virtual arm movement in the X and Y dimensions, subject 2 would be responsible for controlling the Y and Z dimensions of the movement, and subject 3 would account for the X and Z dimensions. A computer would combine the brain-derived inputs in order to guide the virtual arm through all three dimensions simultaneously. This meant that the only way our B3-brainet could move the avatar arm to the center of the spherical target was if at least two of these three subjects could perfectly synchronize their collective motor cortical electrical signals. If they were not synchronized, the arm wouldn't move. If their motor cortices got in sync, however, the computer would generate a continuous three-dimensional kinematic signal that would move the arm toward the target. And all this had to happen while the three subjects remained unaware of each other's participation in the experiment. To check on the accuracy of their own mental work, each subject would receive visual feedback, displayed on a computer screen, depicting the movements of the virtual arm in the two dimensions controlled by his individual brain. Finally, in each trial of this task, if the virtual arm intercepted the target in less than a predetermined period of time, each of the three subjects would receive a very enjoyable fruit juice reward.

Using these simple rules, we began to train the three subjects in earnest. The subjects easily succeeded in doing their individual jobs, controlling two-dimensional arm movement coordinates, but in the vast majority of the trials, despite doing their individual jobs correctly, they were not all in sync, so no correct three-dimensional avatar arm movement was produced. That meant no one got any juice. That was no fun, neither for the subjects nor for us. But even during those early training sessions, out of nowhere, two or even all three subjects would sometimes achieve perfect motor cortical synchronization and the arm would move and reach the target.

After three weeks observing this dynamic mental dance, spontaneously searching for the elusive sweet spot of the three-brain temporal synchrony, we began to notice a few hints that things seemed about to click, so to speak. With more training sessions logged, the number of trials in which two or even all three brains had displayed brief periods of motor cortical synchronization started to increase, slowly but surely. When that happened, all three subjects experienced a fleeting but consistent taste of victory

Three weeks after the first experiment with the B3-brainet was attempted, we reached one particular afternoon with a great feeling of anticipation. As the three subjects decided to really go for it, now in the eleventh testing session, initially the situation looked very familiar, which is a polite way of saying that nothing worked. But then it suddenly happened: everyone in the lab began hearing a much-anticipated metallic song: the coherent beating rhythm of three solenoid valves, one per room, that signaled success, represented by the synchronous delivery of reward to all three participants. As those bursts of solenoid synchronicity built up and became almost continuous, everybody present realized that a spectacular deed was under way: the motor cortices of our threesome brainet had learned how to get in sync and work together in perfect temporal harmony. Indeed, by the end of that day close to 80 percent of the attempts made by the three subjects were synchronized. Despite being located in separate heads, and without being physically connected in any way, these brains were now contributing to a single, distributed organic computational unit in which they took advantage of a digital computer, mixing the electrical signals produced by a meager 775 neurons, to compute a motor program capable of moving a virtual arm to its intended target.

If the original demonstration of a brain-machine interface in our lab twenty years earlier caused a stir and launched the field of modern brain-machine interface research in earnest, what would the first demonstration that multiple brains could synchronize their electrical storms to achieve a common motor goal do? We had no idea at the time. All we wanted was to dig into the terabytes of data collected during those three weeks and see what had really happened during the time it took those three subjects to learn to collaborate mentally to generate a coherent movement. By the end of this analysis, however, a large variety of behavioral and neurophysiological findings shed further light on what had happened during the eleven days in which the B3-brainet was fully operational. First, we confirmed that, as a whole, the B3-brainet had increased its success rate from 20 percent (day 1) to 78 percent (day 11). As originally predicted, the highest rate of success was obtained when all three subjects were fully engaged and capable of synchronizing their cortical activity properly (figure 7.2). Indeed, when we analyzed the simultaneous cortical recordings obtained from the three subjects operating the B3-brainet, adding some data obtained from a two-brain system, or B2-brainet, we found that correct trials were significantly correlated with the transient production of high levels of cortical synchronization across the three subjects: that is, groups of cortical neurons located in one of the individual brains began to fire their electrical pulses at the same time that clusters of cortical neurons in the other two brains did the same.

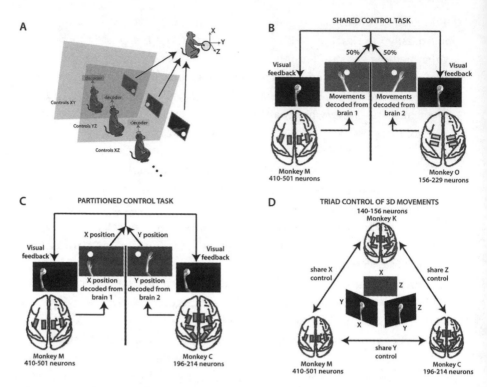

Figure 7.2. Different configurations of monkey brainet. *A:* General arrangement of a monkey brainet used to perform a shared motor task. Monkeys were located in separate rooms. Each monkey faced a computer screen that displayed a virtual avatar arm. The behavioral task consisted of using three-dimensional movements of an avatar arm to reach virtual targets on a screen. The 3-D avatar arm movements were produced by the combination of the cortical electrical activity produced simultaneously by the group of monkeys forming a given brainet. *B:* An example of a shared motor control task in which each of two participating monkeys contributed 50% to the (X, Y) position of the virtual arm. Cortical locations of the implanted microelectrode arrays are shown below the task diagram. *C:* Partitioned control task, in which one monkey contributed to X position of the avatar arm and the other to Y position. *D:* Detailed representation of a three-monkey brainet task. Each monkey performed a two-dimensional task, and all three together controlled three-dimensional movements of the avatar arm. (Originally published in A. Ramakrishnan, P. J. Ifft, M. Pais-Vieira, Y. W. Byun, K. Z. Zhuang, M. A. Lebedev, and M. A. Nicolelis, "Computing Arm Movements with a Monkey Brainet," *Scientific Reports* 5 [July 2015]: 10767.)

Several other results got our attention, too. For example, in instances when one of the subjects slacked off and stopped participating in the game for a while, the remaining two members more than compensated for the temporary loss in brain power. They simply increased the electrical firing rate of their motor cortical neurons, raised the level of cortical synchronization between them, and got the virtual arm to the target as needed, without the third member of the brainet contributing to the effort. Since the slacker did not get any juice when he chose to take a nap, his dropping out of the game was not rewarded. That ensured that he got back into action shortly after.

After waiting so many years to run these experiments, I was tempted to run back to my participants to celebrate—after all, we had just demonstrated successfully the operation of the first-ever shared brain-machine interface built in a laboratory. Unfortunately, extracting anything beyond a few grunts and chirps from our collaborators was simply impossible. Not because they were shy, which they certainly were not, but because, being separated by about 25 million years of evolution, their brains played to a different tune. You see, Mango, Cherry, and Ophelia were three gorgeous rhesus monkeys. And their brains could not synchronize in response to human language and high-fives!

After this first success, there was no going back in our experimenting with brainets. In the next major development using this new approach, our monkey subjects were known simply as Passenger and Observer. Inside the ten-by-ten-foot laboratory that served as their mental playground, they quickly learned their roles: the Passenger used his brain's electrical activity to drive an adapted electronic wheelchair (or was driven by a computer controlling the wheelchair) to grab the grapes he craved, while the Observer had to sit on his own chair and follow the peculiar driving exercise that took place right before his eyes. And although playing Observer might sound like a minor sidekick's role, there was a good payoff associated with the job: if the Passenger got to the grapes before the trial time expired, the Observer would also be rewarded for that milestone with a sip of his favorite fruit juice. It is important to mention that, as has been the case for twenty years in our lab, this experiment could be run only because Gary Lehew, also known as "the magician," had adapted a pair of secondhand electronic wheelchairs so that they could be steered and driven by electrical brain activity alone.

It took several days, but eventually the Passenger-Observer pair mastered the task to the point at which, no matter the location of the wheelchair's departure position, the Passenger got the grapes in almost every attempt. Thus, in every trial, once given the command to go for it, the Passenger's brain at once remapped the new location of the grape dispenser pod in the room as well as its relative spatial position to the wheelchair under the Passenger's brain control. That allowed the Passenger's brain to define the best trajectory through which it could reach the sweet grapes. After a swift mental calculation, the Passenger's brain generated the motor program needed to steer the wheelchair to the target. A few hundred milliseconds later, off the Passenger went, driving away while comfortably seated in his wheelchair cockpit.

While the two were learning to enact this rather unique social interaction, Po-He Tseng, a talented Taiwanese engineer and a postdoctoral fellow in my lab, was busy recording the electrical sparks simultaneously produced by hundreds of neurons located in multiple cortical areas of both the Passenger's and the Observer's brains. That marked the first time in more than a half-century history of intracortical primate neurophysiology that someone was recording large-scale neuronal electrical commands generated simultaneously by two interacting rhesus monkey brains. To make the feat even more astounding, the brain signals of each monkey were obtained through a 128-channel wireless interface that was capable of sampling and broadcasting the action potentials produced by up to 256 cortical neurons per animal. In fact, the invention of this new multichannel wireless interface by a team of Duke neuroengineers working in my lab had become an essential component in the design and execution of the particular social task that kept Passenger, Observer, Po-He, and the rest of our team very busy for many months to come.

While Po-He listened to the daily afternoon broadcasts of his private Radio NeuroMonkey, Passenger and Observer went about their business, day after day, week after week, coalescing into a very efficient playing team. In the words of the most famous Brazilian soccer announcer of our times—my friend Oscar Ulisses—those two had "developed the right kind of team chemistry," like a pair of soccer strikers or a tennis doubles team who play together often. As Po-He was soon able to verify, as this dyad performed the task together— that is, in full view of each other—these two monkeys began to exhibit much higher levels of cortical electrical synchrony than one would expect by chance alone (figure 7.3). In other words, as Passenger drove and Observer watched his driving, a higher fraction of the electrical signals produced by hundreds of neurons in the motor cortex of those two animals began to occur at the same moment in time, despite the fact that these neurons were located in two distinct brains. Indeed, in many instances, this level of interbrain correlation

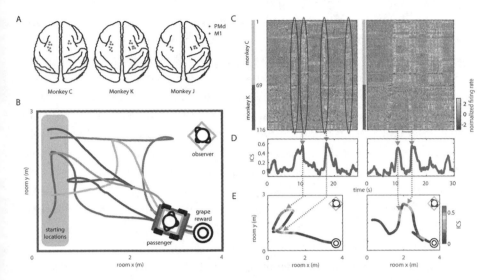

Figure 7.3. Interbrain cortical synchronization (ICS) during a primate social task. *A:* Locations of cortical implants in three monkeys (C, J, and K). Wireless multisite, multichannel neuronal-ensemble recordings were conducted in the primary (M1) and premotor dorsal (PMd) cortices in both hemispheres. *B:* Two monkeys (Passenger and Observer) were placed in a 5.0-by-3.9-meter room. During each trial, the Passenger navigated from a starting location to a stationary grape dispenser. Five representative routes of the wheelchair are plotted. *C:* Neuronal-ensemble activity for two representative trials. Each horizontal line corresponds to a neuron. Individual neuronal action potentials are represented by white vertical bars. Episodes of interbrain cortical synchrony between monkeys C (Observer) and K (Passenger) are highlighted by vertical ellipses in the leftmost plot. *D:* Quantification of ICS for the trials shown in *C*. Instantaneous values of the distance correlation were computed with a sliding window of the same 3-s width as the gray bars in *C*. Correlation peaks are marked by arrows. *E:* Wheelchair routes for the same trials as in *C* and *D*. (Originally published in P. Tseng, S. Rajangam, G. Lehew, M. A. Lebedev, and M. A. L. Nicolelis, "Interbrain Cortical Synchronization Encodes Multiple Aspects of Social Interactions in Monkey Pairs," *Scientific Reports* 8, no. 1 [March 2018]: 4699.)

reached around 60 percent, while the value expected by chance alone was close to zero.

Initially, we thought that such simultaneous neuronal firing could result from both animals being exposed to a common sensory input, like visual signals describing the room, which reached both brains at the same time. As it turned out, however, things were much more interesting than that trivial potential explanation. Further analysis revealed that neurons located in the motor cortical areas of both monkey brains were firing together because the two animals' motor cortices were calculating concurrently the pair of velocity vectors needed to steer the Passenger's wheelchair toward the grape dispenser. Such a mental calculation coming from the Passenger's brain was more than expected since this monkey was in charge of driving the wheelchair. But at the same time, even in its passive observing role, the Observer's brain was also busy calculating, from its own viewpoint, what was needed to take the wheelchair to the right location—essentially some of his motor cortical neurons were carefully monitoring the wheelchair movements toward the target location, a goal that ensured the Observer some juice! Clearly, the brains of both monkeys had become tuned to generate the same type of unnatural motor signals—rotation and translational velocity vectors to guide a wheelchair—needed to get them the rewards they both craved.

Digging even deeper into the data, Po-He found that the synchronization of motor cortical neurons in these two monkey brains also increased when a particular combination of monkeys playing Passenger and Observer got closer to one another, especially when this distance got to be around one meter. That meant that the synchronous neuronal activity recorded across these two animals' motor cortices was correlated with the intersubject distance between the Passenger and the Observer during their social interaction. Further analysis revealed that this increase in cortical synchrony happened when the dominant monkey in the pair was playing Passenger and a lower-ranking animal served as the Observer. When their roles were reversed, meaning that the dominant monkey became the Observer and the lower subordinate assumed the Passenger role, such an increase in cortical synchrony did not materialize or was not as high. Suddenly, we realized that the magnitude of this interbrain cortical synchrony could predict the relative social ranking of the monkeys involved in our experiment.

Interestingly enough, the intersubject separation distance at which we observed maximum interbrain synchronization—about one meter—is approximately equivalent to the reach limit of a rhesus monkey's arm. That means that starting at this distance, a monkey could use his hands to either groom or attack another member of his group. This further suggests that the increase in interbrain synchronization observed when the dominant monkey, while playing Passenger, approaches a lower-ranked animal conveys an important clue about their social relationship.

But this was not the end of the story. By focusing only on what happened in the motor cortices of monkeys that used our wireless brain-machine interface to drive the wheelchair, Allen Yin, a biomedical engineering graduate student in my lab, discovered that a high percentage of the recorded cortical motor neurons varied the magnitude of their firing activity according to the relative spatial location of the wheelchair in relation to the reward pod. Some neurons fired more when the wheelchair was closer to the grapes, while others were more active when the vehicle was farther away. All of a sudden, as a result of this finding, we could accurately predict the spatial trajectory made by the wheelchair, from its starting position all the way to the grape dispenser, just by looking at the Passenger's motor cortex. And, as predicted from the degeneration ensemble principle of the relativistic brain theory, this trajectory could be estimated in different trials by the collective firing of different cortical neuronal samples.

At this point, we realized that taken together, the results obtained by the Passenger-Observer experiments revealed a very different view of the primary motor cortex—one that would prove essential to my understanding of how individuals can become synchronized into brainets while also providing more evidence for my theory of how individual brains work. For starters, it became evident that in addition to being able to code body movements—the classic function neuroscientists have assigned to this cortical area for over a century—neuronal circuits in the primary motor cortex could quickly learn to encode the movements needed to drive artificial devices that require completely different motor programs than those needed to move the limbs of a primate body. On top of that, the primary motor cortex neurons were able to simultaneously encode a relative representation of space between the subject's body and the end target of the planned whole-body movement as well as the distance between co-specific individuals. To this list, we also have to include the fact that about half of the monkey motor cortical neurons that participate in the planning of limb movements are also capable of modulating their electrical firing during both anticipation of a reward or according to whether the animal's reward expectation was matched or not in a given trial. No modern theory dealing with the primate motor cortex had anticipated that all these functions could be carried out simultaneously by neuronal circuits in this cortical area. Nor that very relevant social parameters could be encoded by enhanced levels of cortical motor neuron synchrony across animals. But that was exactly what the multitasking principle of the relativistic brain theory predicted: a given cortical area—like the primary motor cortex—participating in multiple concurrent functional tasks.

The first thought that came to my mind when I tried to make sense of these very interesting findings was that Po-He had stumbled upon a particular type of cortical cell known as mirror neurons. Originally described by the renowned Italian neurophysiologist Giacomo Rizzolatti, a professor at the University of Parma, in experiments conducted with rhesus monkeys in the 1990s, these neurons received their name because of a peculiar physiological behavior: in addition to modulating their firing rate—up or down—when the monkey was preparing or executing a hand movement, these cells also fired when the same animal was simply observing another monkey or a researcher making the same type of movements. A few years after the original discovery in rhesus monkeys, mirror neuron–like activity was also observed in humans, thanks to the employment of modern brain-imaging techniques such as magnetic resonance imaging.

In the original report on monkeys, Professor Rizzolatti reported the existence of mirror neurons only in a higher-order motor cortical area located in the more lateral region of the frontal cortex. Rizzolatti likes to identify this cortical region as F5, a term derived from an old nomenclature. For most of us cortical neurophysiologists, that region is known as the ventral division of the premotor cortex. Soon, however, it became clear that mirror neurons were not restricted to the premotor cortex. As Stefano Rozzi points out in a comprehensive review of the field, subsequent studies in both humans and monkeys identified the presence of mirror neurons in many other cortical regions of both the frontal and parietal cortices, suggesting that this type of motor activity is produced by a highly distributed frontoparietal neuronal circuit. As such, the mirror neuron circuit includes multiple cortical areas involved in the generation of hand, mouth, and eye movements. Interestingly, Rozzi also indicates that in songbirds, mirror neuron activity has been observed in brain structures involved in song production and learning.

All in all, the discovery and widespread presence of mirror neuron activity in the frontal and parietal cortices of monkeys and humans suggest that this system plays a very fundamental role in mediating social interactions both among animals and in human groups. This can be easily understood when one realizes that the electrical activity generated by mirror neurons reflects not only the preparation and execution of movements by the individual but also the representation of similar movements being performed by other members of her immediate social group, or even from other primates (like the experimenter in the case of lab monkeys). Researchers have discovered that mirror neurons can also signal the particular viewpoint of the individual observing the movement of another subject as well as the reward value of this action. Altogether, these results suggest that the classic mirror neuron designation may not do full justice to the many functions carried out by frontoparietal networks of these cells.

In practice, the discovery of mirror neurons revealed that motor cortical areas have continuous access to visual information. One interesting aspect is that visual input reaches the motor cortex via different paths through the brain. Among those, one of the most interesting is the connection that allows visual signals from the inferior temporal cortex, a component of the primate visual system, to reach the ventral premotor cortex of the frontal lobe via a relay station in the parietal lobe. The inferior temporal cortex neurons tend to respond when monkeys and humans look at complex and elaborate objects. Moreover, a subset of these neurons is known to increase its firing rate in both monkeys and humans when faces of other co-specific individuals are presented to the subjects.

There were several similarities between the classical properties of mirror neurons and the results obtained by Po-He. Yet there was an apparent mismatch: our findings derived from neuronal recordings obtained from the primary motor cortex and in the dorsal division of the premotor cortex of the monkey frontal lobe, not in the ventral portion of the premotor cortex where Rizzolatti had originally identified these cells. This mismatch was compounded by the fact that several brain-imaging studies in humans had not identified mirror neuron activity in the primary motor cortex either. After a careful literature search, however, I discovered at least two monkey studies that recorded mirror neuron–like activity in the monkey primary motor cortex. In one of these studies, neurophysiologists observed that most of the mirror neurons increased their firing rate while observing someone else performing a movement, while a smaller percentage of these, the primary motor cortex neurons, responded by reducing their firing, a phenomenon that had also been reported in the premotor cortex. The same study also showed that mirror neurons in the primary motor cortex tended to produce much higher increases of firing rate during the execution of movements by the monkey, rather than during the observation of someone else's movements. That smaller modulation of firing rate during movement observation may explain why many imaging studies in humans were unable to detect mirror neuron activity in the human primary motor cortex. That magnetic resonance imaging missed the occurrence of these neurons in the primary motor cortex became almost a certainty when a new method, magnetoencephalography, which records the tiny magnetic fields produced by the cortex, was employed. Researchers who employed magnetoencephalography had no trouble in identifying mirror neurons firing in the primary motor cortex of human subjects. Interestingly, the use of magnetoencephalography also seemed to reveal that autistic children, albeit showing mirror neuron activity in their primary motor cortex, did not seem to take advantage of the presence and activation of these neurons to engage in normal social behaviors.

Based on what we found in the literature about mirror neurons, it became much more plausible to interpret Po-He's observations with the Passenger-Observer experiments as emerging from a type of interaction—whole-body mobility using an artificial device—that had, until now, not been reported as part of the repertoire of mirror neurons located in the primary motor cortex of primates.

But that was not all.

A comprehensive review of the neurophysiological properties that characterize mirror neurons, and the fact that they could be found in both the primary motor and somatosensory cortices, also made me think about a series of previous studies conducted in my lab in which we may have stumbled upon this class of cortical cells almost by chance, without knowing it. Since 2012, as part of the training monkeys have to receive in order to learn to control a brain-machine interface, we had conducted several experiments in which the animals passively observed hundreds of movements generated by an avatar arm projected on a computer screen placed in front of them (figure 7.4).

Figure 7.4. Passive observations. *A:* Monkey was seated in front of screen with both arms gently restrained and covered by an opaque material. *B:* Actual left and right arm X-position (black) compared with predicted X-position (gray) for passive observation sessions. Pearson's correlation, *r*, is indicated. *C:* Performance of monkey C and monkey M quantified as fraction correct trials. Shown separately for monkey C are different decoding model parameter settings (light gray, dark gray markers) as well as sessions during which the animal moved the avatar using only brain control of this virtual actuator without the use of the hands (BCWOH) (black, both monkeys). *D:* Fraction of trials in which both left (gray circles) and right arms (black circles) acquired their respective target during brain control. Linear fit for learning trends of each paradigm shown in *A-B. E-F:* Fraction of correct prediction by k-NN of target location for right (black) and left arms (gray) over the trial period during both PO (*E*) and BCWOH (*F*) in both monkey C and monkey M. (*G:* Mean k-NN target prediction FC neuron dropping curves separated by cortical area for each monkey (same columns as in *E-F*). (Reproduced with permission. Originally published in P. Ifft, S. Shokur, Z. Li, M. A. Lebedev, and M. A. Nicolelis, "A Brain-Machine Interface Enables Bimanual Arm Movements in Monkeys," *Science Translational Medicine* 5, no. 210 [November 2013]: 210ra154.)

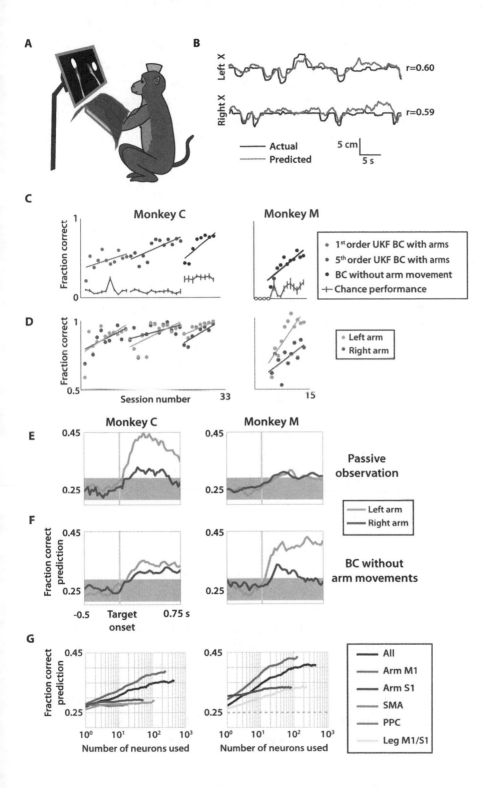

A

B

Left X — r=0.60

Right X — r=0.59

— Actual
— Predicted

5 cm | 5 s

C

Monkey C

Monkey M

Fraction correct

1

• 1st order UKF BC with arms
• 5th order UKF BC with arms
• BC without arm movement
+ Chance performance

D

Fraction correct

1

0.5

Session number   33   15

• Left arm
• Right arm

E

Monkey C   Monkey M

0.45   0.45

0.25   0.25

Passive observation

— Left arm
— Right arm

F

Fraction correct prediction

0.45   0.45

0.25   0.25

-0.5   Target onset   0.75 s

BC without arm movements

G

Fraction correct prediction

0.45   0.45

0.25   0.25

10^0 10^1 10^2 10^3 10^0 10^1 10^2 10^3

Number of neurons used   Number of neurons used

— All
— Arm M1
— Arm S1
— SMA
— PPC
— Leg M1/S1

During these passive observation sessions, we simultaneously recorded the electrical activity of hundreds of neurons located in both the primary motor and the somatosensory cortex. Invariably, a high percentage of these neurons became tuned to the various movements made by the avatar arm, their firing rates modulated in response to those movements. Once the monkeys were connected to the brain-machine interface, these entrained neurons enabled these subjects to rapidly become proficient in controlling the movements of avatar arms using only their brain activity. Simply put: passive observation of a virtual arm was enough for the monkey to acquire such a motor proficiency.

Looking back at this finding, I realized that a large fraction of neurons in the primary motor and somatosensory cortex could exhibit physiological properties compatible with the classic definition of mirror neurons. In fact, this may be the key reason why these animals can learn to use a brain-machine interface to move these surrogate virtual appendices. Curiously, as monkeys went through more passive observation sessions, more neurons in those two regions began to modulate their firing rates. This observation raises the very interesting hypothesis, not discussed in great detail in the mirror neuron literature, that the particular physiological properties of these neurons can be acquired through learning a motor task simply by observing movements made by someone else. This hypothesis, if confirmed, could have a profound impact on the future of neuro-rehabilitation, but also for other practical applications of the brainet concept. For example, in human social activities that aim at achieving a high level of collective perceptual-motor performance—for instance, playing a team sport—practicing in a virtual environment may enhance mirror neuron activity in the brains of the interacting players. I can see how, as a result of reaching a high level of mirror neuron recruitment, teammates could become capable of easily anticipating the motor intentions of one another with very little overt production of movement by any of them. Such a demonstration would basically imply that any training that enhances the players' collective mirror neuron activity may contribute to the enhancement of the team's collective motor performance.

Although we apparently solved the first mismatch—the existence of mirror neurons in the primary motor cortex—there was something else that puzzled us: the primary motor cortex and premotor cortical neurons recorded in our Passenger-Observer experiments were not increasing their firing rate by observing the other monkey's hand, mouth, or eye movements. Rather, they were increasing their electrical activity as a result of whole-body movements, mediated by an artificial actuator, the electronic wheelchair, as the Passenger drove it through the room. Different from the classical setups used until now to investigate mirror neuron activity, in which a single monkey remains immobile in a chair observing the action of another subject (usually the experimenter), we now had a pair of monkeys interacting in a task that always guaranteed that at least one monkey (the Passenger) was continuously moving in the room. Moreover, we were recording from the motor cortices of the two monkeys simultaneously. Because of this particular task design, our experiments offered for the first time the opportunity to simultaneously record the activity of hundreds of cortical neurons from the brains of two monkeys fully engaged in a social task. It turned out that this was also the first time that presumptive mirror neuron activity was simultaneously sampled from the two brains of a pair of monkeys engaged in a collective motor task that involved direct social interactions.

The type of interbrain synchronous cortical activity observed in the Passenger-Observer experiments is also known in modern neuroscience by the term brain-to-brain coupling. During the past decade, the potential relevance of brain-to-brain coupling for the establishment and maintenance of animal social behaviors has begun to get traction among neuroscientists as a true paradigm shift in brain research. Basically, this view proposes that signals generated by the brain of a subject and received by another can functionally couple the two central nervous systems in time and space. In a nice comprehensive review article about this emergent new field, Uri Hasson, a professor at Princeton University, and his colleagues describe a series of examples of key social behaviors that involve brain-to-brain-coupling in both animals and human subjects. For example, in the wild, songbirds tend to learn a new song as a result of their social interactions. Hasson and colleagues highlight this fact by describing the typical courtship behavior of cowbirds. In this species, male birds learn songs that can elicit strong responses in female birds, which cannot sing themselves. Instead, females signal their appreciation for a good serenade by producing slight movements of their wings. These tiny wing movements serve as a powerful synchronizing signal—or reinforcement—to the male singer, likely through mirror neuron activity. Encouraged by the female's positive motor response, the avian Pavarotti goes on overdrive and starts repeating the particular song components that entice the female. Not satisfied with that, he also starts producing more elaborate songs, hoping to attract the attention of other females. This is key because the target female bird seems to select the male she wants to mate with by judging how other females react to his singing (apparently not much has changed in terms of courtship strategies in the animal kingdom since birds learned to sing).

The interaction between two adult human beings who are using language to communicate face-to-face offers yet another fundamental example of brain-to-brain coupling and the tremendous impact it has had on human social interactions. Although there are many fascinating aspects one could talk about involving human speech, here I want to focus exclusively on the key neurophysiological attribute involved in a human dyad communicating via the production and reception of speech. Like any other motor behavior, speech is produced as a consequence of a motor program, originally generated in the motor cortex of the frontal lobe. Once downloaded to the brainstem neurons that control the muscles of the larynx, vocal cords, and tongue, this motor program generates an acoustic signal that, in addition to having its amplitude modulated up and down, cycles at a basic frequency band of 3–8 hertz. This oscillatory envelope basically defines the essential rhythm or frequency at which syllables can be generated by human language, which ranges from three to eight syllables per second. This corresponds well with the brain's theta rhythm, oscillations in neuronal activity in the 3–10 hertz range. Furthermore, groups of neurons located in the human auditory cortex, while receiving inputs from speech a subject is hearing, generate theta-like oscillations in the 3–8 hertz range. As Hasson and his colleagues point out, the presence of similar oscillatory activity in the brain production and reception systems for language has led many theorists to propose that this match in brain rhythms could play a pivotal role in oral communication in humans. Basically, by utilizing a similar frequency envelope to produce, transmit, and process oral language, the human brain ensures that the sounds that define speech can be transmitted optimally, and even be amplified, to enhance the signal-to-noise ratio, which often can be disturbed by environmental interference. The importance of the 3–8 hertz rhythm for speech comprehension can be further stressed by the fact that if human listeners are exposed to language signals that oscillate at higher than 3–8 hertz, they exhibit difficulty in understanding the content of these signals.

Obviously, there is more than sound processing involved in language communication. Face-to-face contact also enhances speech comprehension in adult humans. This happens because the typical mouth movements we employ to produce speech also follow more or less the theta rhythm frequency band. Basically, that means that when we are facing someone who is speaking to us, our brains are receiving two streams of 3–8 hertz signals, one auditory and one visual. The visual stream reinforces the acoustic signal that, once translated into electrical signals in the inner ear, reaches the auditory cortex where the process of speech interpretation begins. As such, according to Hassan and colleagues, facing someone who is talking to us is equivalent to ramping up the amplitude of the speech signal by fifteen decibels.

Overall, these results strongly suggest that through the use of language, brain-to-brain coupling is established by coherent synchrony between the brains of the speaker and the listener. Using my jargon, a language-based brainet is established initially because the production, transmission, and interpretation of analog speech signals are mediated by brain signals occurring at the same frequency range in both the speaker's and listener's brains. For all intents and purposes, therefore, this overlap in frequency is the first step needed to establish the bonding of brains into a distributed organic computer, the brainet.

In addition to offering a canonical example of brain-to-brain coupling, interpersonal human communication via oral language also allows me to describe how the relativistic brain theory accounts for the establishment of brainets. As we saw in chapter 5, the theory proposes that binding of multiple cortical areas within a single brain is mediated by neuronal electromagnetic fields. By using this capability, relativistic brains can also take advantage of such analog neuronal signals to quickly establish and maintain stable brainets. In the case of oral language, for example, electromagnetic fields would ensure the concurrent activation of the many cortical (and subcortical) regions needed for both the emitter's brain to generate a language message and the recipient's brain to process and interpret it upon reception. In the case of the recipient, such instantaneous cortical binding would then allow for the fast decoding and comprehension of both the syntactic and semantic contents of the message sent by the emitter. As a result, like in the case of the Passenger-Observer experiment, interbrain cortical synchronization between the subjects involved in an oral dialogue would emerge quickly and lead to functional coupling of their brains. Thus, for the relativistic brain theory, brainets are formed through a process of analog rather than digital neuronal synchronization mediated by a communicating signal. In humans, oral language plays this pivotal role often. Indeed, it is conceivable that in its primitive stages, the kind of language used by our hominid ancestors worked primarily as an interbrain synchronization signal rather than the elaborate communication medium known to us today.

There are several advantages to relying on analog synchronization rather than digital to create brainets. For starters, analog synchronization is faster, easier to establish, and more malleable than digital, since the latter requires much more temporal precision between participating signals to occur.

Furthermore, analog synchronization can work without predefined subsystems, meaning that it does not require extra information about the underlying signals to be enacted. It can simply happen when two continuous signals get entrained to each other.

In addition to emphasizing the employment of analog synchronization, the relativistic brain theory postulates that a classic Hebbian learning principle is involved in the establishment and long-term maintenance of brainets. In this case, instead of talking about two interacting neurons that share a synapse, as in the original formulation introduced by Donald Hebb in 1949, we would be talking about two (or more) brains that are connected by a communicating signal or message. The Hebbian principle states that when two neurons that share a synapse fire together in close succession, the strength of their synapse is enhanced. Figure 7.5 illustrates this principle by showing two neurons, 1 and 3, and the direct synapse that neuron 1 makes with neuron 3. If an action potential produced by neuron 1 (known as the presynaptic neuron) drives neuron 3 (or the postsynaptic neuron) to fire in sequence, the synapse between

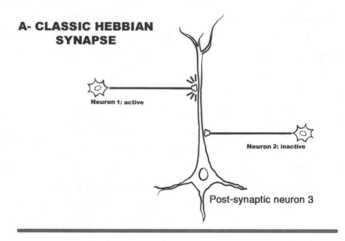

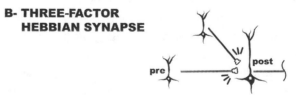

Figure 7.5. Schematic representation of a classic Hebbian synapse (*A*) and a three-factor Hebbian synapse (*B*). (Image credit to Custódio Rosa.)

these two neurons is enhanced. By the same token, I propose that when two people engage in a conversation, their brains could become functionally coupled via a Hebbian learning mechanism that enhances the level of inter-brain synchronization. While in the case of synapses, a chemical—known as a neurotransmitter—is responsible for the delayed communication between the pre- and postsynaptic neurons, in the case of brain-to-brain coupling during a conversation, language (in addition to other communicating signals) plays the role of the coupling signal. As we will see, such a simple "wireless" analog coupling mechanism could explain why humans tend to build brainets capable of synchronizing a large number of individual brains into participating in social groups, which thrive on the exchange of much more abstract constructs, such as a common set of beliefs, culture, and knowledge, across vast spans of time and space, throughout human history.

But there is much more occurring within the brains of a speaker and listener as they become part of a human brainet. Thanks to the Hassan article, I was introduced to a very interesting set of experiments carried out by Greg Stephens and collaborators that highlight some other aspects involved in establishing a language-based brainet. In this study, functional magnetic resonance imaging was employed to map the activation of the brain areas of a person reading aloud, without any prior practice, a real-life story. The audio recording was then played to a listener, whose brain activation pattern was also obtained using functional MRI. The patterns of brain activation obtained from the speaker and listener were then analyzed to search for potential correlations. The authors observed that the brain patterns of the speaker and listener exhibited clear signs of temporal synchronization. To show that this temporal brain-to-brain coupling was indeed meaningful for the communication of a message between the speaker and the listener, the researchers ran control experiments in which the speaker produced a narrative using a language that the listener could not understand. When the brain patterns of the pair were re-analyzed, a significant reduction in the brain-to-brain synchronization was observed, indicating that the brainet had not formed properly in this condition.

How the human central nervous system comprehends language also presents good evidence for the importance of the brain's own point of view, a key tenet of the relativistic brain theory. In their review paper, Hassan and collaborators describe that in a speaker-listener language interaction, a series of particular cortical and subcortical structures are actively engaged in anticipating the next utterances that the speaker is about to produce; as far as language is concerned, our brains listen before they really hear.

Although language has mediated, since time immemorial, a great many human social interactions, it is not the only way in which human brains can become entrained and form a brainet. Hand gestures can do the trick as well, as likely can mutual tactile stimulation and certain hormones, like oxytocin, which is liberated when mothers breast-feed their newborn babies or when people fall in love. In both cases, oxytocin seems to mediate the establishment of a strong pair bonding between subjects, which may involve an increase in analog cortical synchronization within each of their brains.

In another very elegant study, Hasson and his colleagues show that by presenting complex video clips—extracted from real action movies—sequentially to multiple subjects, one can induce a surprising level of brain coupling across individual viewers, as measured by brain magnetic resonance imaging. In a departure from previous studies that focused only on visual areas—again, a finding that could be considered trivial when multiple subjects receive the same visual input—Hassan indicated that this interbrain synchrony was established by recruiting in addition a series of other cortical regions, normally known as association areas. Part of this global cortical recruitment and interbrain synchronization was generated by what Hasson defines as a more general component, which contributed to the widely distributed representation in the cortex of the complex visual images presented to the subjects. A second source of this general response may be the increase in arousal and attention load produced in viewers by some of the more emotionally charged scenes contained in the video clips selected for the experiments. Aside from the basic science interest of this finding, this observation raises the more practical possibility that interbrain synchrony could be employed in the future to assess in a quantitative way how well audiences become engaged (or not), from an attentive and emotional point of view, by the delivery of visual or auditory messages such as those originating in a movie scene, a TV commercial, or a political rally, just to mention a few examples.

In addition to these general components, the authors also suggest the existence of a very selective mechanism in processing the scenes contained in the movies that affects interbrain synchrony across subjects. Basically, they found that the processing that goes on to identify complex objects does not seem to be very dependent on the placement of the particular visual input—a human face, for instance—on a given region of the viewer's retina. No matter the angle in which a face was shown in the video clip, inferior temporal cortex neurons involved in face recognition seemed to be able to respond to it without much trouble. This particular finding illustrates well why relying on a more malleable, analog-based mechanism of synchronization, rather than a precise digital one, can be advantageous for brainet formation. Basically, different from the digital case, there is no need for strictly identical signals to produce synchronization when analog signals are used. Translated to the case in question, despite viewing the same face from different angles, the brains of multiple subjects would be able to synchronize among themselves into a brainet rather well and quickly.

Another aspect that got my attention in this study was that whenever the video clip showed a precise hand movement, neurons in the somatosensory cortex tended to fire in all viewers, contributing to the brain-to-brain coupling measured by Hassan and collaborators. Once again, we are back to talking about the mirror neuron circuit being recruited across multiple subjects, even when these individuals are engaged in what one would consider a few years ago as a typical and pristine—with some intermittent tactile exploratory exceptions—visual discrimination task: going to the movies.

Suddenly, going to the movies acquired a completely different dimension, at least for me!

If, by any chance, you were not very impressed by this movie theater example, basically because you reasoned that there is nothing extraordinary about the brains of people who are watching the same movie together producing similar patterns of activity at the same time, I owe you an explanation. I did not bring up this experiment because a common visual signal can transiently synchronize many brains. Indeed, that would be uninteresting. Instead, I am much more interested in the effect that watching the movie has afterward, when audiences leave the theater. To understand where I am going, one just needs to compare what would happen if we had decided to run a control experiment involving two groups of viewers: one formed only by humans and another one containing only chimpanzees. In this experiment, each group of viewers would be placed in a separate room so that they both could watch the same show, let's say an episode of the original TV series *Star Trek*, broadcast in the mid-1960s. During the showing, the electrical activity produced by the brains of all individuals in each group would be measured by wireless electro-encephalogram recordings. As reported by Hassan and colleagues, this would reveal the occurrence of brain synchronization developing across subjects in both audiences. Trivial again, you would say. Sure, I would have to agree. But the best part of this experiment would happen if we then decided to follow these two audiences and see what they did after experiencing this brief period of synchronous brain activity driven by a common visual input. While observation of the chimpanzee audience would reveal that nothing substantial happened after the show was over—that is, the chimp group went back to its usual business of daily living without exhibiting any explicit sign that watching the episode together had affected their routine—a much different collective social behavior would have taken place with the human audience. Following their theatrical experience, this human group became busy talking to each other, and also to members of their social group who did not attend the showing, about the incredible adventures of Captain Kirk and Mr. Spock. As a result, viewers and some nonviewers decided to create a *Star Trek* fan club, dress like Captain Kirk or Mr. Spock to go to school, and even propagate to their friends and family the reasons one should mistrust the Romulans. Some went as far as to become fluent speakers of Klingon and attend annual *Star Trek* conventions to seek autographs and have their pictures taken next to the show's main actors.

Essentially, what I am trying to portray is that after they had been confined together in a dark room and fed a fictional narrative, embellished with extraordinary visual inputs and captivating music that appealed deeply to their emotions, expectations, desires, beliefs, and worldviews, members of this human audience became integrated into a new mental abstract framework, a fantasy that pretty much dictated the very way they behaved afterward. Granted, my example may sound a bit cartoonish, but it still allows me to describe the potential neurophysiological mechanisms that allow the transformation of an initial transient period of brain-to-brain coupling, as generated by a common visual input, into a very cohesive human brainet driven by the sense of belonging to a social group bound by a new set of abstract beliefs—in this case, being part of a sci-fi adventure. According to the relativistic brain theory, that may happen because such a transient interbrain synchronization period may be followed by a crystallization phase mediated by the release of powerful neurochemical modulators, among which is the neurotransmitter dopamine, all over the cortex. Because the initial transient synchronization generates strong hedonic sensations shared by a large number of human subjects targeted by the common visual input, it likely activates the extensive brain circuit that mediates our intense drive to seek reward and pleasure experiences. In addition to its role in motor behaviors, dopamine is the key neurotransmitter employed by the neuronal circuits that mediate natural reward-seeking behaviors, like sex and the search for palatable foods, but also a series of addictive behaviors, such as drug addiction and compulsive gambling.

Here, I propose that following an initial phase of transient interbrain syn-chrony, the concurrent release of dopamine in many brains as a result of a common experience could contribute to produce a much longer-term bonding of a social group into a brainet. At the level of neurons, dopamine is known to induce changes in synaptic strength. That can be seen in the bottom graph of figure 7.5, which illustrates what is known as a three-factor Hebbian synapse. According to this mechanism, a neuron that uses dopamine (or other neuro-modulators) can exert an important modulatory effect on a classic Hebbian neuronal synaptic interaction, effectively creating a supervised mechanism of synaptic plasticity. Essentially, this third neuronal contribution can provide an error signal or report the magnitude of a reward in relation to what was origi-nally expected, or even offer a measurement of the overall attention level or global awareness state of the brain. Dopamine is well known to signal reward and as such could be used to modulate synaptic plasticity and hence Hebbian learning. In the case of social behaviors, concurrent release of dopamine in multiple individual brains could potentiate the Hebbian-like mechanism that allows brain-to-brain coupling to emerge via interbrain synchronization. This dopamine-mediated modulatory effect could guarantee that what began as a transient state of interbrain synchrony through a common visual input can be maintained for much longer periods of time. That would be particularly relevant if one considers that once formed, a social group tends to generate self-reinforcing hedonic signals by its frequent interactions. Essentially, the relativistic brain theory raises the hypothesis that mechanisms that operate at the synaptic level—like Hebbian learning and reward-based neuromodula-tion—could also manifest themselves at the scale of brain-to-brain interac-tions and contribute to the formation and maintenance of brainets that under-lie the establishment of social groups in animals and humans.

By now you may be asking why such an evolution from transient to long-term brain-to-brain coupling does not happen in chimpanzees. As we saw in chapter 2, despite their clear capacity to imitate, chimpanzees do that much less often than humans. Essentially, this means that chimpanzees still tend to focus more on emulating—copying the end result of a goal—while humans are much better imitators—focusing primarily on reproducing the process through which one achieves a motor goal. Moreover, thanks to the dramatic enhancement in communication bonds provided by language, humans are much better at teaching new skills and disseminating new ideas to others. In other words, in humans a mental insight or abstraction can spread rather quickly and efficiently among social groups through gossiping and, in the long run, by the establishment of cultural instruments.

To bring this unique ability to form brainets into the perspective of the relativistic brain theory, let's focus on a key component of this argument: how the brains of different primates react during observation of a motor act. When the spatial patterns of cortical activation during motor resonance are compared between chimpanzees and humans, one can identify immediately a striking difference, reflecting the fact that chimpanzees devote relatively more emphasis to emulation, whereas humans are much more skillful imitators. Figure 7.6 reproduces this comparison by showing the distribution of cortical activation when either chimpanzees or human beings observed the same motor behavior being produced by an experimenter. Immediately, one notices that while in chimpanzees the cortical activation pattern was primarily restricted to the frontal lobe, with great recruitment of the prefrontal cortex and a much smaller recruitment of the parietal lobe, human observation of a motor act generated a pattern of cortical activation that is widely spread across the frontal, parietal, and occipitotemporal cortex. Among this vast cortical territory, humans tended to exhibit a higher activation of four interconnected regions: the ventral prefrontal cortex, the ventral premotor cortex, the inferior parietal lobe, and the inferotemporal cortex. In analyzing these results, Erin Hecht and Lisa Parr concluded that the pattern of chimpanzee cortical activation more closely resembles the one found in the rhesus monkey rather than the one measured in humans during motor resonance. Indeed, when examined in detail, the pattern of cortical activation in humans during motor resonance relies heavily on connecting cortical regions that, on one side, are more related to the representation of intention, context, and goal outcome, like the ventral prefrontal cortex, with cortical areas that are primarily devoted to the nitty-gritty sensory and motor integration aspects required for planning the execution of sequences of precise movements needed to mimic an action. This latter circuit includes the ventral premotor area of the frontal lobe—the region where mirror neurons were first identified—and multiple regions in the parietal and occipitotemporal regions. Hecht and Parr speculate that these differences in the pattern of cortical activation may explain why "despite the fact that chimpanzees can imitate, they normally do not."

While the studies described above focused primarily on the patterns of cortical gray matter activation, comparative analysis of white matter distribution across the frontoparietal-temporal circuit of monkeys, chimpanzees, and humans matches the functional data very nicely. Three major cortical white

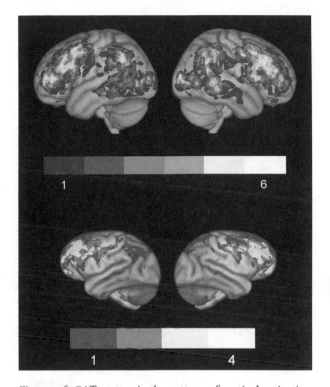

Figure 7.6. Differences in the pattern of cortical activation between humans (top) and chimpanzees (bottom) during observation of grasping gestures by a third subject. (Modified with permission of the *Journal of Neuroscience*, from E. E. Hecht, L. E. Murphy, D. A. Gutman, J. R. Votaw, D. M. Schuster, T. M. Preuss, G. A. Orban, D. Stout, and L. A. Parr, "Differences in Neural Activation for Object-Directed Grasping in Chimpanzees and Humans," *Journal of Neuroscience* 33, no. 35 [August 2013]: 14117–34; permission conveyed through Copyright Clearance Center, Inc.)

matter bundles are involved in this analysis. The first one is the so-called extreme capsule that connects key regions in the temporal lobe—like those located in the superior temporal sulcus (STS) and the inferior temporal cortex—to the inferior prefrontal cortex (figure 7.7). The second one, which connects the STS and a region of the mirror neurons located in the parietal cortex, is formed by the bundles of the so-called inferior and middle longitudinal fasciculus. Finally, there is the superior longitudinal fasciculus that mediates the

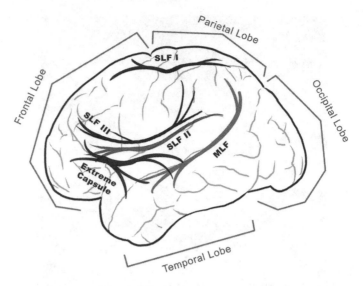

Figure 7.7. Lateral view of the human brain showing the main lobes (frontal, parietal, temporal, and occipital). The detailed organization and subdivisions of the superior longitudinal fasciculus (SLF I, II, and III), one of the main white matter tracts connecting multiple cortical areas, is also illustrated, as are the extreme capsule and the middle longitudinal fasciculus (MLF). (Image credit to Custódio Rosa.)

communication between the pools of mirror neurons located in the parietal and frontal lobes.

Comparative analysis of these three white matter structures revealed that in rhesus monkeys the connection linking the temporal lobe structures to the frontal cortex—the so-called ventral component—far outweigh the dorsal frontoparietal pathway, mediated by the superior longitudinal fasciculus, and the temporal-parietal linkage. As such, the superior temporal sulcus was the node providing most of the connectivity in the monkey's brain-wiring diagram. In chimpanzees, the dorsal frontoparietal connectivity increased somewhat, but it is still no match for the ventral component. As a result, no cortical area dominated the traffic of nerves in the mirror neuron circuit (figure 7.8).

The situation changed considerably when the human brain was considered because the density of dorsal and ventral connectivity became much more equilibrated, and the parietal area that concentrated mirror neurons assumed the role as the key connectivity hub in the circuit that links temporal, parietal,

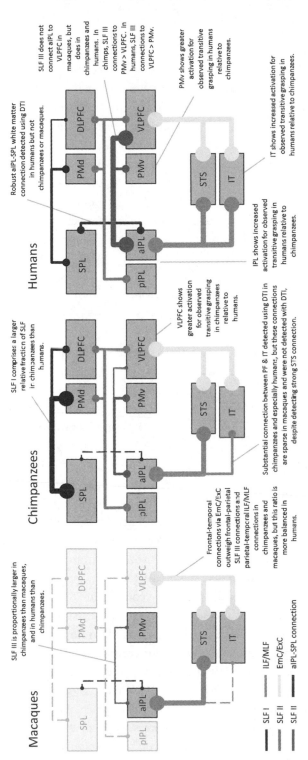

Figure 7.8. Schematic summary of the differences in cortical white matter connectivity between macaque monkeys, chimpanzees, and humans. AIP = anterior intraparietal area; aIPL = anterior inferior parietal lobe; DLPFC = dorsolateral prefrontal cortex; EmC/ExC = extreme/external capsules; ILF/MLF = inferior/middle longitudinal fasciculi; IT = inferotemporal cortex; pIPL = posterior inferior parietal lobe; PMd = dorsal premotor cortex; PMv = ventrolateral prefrontal cortex; SLF = superior longitudinal fasciculus; SPL = superior parietal lobe; STS = superior temporal sulcus; VLPFC = ventrolateral prefrontal cortex. (Originally published in Erin E. Hecht and Lisa Parr, "The Chimpanzee Mirror System and the Evolution of Frontoparietal Circuits for Action Observation and Social Learning," in *New Frontiers in Mirror Neurons Research*, edited by Ferrari and Rizzolatti [2015] Figure 9.4. By permission of Oxford University Press.)

and frontal lobes (figure 7.8). This happened thanks to an enhancement in both frontoparietal and temporoparietal interactions. According to Hecht and Parr, "The ventral extreme capsule connection within this network offers a route of information transfer that may support copying actions' end results. Conversely, connections through the superior, middle, and inferior longitudinal fasciculi might offer a route of information transfer that supports copying action kinematics. Thus, greater ventral connectivity within this network might be related to greater frontal activation during the observation of observed action and to a greater propensity for copying actions' end results, whereas greater dorsal connectivity might be related to greater occipitotemporal and parietal activation during observed action and to a greater propensity for copying actions' methods."

What Hecht and Parr are saying is that the particular distribution and density of cortical white matter linking the frontal, parietal, temporal lobes (and part of the occipital lobe too) play a pivotal role in defining the different mental strategies through which other primates and humans engage in observing and copying motor actions that they watch as part of a social interaction.

Hecht and colleagues did a more detailed analysis of the superior longitudinal fasciculus (figure 7.7) and its subcomponents, revealing that since our human ancestors diverged from chimpanzees, the inferior branch of the superior longitudinal fasciculus, also known as SLF III, increased significantly in size, likely at the expense of the superior branch, the SLF I, which is the larger component in chimpanzees. The SLF III is responsible for connecting the inferior prefrontal cortex, the ventral premotor area, and the anterior part of the inferior parietal cortex. In humans, the SLF III shows a significant increase in projections that terminate in the inferior frontal gyrus. Thus, since the first populations of *Homo sapiens* emerged from Africa, they carried in their brains a clear expansion in connectivity of the mirror neuron system to include not only the classical ventral premotor area and the parietal and occipitotemporal regions, but also a key component of the prefrontal cortex.

For the relativistic brain theory, all these changes in the human white matter configuration, including the enhancement of the dorsal frontoparietal connectivity and the differential growth of the inferior component of the superior longitudinal fasiculus, led to profound modifications in the pattern of electromagnetic fields generated by these biological solenoids. As a result, a completely distinct pattern of cortical amalgamation takes place in the human brain when compared to chimpanzees and monkeys. Indeed, this dramatic change in the cortical neuronal continuum may even help explain differences between our species and our close hominid ancestors, not only by clarifying why our individual brains are capable of generating much more elaborate behaviors, such as language and tool making, but also by elucidating why we are prone to establish much more cohesive and creative social groups than our ancestors.

Several reasons underlie my lab's motivation to perform experiments with brainets. First, we just wanted to see if it was possible to build a brainet and demonstrate that multiple brains could work together to generate a coherent motor behavior without overt body movements or communication by the subjects. Beyond this initial proof of concept, we also wanted to reassure ourselves about the feasibility of building brainets that could pair paralyzed patients, like those who would participate in the Walk Again Project, with healthy individuals, such as physical therapists, in the hope that the paralyzed patients could take advantage of the collective mental power of the resulting brainet to learn more quickly how to operate a brain-machine interface that could restore mobility. If this idea proves feasible, I can imagine a future when a single physical therapist or physician could lend her brain activity to simultaneously help train thousands of paraplegic patients around the world to operate a shared brain-machine interface capable of improving their clinical condition. As it turns out, as I write this paragraph, the first experiments demonstrating just that idea have been successfully completed in the headquarters lab of the Walk Again Project in São Paulo, Brazil. Once again, things happened much faster than originally anticipated.

The third reason for trying those experiments focused on testing my ideas about the relativistic brain: if the theory was going to hold up, I needed to find some mechanism in the brain capable of producing such widespread synchronization. I say that because even though I postulate that neuronal electromagnetic fields may be involved in this process, it is not easy to tease apart all the details that are required for neuronal synchronization to emerge in an intact brain. Thus, by creating brainets of multiple individual brains I thought we would have a better chance to study the requirements for the establishment of such large-scale synchronization. In a brainet setup one can control the sensory feedback and reward signals that are delivered to each subject participating in the experiment, so I reasoned that by measuring how and when neuronal synchronization occurs between multiple brains required to cooperate, I might gain some key insight on how large-scale synchronization can be produced within a single brain. As it turned out, in the case of the B3-brainet we discovered that the combination of common visual feedback and reward was powerful enough to tightly synchronize the electrical brain-storms produced by three individual brains taking part in the experiments. That meant that the combined B3-brainet was capable of controlling the three-dimensional movements of a virtual arm as if the neuronal signals had origi-nated from a single brain. Therefore, I began wondering if this combination could also play a key role in the consolidation of the neuronal continuum in an individual brain. That is exactly what led me to propose that a three-variable Hebbian learning rule, originally described as a mechanism for synaptic plas-ticity, may also underlie our exquisite ability as a species to form and maintain large brainets that are capable of generating a great number of complex social behaviors. That crosses off one more item on my checklist.

The fourth and final reason I wanted very badly to play with brainets was be-cause I would be able to investigate the key principles that allow such organic computers to form in nature and account for all the True Creator of Everything is capable of accomplishing in order to build the human universe. That is why I firmly believe that equating brains and brainets to individual and distrib-uted organic computers may help us understand why beehives have a lot in common with the pyramids in Egypt. According to this working hypothesis, both of them are stunning tangible outputs produced by different examples of such distributed organic computers: one, obviously, made thanks to the brains of working bees, the other created because hundreds of thousands of men, working across decades, designed and built a common architectural goal that emerged and took shape inside someone's nervous system before it could be projected into everlasting Egyptian stone. Granted, the bee organic computer is much less complex since it operates according to simple environmental or biological synchronizing signals, triggering commands genetically imprinted in the brains of the individual bee workers, which act as a very efficient collec-tive but without any individual awareness or true understanding of the task at hand. Conversely, the Egyptian pyramid-building brainet involved the need to learn abstract and manual skills, create new tools, and design a strategy to solve a variety of problems encountered during the construction process, not to mention that each individual participant was aware of the role he was playing and the (alleged) purpose of the task. It is no wonder, therefore, that Lewis Mumford named this Egyptian building effort the "megamachine," the collective human effort that would serve as the original prototype for the age of mechanization that would flourish thousands of years later—as well as the classic example of a human brainet, in my own jargon. Nevertheless, both the beehive and the Egyptian pyramid examples illustrate the fact that specific outcomes require the collaborative interactions of a large number of individual brains synchronized to produce a common tangible goal. My hypothesis is that in both cases this is accomplished through analog-based synchronization that leads to brain-to-brain coupling.

At this point I can declare my main reason for defending the existence of a common thread linking animal swarms, like those formed by bacteria, ants and bees, fishes and birds, and the large brainets created by humans. Using the thermodynamics framework introduced in chapter 3, I can link all these examples by the conclusion that since the humble origins of life on our planet, individual organisms formed swarms and synchronized themselves together in order to maximize the amount of useful work they could produce by the flow of energy and information they exchanged with their surrounding environment. Essentially, swarming and brainets share the solution for producing the most self-organization and entropy reduction—not to mention maximum Gödelian information embedding—per unit of energy/information flow exchanged with the outside world. For the vast majority of living things out there, that means a better chance to acquire more solar energy and extend their always-on-the-brink lives. In our own case, that provides the fuel needed to generate an exquisite range of mental abstractions, allowing us to suck up potential information from the cosmos and transform it into knowledge.

Months after our brainet experiments were run, I came across some video footage of the facial expressions made by two of our subjects as they controlled a B2-brainet. Watching just a few minutes of the footage brought the eerie feeling that I had seen this look before in my life—not in a lab like ours but out there, in the world, innumerous times, in a variety of circumstances. In a movie theater during a particular scene that captivated the collective emotions, memories, hopes, and desires of an entire audience; in public rallies where the speaker's voice and words mesmerized hundreds of thousands of people who shared the same political ideal; in soccer matches where fans rooted for their teams, singing together as if that kids' game was bigger than life itself. In all these cases, people seemed to merge as part of a collective entity and behave not like individuals but as part of a whole

Now, after running the brainet experiments in my lab, all of a sudden, I had a firm hypothesis to explain why that happened: each of these human social groups—in the movie theater, at the public rally, in the soccer stadium—basically represented multiple examples of distributed organic computers assembled in the nick of time.

In the beginning this idea seemed so foreign, even to a systems neuroscientist like myself, that I tried to forget about it. Yet the more I thought about it and the more I read about animal and human social behaviors and their ancient origins, the more my notion of distributed organic computing seemed to be consistent with a variety of anecdotal evidence known to all of us. As a soccer fanatic, immediately I found myself thinking about the well-known soccer wisdom that no matter how many stellar individual players a team may have—think the "more than 100 gol" Palmeiras team of 1996 or the Real Madrid Galáticos of the mid-2000s—if these star players do not "gel" as a team, you better forget about them winning anything significant. Well, the gelling of a team, the development of the so-called team chemistry, a jargon known to every sports fan, offers a very nice analogy to describe what I think a human distributed organic computer is, what it is capable of doing, and why it may require a lot of training to reach large-scale synchronization, across multiple individual brains, to assemble it. Once it does, however, whether for a short period of time or for decades of ongoing group work, such a distributed organic computer can achieve incredible feats, manifested either by concrete physical indicators or, in the case of humans, even more elaborate intellectual treasures that together define our culture and legacy as a species.

Following my epiphany of the soccer analogy, things got even worse. Now, suddenly, I would find myself thinking about a symphonic orchestra like the Berlin Philharmonic playing my favorite opera overture, Tannhäuser, not as a unique collection of highly skilled musicians but as another exquisite example of how, thanks to years of training, a handful of baton gestures from a conductor, and some stunning real-time auditory feedback, the motor cortices of dozens of human brains could become entrained or synchronized at millisecond scale to participate in such a mesmerizing example of collective sound sculpturing.

Given all these examples, I can now propose my operational definition of a brainet: basically, a brainet is a distributed organic computer composed of multiple individual brains that become synchronized—in the analog domain—by an external signal such as light, sound, language, chemicals, or radio or electromagnetic waves and, as a result, is capable of producing emergent collective social behaviors. Like individual brains, such distributed organic computers utilize organic memory storage to hold Gödelian information while transmitting Shannon information, and are capable of collective learning through a mechanism similar to Hebbian plasticity, scaled to the level of entire brains that interact with one another. As such, brainets also exhibit self-adaptation capabilities. Moreover, such a human distributed organic computer, due to its immense complexity, is also capable of a wondrous type of computing operation, something that, so far at least, defines a unique trait in the universe out there; it is capable of taking potential information provided by the universe and shaping it into knowledge that can then be packed and transmitted to future generations so they can continue our species' main existential mission: universe building.

8 • The Case for a Braincentric Cosmology

Despite the bitterly cold night, they came in surprisingly large numbers. Scattered in many small groups, they emerged from all corners of the dark frozen forest, ready to mingle into the tightly packed procession that gradually acquired its own marching rhythm. Once there, men, women, children, and all the elderly who could still walk by themselves instinctively welded their freezing bodies into a human spear of sorts, trying to pierce the blizzard weather and continue to slog forward, looking pretty much like a new emergent force of nature. Walking in silence, that human harpoon tracked a single dim cone of light, generated by the ceremonial torch carried by their most venerated shaman who, as the procession's leader, was in charge of taking them to their new underground temple.

Walking in perfect synchrony, they kept their heads low, bending their torsos forward to reduce the brutal impact of the swirling, bone-licking frigid wind. The trail of footprints they left in the deep snow, their steadfast resolve regardless of the terrifying sounds made by the predators that eagerly tracked them at close range, and their sluggish but insistent progress through the night offered an unmistakable testimony of the unlimited sacrifice they were willing to endure to demonstrate their devotion. Locked in the midst of that perilous march, they were all willing prisoners: bewitched and converted, since early life, by an irresistible, intangible urge no other animal, nor even their close ape and hominid ancestors, had ever experienced before. Without even realizing it, they were the first of a kind. Relentlessly marching, determined to reach, by any deed or cost, first the entrance and then the depths of their latest shrine, these pilgrims were defiantly following, despite all well-recognized risks, nothing but a mental mirage. Yet nothing on Earth would impede the unyielding advance of what one may as well consider the first human brainet ever assembled by pure belief.

Approximately forty thousand years ago, probably in the midst of a freezing glacial night like the one portrayed in the previous paragraph, men and women sporting bodies and brains like ours inaugurated one of the most enduring traits of our species: the unique ability to create and widely disseminate mental abstractions that, despite having their true origins in the biological coils of our own minds, are projected onto the outside world as if they represented the most unquestionable and irrefutable truths worthy of blind worship.

Since there were no clocks to be consulted, theirs was a life without many shades of time—only day and night and the transition between them, the moon cycle, Earth's natural seasons, and the patterns of animal migrations stamped clear distinctions in their life's routine. During the day they mainly hunted and gathered; during the night, around the fire, they likely shared stories that they later dreamed about. But on special occasions, they marched together, as a proud and cohesive band, toward the very underground places where, forty thousand years later, their descendants would spend a considerable amount of time debating what they had really meant when they decided to descend deep into caves and either paint the bare rock walls with elaborate scenes or simply worship the paintings of others who came before them. Their rock art was signed with their own handprints. Yet, while the richly ornate scenes were full of animals they hunted or that hunted them, there were surprisingly few images of themselves and their kin. Clearly, that was not due to a lack of artistic skill. Instead, the conspicuous absence of human renditions on the painted rock seems to indicate the desire of these ancient generations of human artists to leave behind a permanent record of the mental imagery produced in their brains—a result, as we have seen already, of the collision between the natural world and how the True Creator of Everything perceives it.

To this day, when confronted with the incredible beauty and strength of such underground cave paintings and the realization of what it took for our prehistoric ancestors to produce them in the Upper Paleolithic period, one may wonder what kind of material reward or unique indulgence these early pilgrims sought so eagerly, to the point of putting their lives and those of their loved ones at mortal risk. What kind of riches enticed such a journey through the dangerous forest in search of the entrance to a new underground cave?

Why did they all so blindly trust their lives to an elderly shaman and the dim light that emanated from both his vision and his torch?

Before moving on, I should point out that at the time I write this chapter, new evidence has appeared suggesting that Neanderthals produced similar cave paintings about sixty-five thousand years ago, or twenty-five thousand years before *Homo sapiens* did the same. If this is confirmed, then Neanderthals deserve the distinction of being the first artists of our lineage.

Surprising as it may sound in the second decade of the twenty-first century, at a time in which our consumer society and self-indulgent culture have reached their pinnacle, all the archeological evidence indicates that these men and women were not after any precious goods, power, or delicacy on their excursions. Having satisfied their immediate survival needs through hunting and gathering, the men and women of the Upper Paleolithic began to roam through frozen forests in sizeable groups in search of underground caves they could occupy, richly decorate with their own hands, and then visit routinely. Indeed, somewhere in the midst of a glacial landscape that dominated the southwest and northeast of the Pyrenees, in what today is southern France and northern Spain, Upper Paleolithic human nomads left behind richly ornate historical records in the shape of elaborate colorful paintings on the rock walls and ceilings of deep and convoluted underground caves. Overall, these artistic relics constitute the surviving fragments of fundamental aspects that defined the physical and mental lives of ancient members of our species who acquired the will and the skill to leave reports of their experiences and thoughts in a medium other than their own memories. Thus, to properly characterize the epic nature of the Upper Paleolithic people's achievements, it is essential to emphasize that until they began painting, engraving, and sculpturing the walls of underground caves, for thousands and thousands of years, the only medium available for members of our species to communicate their experiences was oral language. Likewise, the only medium available for long-term storage of such reports was human memory. Thus, until thirty thousand to forty thousand years ago, the actual neuronal substrate of the human brain served as the primary repository of both the individual life story and the accumulated history of our species. As such, it was only through language that both these historical records could be transmitted to present and future generations. When our ancestors went underground and began to paint the walls and ceilings of caves, they triggered a major communication revolution in the way in which human history was recorded and stored. On sheer rock, they suddenly became able to project their most intimate feelings and representations of the world around them and, in some cases, create lasting records of the most inner human emotions and thoughts that, to this day, no spoken or written language can properly reproduce. In this context, one may even say that by learning to paint, our ancestors shattered wide open the last doors that kept the human brain prisoner of its cranial cell. Indeed, in line with the thinking of the Austrian philosopher Ludwig Wittgenstein, the Magdalenians inaugurated the human tradition of showing with their own hands what could not be spoken about using language alone. Employing the terms of the relativistic brain theory, our Upper Paleolithic ancestors used painting instead of talking to better depict mental manifestations of high-dimensional Gödelian information—emotions, abstractions, thoughts—that cannot be conveyed well by low-dimension Shannon channels such as language.

Once fully liberated into the world, there was no way back for it, because this initial humble transfer of raw human mental imagery—derived, like all other brain products, from large-scale neuronal electromagnetic activity—to an artificial medium, in this case rock, not only allowed humans to express and communicate the way they represented and interpreted the natural world, the basis of their life's philosophy, their ethical and moral codes, and their cosmological views, it also launched in earnest the enduring human quest, which continues undeterred to this day, of identifying new forms of media and new communication channels to store and disseminate human thoughts, views, opinions, and knowledge as widely and as fast as possible throughout human civilization. During the last thirty thousand to forty thousand years, this quest has evolved from painting some mental images on rocks to the current ability to download, in real time, electrical brain activity underlying sensory and motor behaviors directly into digital media, as we do in our brain-machine interface experiments.

Not too shabby at all.

All in all, the *Homo sapiens* of the Upper Paleolithic—and perhaps the Neanderthals before them—pioneered the expression of a dominant trait of the human ethos, often manifested as if it were some kind of atavistic curse, that can be clearly identified in a variety of ways throughout the history of all major human civilizations. I am referring to the apparently innate human obsession of fully committing one's allegiance, of gambling one's present and future life, and establishing rigid codes of ethical and moral conduct, based on nothing more than an intangible mental abstraction.

As we do today, the Magdalenians (a name derived from Le Madeleine, a cave in the region of Dordogne, France), as the western European cave-painting people of the Upper Paleolithic period are commonly referred to,

lived and died under the spell of powerful mental abstractions: primordial mythologies created, disseminated, and assimilated as nothing less than tangible
reality by the True Creator of Everything. According to the theory I am advancing in this book, back then—and throughout the history of our species—such
worldviews were brewed initially in the confines of the tangled brain circuits
of a single individual or a restricted group. Soon enough, however, individual
mental abstractions spread across entire human communities like fire in dry
bushes, acquiring a life of their own and a dimension so powerful, so influential, and so irresistible that, invariably, each of them rose to become the dominate theology, creed, cosmology, ideology, or scientific theory—the names vary
but their true neurobiological origins are most likely the same—determining
individual and collective behaviors, not to mention the overall culture that
defines the core principles guiding entire human civilizations.

In this sweeping process of social takeover, each of these dominant mental
abstractions, at any given moment in our history, suddenly imposed out of
thin air what was legal or illegal, acceptable or unacceptable, proper or improper in terms of human conduct in all aspects of life, as a result casting
an omnipresent and often overbearing shadow over all aspects of human
existence. Accordingly, during the entire course of human history, as each
new mental abstraction managed to ascend and defeat the previous dominant
neural mirage, it was able, time after time, to dictate its dogmas and canons,
even when they flagrantly contradicted grounded reason and established facts
about the surrounding natural world.

Based on this premise—that mental abstractions played an essential role
in dictating our species' entire history—I propose that the cosmological description of the approximately one hundred thousand years needed to build
the human universe (that is, the totality of all intellectual and tangible material achievements made by *Homo sapiens*) can be radically reframed using a
very different viewpoint: one that has its epicenter within the human brain
working by itself or as part of human brainets. According to this cosmological reshuffling, the so-called human universe was gradually erected as distinct mental abstractions—and the social groups that pledged allegiance to
them—competed among themselves in a grand struggle for the domination
of humankind's collective mind, with the goal of achieving the sort of hegemonic position that granted the winners, at each crucial bifurcation in human
history, the power to plot the main course to be followed.

During this never-ending mental battle to become the ghost writer of our
species' history, the first step in the transition, from an old to a new dominant

mental abstraction, likely took place when a new mental construct, introduced as the result of a novel mental insight by an individual or a small group, managed to disseminate freely among a community and, at last, take hold of the minds of large numbers of people. I propose that this entire process can only happen because of both the exquisite neurophysiological properties of an individual human brain—namely, its reliance on tiny neural electromagnetic fields to fuse neuronal space and time into a continuum—and the unique capability developed by our species to synchronize large numbers of human brains in order to form very cohesive human social groups, or brainets. Based on this new view, I suggest that since the dawn of our species, tightly knitted human brainets, formed by the dissemination of particular mental abstractions, competed among each other for power and eventually for determining our species' fate.

In this brain-based framework, the entire course of human history was influenced by the outcome of such social disputes, while the self-organizing processes that emerged during these clashes gave rise to the distinct cultural, religious, political, and economic systems experienced throughout history. All in all, I dare to say that this brain-centered cosmology does a bit more justice to the truly unique legacy of our species to a cosmos that, although it existed billions of years before we emerged as an animal species on Earth, as far as we can say, depended on an obsessive observer who was available and willing to make an attempt at reconstructing its history, using his own brain-centered point of view as a frame of reference.

Although the idea of a brain-centered cosmology may sound extravagant and even outlandish at first, primarily due to the circularity involved in it—a universe leading to the emergence of a brain that devotes itself to reconstructing the history of the very universe from which it came—it is greatly reassuring to discover that many people of great intellectual stature over the centuries have proposed a similar reframing of our brain's position in the universe. For instance, in 1734, the Italian scholar Giambattista Vico, in *Principles of a New Science*, suggested that the time was ripe for the creation of a "new science," one that focused primarily on the investigation of the principles of human society. According to a citation originally reproduced by J. David Lewis-Williams in his book *The Mind in the Cave*: "[Vico] argued that the human mind gives shape to the material world, and it is this shape, or coherence, that allows people to understand and relate to the world in effective ways. The world is shaped by, and in the shape of, the human mind, despite the fact that people see the world as 'natural' or 'given.' In performing this task of shaping the world, humanity created itself. This being so, there must be a universal 'language of the mind,' common to all communities. Structuring, making something coherent out of the chaos of the natural world, is the essence of being human."

Echoing Vico, the great American mythologist Joseph Campbell stated in *Myths to Live By*, "It is a curious characteristic of our unformed species that we live and model our lives through acts of make-believe." Elaborating on this thought, he notes,

> Monkeyshines of this kind still have an effect. They represent the projection into the daylight world—in forms of human flesh, ceremonial costume, and architectural stone—of dreamlike mythic images derived not from any actual daylight-life experience, but from depths of what we now are calling the unconscious. And, as such, they arouse and inspire in the beholder dreamlike, unreasonable responses. The characteristic effect of mythic themes and motifs translated into ritual, consequently, is that they link the individual to trans-individual purposes and forces. Already in the biosphere it has been observed by students of animal behavior that where species-concerns become dominant—as in situations of courtship or of courtship combat—patterns of stereotyped, ritualized behavior move the individual creatures according to programmed orders of action common to the species. Likewise, in all areas of human social intercourse, ritualized procedures depersonalize the protagonists, drop or lift them out of themselves, so that their conduct now is not their own but of the species, the society, the caste, or the profession.

To leave no doubt, Campbell concludes, "For it is simply a fact—as I believe we have all now got to concede—that mythologies and their deities are productions and projections of the psyche [that is, the human brain]. What gods are there, what gods have there ever been, that were not from man's imagination?"

As we will see in this and in the next chapter, many other scientists, philosophers, and artists shared the same viewpoint, although this general idea was rarely referred to as a braincentric cosmology, the name I decided to use to baptize this theory. In light of that, by building upon Campbell's argument and those of many other thinkers who have supported the same view in the past, I believe that presently we are in a much better position to advance and back up scientifically the adoption of a braincentric cosmology as a new epistemic model to describe the human universe. I say that because, unlike previous attempts that involved primarily rhetorical and philosophical arguments, we can now rely on a comprehensive and cohesive neurophysiological argument to defend such a braincentric framework. Indeed, after introducing the main tenets of the relativistic brain theory in the previous chapters, my next goal here is to combine them all to build a formal case for why talking about a cosmology centered on the human brain makes so much sense. In fact, knowing what I know now, I do not see how such a viewpoint can be avoided.

Before I begin, however, I would like to emphasize that a braincentric cosmology does not imply supporting any anthropocentric definition of the universe. Indeed, nothing in this new cosmology presupposes that humankind occupies or plays any exceptional role in the cosmos. Furthermore, this braincentric cosmology is not equivalent to and hence cannot be simply dismissed as a different manifestation of solipsism or Kantian's idealism. Such a braincentric view does not at all negate the existence of an external natural world. Rather the opposite: it simply proposes that the universe provides the pool of potential information used by our human brains to generate mental representations of it. Thus, by definition, the braincentric cosmology that I propose assures the existence of a tangible universe out there.

The sequence of my argument will follow the bottom-up inverted pyramid depicted in figure 8.1. Initially, my goal is to discuss how the relativistic brain theory accounts for the exquisite ability of the human brain to generate and spread mental abstractions. So far we have encountered several examples of this peculiar human attribute when we discussed phenomena such as the body schema, the sense of self, pain, and the phantom limb sensation. All in all, those are clear examples of how the human brain creates self-referred mental constructs that define its internal neural rendition of the very body it inhabits. But the human brain is capable of generating much more elaborate mental constructs than those. In fact, I will argue that thanks to this incredible property, our brains actually build the only comprehensive definition of reality we humans can ever experience.

But before I get that far, let's build the argument one step at the time.

Let's start by discussing how the human brain handles what the outside world has to offer to it. According to my braincentric model, the stuff the universe offers to us or any intelligent observer out there is only potential information. Indeed, this view is very similar to the classic Copenhagen interpretation of quantum mechanics that proposes that before an observation or a measurement is made, one can talk about the external world only in terms of probabilities. Put in other words, before a measurement is made, whatever

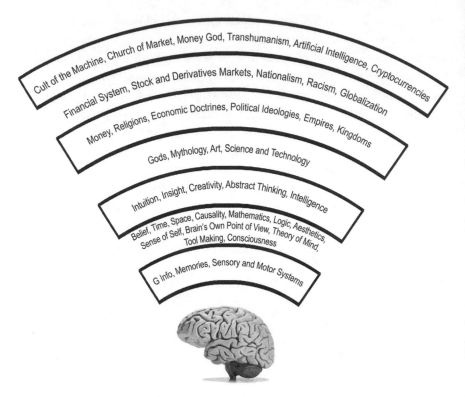

Figure 8.1. The braincentric cosmology: different levels of mental abstractions created by the human brain. (Image credit to Custódio Rosa.)

is out there in the world remains indefinable, meaning that although there is something out there—and I have no doubt about that whatsoever—it is meaningless to talk about what it is until it is witnessed or measured by an intelligent observer.

Instead of probabilities, I prefer to use the term *potential information* to describe this indefinable quantity because, in my view, without intelligent life-forms like us playing the role of avid observers and interpreters, nothing out there can cross the crucial threshold needed to become information. As such, like the distinguished American physicist John Archibald Wheeler, I subscribe to the notion that the universe can be defined or described only by the accumulated observations generated by all the intelligent life-forms that are capable of creating a coherent description of the very cosmos they inhabit. Given that, so far, we can vouch for the existence of only one such observer— *Homo sapiens*—the relativistic brain theory proposes that the human brain is responsible for the key operation of sampling the potential information that exists in the vast cosmos that surrounds us and transforming it first into Shannon and then into Gödelian information that can be used to build the brain's internal description of reality (see figure 3.2). Such a transduction, therefore, is the initial step toward building a brain-made version of the cosmos, the human universe that I have been talking about throughout this book.

Let's now follow the inverted pyramid of figure 8.1 to unveil the entire case supporting a braincentric cosmology. The first layer of the figure means only to remind us of the key anatomical and physiological properties that define the operation of the organic computer known as the human brain. As we saw before, the key attributes of the human brain include having at its disposal a large mass of neurons connected in a particular way so that complex electromagnetic fields can be created. Such analog fields support many functions, among which I include the fusing of the brain into a continuum but also providing the analog substrate through which large numbers of brains can become synchronized into brainets. At this first level, one can also include the diversified multichannel sensory apparatus that allows continuous sampling and transducing of inputs from the outside world into multiple incoming streams of Shannon information. Once this process of transduction takes place by specialized sensory receptors at the body's periphery (eyes, skin, ear, tongue), the resulting streams of Shannon information—in the form of sequences of action potentials—are quickly transmitted by peripheral nerves and the subcortical structures that define the brain's sensory pathways to the cortex. Once there, another fundamental transduction operation takes place, now at the level of neuronal circuits: the generation of electromagnetic fields, derived from neuronal electrical currents, responsible for mediating the conversion of digital Shannon into analog Gödelian information (see figure 3.2). As we saw in chapter 3, Gödelian information reshapes the micro/macro structure of brain tissue through the process of neuronal plasticity, as it is continuously embedded in neuronal tissue as long-term memories. Thanks to this latter mechanism, the human brain can gradually develop and refine its own internal point of view throughout one's life. Thus, every time new incoming sensory information is acquired, it is compared to the content of that brain's internal point of view, in order to both update it and to define one's perceptual experience at each given moment. This first level of figure 8.1 also reminds us that a series of neuronal ensemble principles constrains the operation of our brains (see chapter 4).

The second level of figure 8.1 indicates that, thanks to these basic attributes, the human brain, working in isolation or as part of brainets, can transform snippets of potential information it collects from the external world into a broad range of mental constructs that, when combined, define the brain-generated rendition of material reality. If one follows figure 8.1 from the second layer up, one can identify a hierarchical progression of such mental abstractions, from the most basic to the most elaborate. According to my own hierarchy, at the lowest level, this list includes primitive concepts such as time and space, the isolation and naming of individual objects, a comprehensive internal representation of cause-effect relationships, and the emergence of our rich perceptual experiences. At this level I also place the brain's capacity to generate meaning and semantics. Furthermore, this tier also encompasses the brain's own point of view and its main contributor, the unique human mental attribute commonly known as belief. This second level also incorporates our ability to create mathematics and logic to account for natural phenomena.

For me the elucidation of the neurophysiological mechanisms that explain how our brains generate and rely on pure belief to guide so much of human behavior is of paramount importance. I say that because usually it is through naked belief, and nothing else, that we humans create or subscribe to a wide and often disparate spectrum of mental abstractions in an attempt to elucidate primordial existential questions: things like the origin of the universe and the meaning of our lives. Although neuroscientists do not usually discuss the potential neurophysiological mechanism of belief, the relativistic brain theory proposes that belief can be defined as a "Gödelian operator." What I mean by that is that in our brains belief defines a mechanism that modulates Gödelian information somewhat like a typical mathematical operator (multiplier or divider, for example) does for numbers. By doing so, belief can affect (amplify, multiply, diminish, create, erase, maximize, minimize) human perception, emotions, expectations, attention, the readout of our memories, and many other essential mental functions. Essentially, as a whole, belief has the power to shape most, if not all, the content of the brain's own point of view. Therefore, it is no wonder that humans are exquisitely proficient in creating a huge range of mythological and religious descriptions, not to mention a vast list of gods, goddesses, heroes, and villains, in an attempt to explain, without any further requirement or need for any sort of empirical validation, all sorts of natural phenomena that, on a first and superficial inspection, totally defy one's comprehension. Indeed, one could argue that it is thanks to the pervasive and seductive power of sheer belief in the existence of supranatural causes that most of humanity has endured for millennia, with only sporadic manifestations of protest, the enormously precarious living conditions that have been routinely imposed on them, either by nature or by man-made political and economic systems.

Although I treat belief as a Gödelian operator whose initial roots are deeply embedded in our brain circuits, as a neuronal deposit of the inheritance transmitted to us by our ancestors, belief can also be acquired throughout one's life and disseminated by typical channels that convey Shannon information, like oral and written language, for example. That means that we all tend to be influenced in our beliefs by social contact, particularly with our families, friends, teachers, and other people who are perceived as being authorities in their fields or play dominant roles in society. The possibility of learning a belief may explain, for instance, medical phenomena like the placebo effect that we discussed before, as well as why so many people can be misled into believing the so-called fake news disseminated by modern technologies of mass media, particularly when it originates from someone most people judge to be credible—like, for instance, a president of a country. As we will see briefly but in more detail in chapter 11, the possibility of influencing people's belief through mass communication plays a decisive role in the formation of belief-based brainets like the one depicted in the opening passage of this chapter.

That beliefs can be learned by supervised instruction also speaks volumes for the importance and potential impact of educational systems in modern societies. I say that because according to the theory described here, a proper humanistic education can be a very powerful tool to shape collective human attitudes toward a huge variety of serious social problems that are widespread nowadays—racism, homophobia, xenophobia, and violence against minorities and women, just to mention a few items in a very long list. We will also return to this vital point in chapter 13.

Moving one step up in figure 8.1, we enter the domain of more complex mental functions such as intuition, insight, creativity, abstract thinking, and intelligence. From here, we can derive a series of complex mental abstractions such as gods, heroes, and mythology but also artistic manifestations, science, and our ability to produce and become very proficient in the use of very elaborate tools to change our surrounding environment and, more recently, even ourselves. Building on those blocks, we can now cross a threshold into a realm where large numbers of individuals begin to organize themselves around complex mental abstractions, leading to the establishment of ever-growing social, economic, religious, and political structures, thanks to the human brain's capability to synchronize into brainets. This is where kingdoms and empires, city-states and nations, political parties and economic philosophies, artistic movements and schools of thoughts come from, according to the braincentric view. It is from the same initial mental substrate that institutions built purely on belief—such as the Catholic Church or the international financial system, just to mention a couple—rose to become accepted as divine creations or tangible reality by billions of people. For me, all these are clear examples of mental abstractions that, ultimately, became larger than human life itself.

Having come this far, I am finally prepared to disclose my operational definition of a mental abstraction. For me, a mental abstraction is an analog brain computation involving the generation of a Gödelian representation that attempts to significantly reduce a large volume of potential information sampled from the outside world after it is compared against the brain's own internal point of view (where belief reigns supreme). The result is a low-dimension, all-encompassing mental model of either portions or the entirety of material reality. According to this definition, mental abstractions are Gödelian-info composites, best guesses or hypotheses that our relativistic brains generate in order to try to make sense of what is out there in the universe, in an effort to acquire an ecological advantage that enhances our chances of survival.

To refine such a definition, I will use a metaphor that may appeal to more mathematically inclined readers. The main downside of using this mathematical analogy, however, is that it is not very accurate on details, it clarifies only the broad gist of my definition. If we can keep this disclaimer in mind, I would say that mental abstractions are generated by a neuronal transformation that is somewhat analogous to the well-known multivariate statistical method known as principal component analysis. Put in a very simplified way, principal component analysis is used when one wants to identify the existence of linear correlations between large numbers of chosen variables to describe a particular phenomenon. Once these correlations are identified, such analysis allows the original multidimensional space defined by these chosen variables to be significantly reduced into a much smaller set of orthogonal components that, taken together, account for all the original variability described by the much larger set of initial variables. This happens because each of the produced principal components is formed by a particular linear combination of the original variables.

Before I move any further, it is important to emphasize that I am not saying that the brain literally engages in principal component analysis to generate mental abstractions. Far from it! If that were the case, any Turing machine would be able to generate mental abstractions galore. As we know, this does not happen now and will not happen in the future. But why is principal component analysis not a perfect analogy? For starters, it is a linear method and the brain, clearly, takes advantage of nonlinear processes to generate its main mental by-products. More important, when the brain generates any mental abstraction as a way to reduce the dimensionality of the variables it has available, it does that by adding or filtering it through its own internal bias, the brain's own internal point of view. Put another way, the brain takes advantage of Gödelian operators, such as belief and other primitive neuronal routines, embedded in our brains as part of the collective inheritance we received over millions of years from our ancestors, to modulate the process through which potential information is integrated and combined into a new mental abstraction. Therefore, using the arguments discussed in chapters 3, 5, and 6, I propose that mental abstractions are analog constructs made of Gödelian information, which are built through noncomputable operations that involve dynamic, nonlinear mixing of neuronal electromagnetic fields. That is why no digital computer will ever come up with a new god or scientific theory by itself. Yet, like belief, our brains can also project mental abstractions into lower-dimension Shannon information and disseminate it through the usual communication channels, such as oral and written language.

A simple example may clarify a bit more my definition of mental abstraction as well as the well-known fact that, given the same original set of potential information and observations describing a particular event in the natural world, two different brains can come up with diametrically opposed mental abstractions to account for it.

Suppose two individuals with very different backgrounds—a deeply religious person and an agnostic meteorologist—are on top of a skyscraper in São Paulo, Brazil, when an imminent tropical thunderstorm begins to manifest in the city's skies. Both these observers can see the clouds turning darker, feel the wind speed picking up violently. Suddenly, out of nowhere, a staccato sequence of silver lightning bolts begins scratching the horizon, followed by ominous deafening thunder announcing the opening of the skies to the descent of true walls of water. Although both observers were exposed to the same information, when asked to explain what caused the natural phenomenon they have just witnessed, their opinions may be very different. In all likelihood, the deeply religious man may simply say that the storm was a creation of a God who, from above the clouds, decided to throw lightning bolts and scream loudly because he was not happy with the way people down below were behaving. The meteorologist, on the other hand, would provide a totally different explanation, one based on his accumulated knowledge of the climate conditions that underlie the production of tropical storms.

In both cases, our observers are taking advantage of very distinct mental abstractions (religion and science)—and their individual beliefs in them—to provide a comprehensive explanation for the complex climatic event they have just observed. One can argue that in both cases a significant dimension reduction of the original variables and observations took place when their individual beliefs were allowed to operate on the resulting collision between incoming Shannon and Gödelian information inside their brains. Put in other words, in the most general sense, both the God and the scientific theory resulted from similar mental operations that collapsed a complex set of raw data and observations into a low-dimension explanation. The added advantage of producing such a reduced but comprehensive account is that, despite being totally disparate, both mental abstractions can be verbalized and disseminated widely so that, depending on the belief of the audience who listens to them, two very distinct brainets may emerge within the social group at large. One can argue that although generated by a similar neuronal apparatus, there is a profound chasm that separates these two mental abstractions. There are many differences in what they can accomplish too. For example, while the God-made thunderstorm explanation offers a satisfactory account only to those who share the same deep belief in that God, the scientific description, because it can be verified independently by anyone through the application of a particular method, does not require such a particular belief to be accepted. Instead, it requires one to accept that the human mind can produce very good approximations of natural phenomena by using mathematics and the scientific method. No doubt one can call this latter acceptance a particular form of belief, but one has to admit that it carries a very important added value. I say that because, while both explanations provide concise accounts of an event, only the one introduced by the meteorologist has any predictive power embedded in it. Claiming that a supranatural God created the thunderstorm does not help us in dealing with future similar events. Conversely, the possibility of using the scientific description to analyze the current thunderstorm and predict new ones substantially improves our chances of enduring such events in the future, by allowing us to seek shelter in advance, for example. Essentially, although both interpretations are brain-made accounts of the natural world, the scientific one gives us a much better shot at surviving the vicissitudes presented by the outside world because it allows us to adapt to, manipulate, and shape them to our own species' ecological advantage.

Overall, I believe that all mental abstractions, from the simplest to the most complex, are generated by the same type of Gödelian dimension reduction mechanism I have just described. Using this belief, I propose that in a braincentric cosmology the integration of all the mental abstractions created by all the human minds that ever lived, are alive today, or will live in the future, until the ultimate demise of our species, offers the best possible definition of the human universe we can come up with. To further support what I mean, let's spend the final part of this chapter going through a brief exercise in which an attempt is made at reconstructing a few important events of our species' recent history under the viewpoint proposed by a braincentric cosmology. The central goal of this limited exercise is simply to show how such a history can be reframed and retold based on the notion that it basically reflects the continuous dynamic struggle between distinct mental abstractions—and the social groups that pledged allegiance to them—for the hegemonic domination of humans' collective mind.

Let's start this braincentric-inspired historical digression by asking what, exactly, these Upper Paleolithic paintings tell us about our ancestors. Although it is pretty difficult to assert exactly what the prehistoric artists intended to communicate, and many potential theories exist, since the first traces of underground art were discovered, several experts have described the Magdalenians' paintings as expressing highly elaborated visual metaphors of the social organization of these prehistoric communities. For example, in *Prehistoric Cave Paintings*, a deeply touching and insightful reconstruction of Paleolithic cave art, the German art historian Max Raphael proposes that the first mental abstraction known to influence all aspects of human life was centered, rather surprisingly, not on people themselves but on the animals that occupied their surrounding natural world and that, through their sacrifice, guaranteed survival by providing food, clothing, and raw materials (for example, bone) for the manufacturing of key tools and hunting weapons.

After analyzing carefully the paintings made by our ancestors on the rock surfaces of multiple European caves, Max Raphael came to the conclusion that the depicted animal scenes were not simple representations of images seen from afar, as some archeologists originally thought. Instead, in total contrast to the paintings of classic antiquity, the Magdalenians portrayed scenes richly decorated with specific groups of animals as seen from close range. As Raphael proposes, "The Paleolithic hunter struggled with the animal at close quarters, body against body . . . [thus] the object of the Paleolithic art is not to picture the individual existence of animals and men, but to depict their group existence, the herd and the horde."

Further testimony that the artistic skills of our ancestors were far from primitive and simplistic was offered by none other than the immortal Pablo Picasso who, after the discovery of these cave paintings, exclaimed, "None of us could paint like that."

As a matter of fact, the discovery of the grandiose Paleolithic paintings in the underground caves of Chauvet, Altamira, Niaux, Lascaux, and many others can be considered a watershed event in our attempts to reconstruct the history of our recent ancestors. Max Raphael understood completely the breadth of such findings and the feeling of awe they inspired since he was one of the first authors to place these cave paintings in their proper historical perspective. In his masterpiece narrative, he points out that they were the first images to be created by the brains of the first people who "emerged from a purely zoological existence, when instead of being dominated by animals [and the countless vicissitudes and hazards of the natural world, they] began to dominate them."

And in this process they experienced, for the first time in the long and eventful history of the human clan—and of all living forms on Earth or even, who knows? in the entire cosmos—the privilege of being able to reflect upon these experiences and, in an act of pure defiance and revolutionary creativity, commit their mental imageries to an enduring medium—solid rock—depicting, in great splendor, their own brain's viewpoint of reality. What they probably did not anticipate is that the "mental snapshots" imprinted in these cave frescos would be preserved for thousands and thousands of years, so that these primordial impressions of their neural awakening, this true big bang of the human mind, could be broadcasted afresh to future generations, offering a glimpse of what it was like to be human at the dawn of the True Creator of Everything. For this and many other reasons, Max Raphael defines the Paleolithic people as "history-making people par excellence: they were in the throes of a continuous [and totally novel] process of transformation because [for the first time in history] they squarely confronted the obstacles and dangers of their environment and tried to master them."

Max Raphael speculated on what the real motives were behind the artists' renditions. Were animals part of an artist's actions, desires, or deeper thoughts? Did they represent the way the artist saw them in nature or, even more provocatively, did the animals really represent the artist, his social group, and competitive human bands? Whatever the answer to these questions may be—and there is no definitive way to know—Raphael confidently offered the conclusion, which he considered indisputable, that "totemism and magic coexisted in the world-view of the Paleolithics." For Raphael, both their almost sacred act of documenting and worshiping their own thoughts, then translating them into an external medium, and the artistic records they left behind as a result of this emergent mind-centered worldview constitute unique testimonies to the rise of the modern human mind.

It struck Raphael that the same instrument used to slay the animals, the human hand, had been employed to depict the mental imagery of the hunters, supplying the deficiencies of oral language that had, in all likelihood, already become apparent to these ancestors of ours. To compensate for the lack of words to fully describe verbally their most intimate thoughts, desires, and fears, men and women used their hands to draw and paint them on sheer rock, inaugurating an artistic tradition that has endured throughout the history of humankind. The only thing that has changed, from time to time, is the medium in which they engrave their innermost feelings and beliefs. Rock, ceramic, paper, canvas, photographs, electromagnetic waves, magnetic tapes, LPs, CDs, DVDs, the internet cloud: each of these mediums has served as the external depository of the contents of the human mind. Some aspects of their mental images people could not talk about. Instead, they discovered that to fully express themselves they had to commit their own hands to fully imprint their own electromagnetic-carved thoughts into some type of external medium. In that sense, it is simply stunning to realize that the hidden reasons that motivated the Upper Paleolithic underground cave painters to produce their art were the same that, tens of thousands of years later, led another distinguished member of our species, Michelangelo Buonarroti, to carve his powerful brain-generated vision of David into a flawless block of Carrara marble and, according to legend, at the end of his struggle, look at his final creation and simply beg: "Parla, David, parla!" (Speak, David, speak!).

For the Paleolithic humans, in addition to serving as the instruments of tool making, weapon handling, and social and intimate interaction, their hands had become essential "instruments of magic."

Evidence in favor of the human hand's novel quintessential role and mystique, according to Raphael, is further confirmed by the fact that, in many caves, like Gargas and Castillo, one can find dozens of handprints, isolated or in groups, next to depictions of the Paleolithic artists' mental construct of their animal universe. There are two forms of such handprints: positive ones, produced by applying paint to the entire hand and then pressing the palm and fingers on the rock's surface; and negative ones that were generated by laying the hand on the rock's surface while ink was blown from the artist's mouth, creating just the contour of a hand.

My own interpretation of these very moving testimonies is that the presence of so many adult and children handprints next to one another may also convey the message that the authorship and veracity of those artistic renditions had to be recognized and upheld by large social human groups during their visits to these underground sanctuaries, accepted as their most accurate cosmological view of the universe—the first ever built by brainets that encoded the collective work of many human minds together.

Another striking fact unearthed by Max Raphael that upholds his view that Upper Paleolithic humans trusted their hands to provide all sorts of benchmarks is that, in a large number of paintings, the height and width of the depicted animals seem to follow the famous golden section (3:5), a scale that can be easily obtained by spreading the palm of one's hand in the most natural way to divide it in half—Vulcan style—the thumb, indicator, and middle finger separated as far as possible from the last two fingers—on the rock surface.

What stuns me the most about the Upper Paleolithic cave paintings is their heroic dimension: what they represent simply by being there, stamped on those rock walls, and signed by the prints of the artists who painted or worshipped them. Although we will never be completely sure about the artists' original intentions in painting them, there is one irrevocable and profound message we can extract from their effort: in the many millions of years that it took for the human brain to be able to generate any minimally credible explanation for all that surrounds and amazes us, during the Upper Paleolithic, a mental abstraction created inside a human head was translated through the voluntary motor commands needed to guide the artist's hands into an enduring medium, allowing many other members of our species to acquire knowledge aimed at accounting for most, if not all, aspects of the tangible reality experienced by those people. Whether this knowledge was factually true or not by modern standards is totally irrelevant at this point. What really matters is that, by introducing a process for generating and disseminating knowledge, these Upper Paleolithic pioneers ignited a profound shift in the usual way of human living, which until then was characterized exclusively by behaviors needed to ensure immediate survival and the perpetuation of the species. In contrast to a prior pure animal existence, as Max Raphael pointed out, the pious human crowds that ventured through the perils of those frigid forests of forty thousand years ago to contemplate and assimilate the hidden messages of the paintings of their underground temples inaugurated the perennial tradition of elevating a mere mental abstraction to the summit of Mount Olympus, and extracting from it the guiding force to drive and endure the entirety of their otherwise mundane human existence.

And thereafter the same phenomenon manifested itself over and over again in the history of major human civilizations. And each time, once the new mental abstraction had taken hold of the individual and collective minds of humans and converted them and their social group into true believers, neither expressed any resistance in surrendering all aspects of their lives to the new dogma, revoking any possible dissenting thought that could still linger in their minds to the inevitability and the enchantment that the new "mental virus" seeded in their brains.

In *Myths to Live By*, Joseph Campbell introduced a very similar point of view, which was also shared by the culture historian Leo Frobenius, who proposed that through "paideumatic" or pedagogical powers "man—the unformed, uncertain animal in whose nervous system the releasing mechanisms are not stereotyped but open to imprinting—has been governed and inspired in the shaping of his cultures throughout history." That would explain why, as Campbell said, "we live and model our lives through acts of make-believe."

In modern neuroscience language, the highly plastic nature of the *Homo sapiens* brain made him easy prey to the immense predatory powers of his own mental abstractions, which could easily take over any rational way to interpret the natural world. Frobenius proposed that, as we saw above, the first mental abstraction to dominate humans' cosmological view was dictated purely by the mysteries they identified in the behavior of animals. About ten thousand years ago, once people settled in fixed communities and began to make their living through agriculture, the Earth's seasonal cycles, the fertility of its soil, and the abundance of its plants became the new center of humans' religious and cosmological views. As in the case of the Upper Paleolithic people, this new belief influenced all aspects of the Neolithic people, from their artistic manifestations to their rituals. As pointed out by David Lewis-Williams and David Pearce in *Inside the Neolithic Mind: Consciousness, Cosmos, and the Realm of the Gods*, unlike their Upper Paleolithic ancestors, Neolithic societies constructed their temples aboveground. Bertrand Russell in *History of Western Philosophy* added, "The religions of Egypt and Babylon, as well as other ancient religions, began as cults of fertility in which the Earth was [represented] as the female and the Sun the male."

This new mental abstraction led to an increase in the incipient social stratification that had already been triggered in the Upper Paleolithic. As a result, as Lewis-Williams and Pearce state, in the first Neolithic permanent settlements—the first cities built by our species—one can observe the emergence of a social elite, a differentiated upper class that had privileged access to esoteric knowledge and was in charge of performing regular ceremonies for larger crowds as well as teaching the canon. This selected priesthood became highly influential and began to play a key role in the political life of these societies.

Lewis-Williams and Pearce attribute this change in shamanism ritualism to the choice made by Neolithic societies to "build large towns and to construct massive monuments." In doing so, these cultures may have launched yet another enduring human tradition, one that involves building sumptuous edifices and monuments that reflect and pay tribute much more to human imaginary worlds than to the reality of their lives. Totems, sculptures, pyramids, temples, and cathedrals are just a few examples of how sculpturing, architecture, and sophisticated engineering techniques were put to the service of consolidating human beliefs: created out of nothing more than pure mental abstractions, these solid structures were meant to survive both the builders' own lifespans and the history of the societies they created.

According to Leo Frobenius, the next stage in humans' borrowing of their own mental abstractions to create social and political norms came when the early astronomers of the Near East—the old Sumerian priests, as Campbell refers to them—succeeded in shifting "the focus of attention . . . to the mathematics of the seven moving cosmic lights," the heavenly sky above Earth. Suddenly, the heavens became the center of human fascination and the fulcrum of our cosmological views. In Joseph Campbell's words, "Wearing symbolic crowns and in solemn costume, the king, his queen, and their courts duplicated in earthly mime the spectacle of celestial lights."

The manifestation of such a devotion to celestial power morphed into the emergence of powerful kingdoms that revered the source of their heaven-given power by erecting some of the most stunning man-made structures ever seen in the history of humanity, like the great pyramids of Giza. It was also in Egypt that Ramses II, the most prolific builder of all Egyptian pharaohs, took the mental-to-celestial connection to its limit by appointing himself as the first god-king.

Around 2000 BC, however, a major shift in the dominant mental abstraction that related humanity to the universe took place. As Campbell puts it: "In the Mesopotamian texts of about 2000 B.C., . . . a distinction is beginning to be made between the king as a mere human being and the god whom he is now to serve. He is no longer a god-king like the pharaoh of Egypt. He is called the 'tenant farmer' of the god. The city of his reign is the god's earthly estate and himself the mere chief steward or man in charge. Furthermore, it was at that time that Mesopotamian myths began to appear of men created by gods to be their slaves. Men had become the mere servants; the Gods, absolute masters. Man was no longer in any sense an incarnation of divine life, but of another nature entirely, an earthly, mortal nature." Campbell refers to this development as the "mythic dissociation" and identifies in it the key features that would, much later, dominate the theology of the three main monotheistic religions that emerged in the Levant and the Arabic peninsula: Judaism, Christianity, and Islam.

It was only with the emergence of ancient Greek civilization that, for the first time in history, humans placed themselves at the center of their own universe. Among other manifestations, expressed in sculpture and architecture, this momentous swing in cosmological view served as the background of the verses of *The Iliad* and *The Odyssey*, the epic poems attributed to Homer. Although the first written versions of these poems have been estimated to date from the eighth century BC, they describe events that took place a few centuries before, likely around the twelfth century BC.

In *The Iliad* and *The Odyssey*, the Greek gods of Mount Olympus, from Zeus to Apollo, despite their power and utter control of all aspects of human destiny, are portrayed as exhibiting clear human attributes, such as vanity, jealousy, hatred, sensuality, and passion. Indeed, they are prone to a multitude of serious character flaws.

The central place afforded to humans in these epic poems can be clearly perceived when, even in the midst of his descriptions of the most gruesome battle scenes, Homer pauses to devote a significant amount of time to illustrating who the individual about to die is, where he comes from, who his parents and wife are, and who the sons, whom he will never again have the opportunity to embrace since he will soon reside in the depths of Hades. Reading and rereading these passages over the past forty years, I cannot avoid feeling how much we as a species have lost in terms of our humanity. To see what I mean, just compare these two descriptions taken from *The Iliad* to a contemporary account of human deaths in a modern battlefield.

Forthwith Ajax, son of Telamon, slew the fair youth Simöeisius, son of Anthemion, whom his mother bore by the banks of the Simois, as she was coming down from Mount Ida, when she had been with her parents to see their flocks. Therefore he was named Simöeisius, but he did not live to pay his parents for his rearing, for he was cut off untimely by the spear of mighty Ajax.

or

Meriones then killed Phereclus, the son of Tecton, who was the son of Hermon, a man whose hand was skilled in all manner of cunning, workmanship, for Pallas Minerva had dearly loved him. It was he that made the ships for Alexandrus, which were the beginning of all mischief, and brought evil alike both on the Trojans and on Alexandrus himself.

For comparison, here you have a 2016 *CNN* report describing human casualties in the ongoing Syrian war:

In the neighboring Idlib province, another 19 were killed in airstrikes Sunday, the Aleppo Media Center group said.

Here is Joseph Campbell's assessment of the Greeks' immense contribution to our human ethos: "It is in the tragedies of the Greeks that one finds the earliest recognition and celebration of this new, immediately human, center of awe. The rites of all other peoples of their time were addressed to the animal, plant, cosmic, and supernatural orders; but in Greece, already in the period of Homer, the world had become man's world, and in the tragedies of the great fifth-century poets the ultimate spiritual implications of this refocusing of concern were for all time announced and unfolded."

But being the first to place humans at the center of their own universe was not the only major mental feat accomplished by the great Greeks. They are also credited with the creation of mathematics, philosophy, and science, a unique triad of mental abstractions. As Bertrand Russell says, it was the combination of passion and the intense desire for the pursuit of intellectual life "that made the Greeks great, as long as their greatness lasted." As with prior civilizations, Greek art, in the form of sculptures and massive buildings, such as the Parthenon in the Acropolis, projected the Greeks' mental constructs into magnificent edifices that defined the benchmark of classic antiquity architecture for centuries, way beyond Greece's borders and into long-lasting history.

The dominance of the Greek way of thinking, the mental choice to opt for a cosmological view that truly centered on humanity itself, as well as the unique innovations it created, were buried deeply by another major mental earthquake in human history that gave birth to many centuries of obscurantism in western Europe. The so-called Dark Ages came to life by the ascendance and wide dissemination of a mental abstraction that projected both a worldview and a cosmology diametrically opposed to those of the Greeks. During the next European millennium, a supranatural mental abstraction reduced humans to mere servants of a never seen, never heard, but always omnipresent and omniscient Almighty. In direct contradiction to the Greeks, during these one thousand years, the converging canons of the three major religions that originated in the Levant and Arabic peninsula demoted humans from the center of the cosmos to a secondary, insignificant and, for the most part, submissive slave role. Once humans were conceived of as sinners, they and their earthly life became corrupt. From now on, the only worthy goal of a mortal existence was to worship God in hopes of earning the privilege of spending the afterlife in his company in paradise.

While the designation of this powerful single divine entity varied according to the particular monotheistic brainet you belonged to in those times—Jehovah, God, or Allah—the devastating effects produced on distinct human societies were equally somber. As Lewis Mumford states about western Europe in *Technics and Civilization*, "During the Middle Ages the external world had no conceptual significance upon the [human] mind. Natural facts were insignificant compared with the divine order and intention, which Christ and his Church had revealed: the visible world was merely a pledge and a symbol of the Eternal World of whose blisses and damnations it gave such a keen foretaste. Whatever significance the items of daily life had was as stage accessories and costumes and rehearsals for the drama of Man's pilgrimage through eternity." Mumford quotes another author, Emile Mâle: "In the Middle Ages, the idea of a thing which a man formed for himself was always more real than the actual thing itself, and we see why these mystical centuries had no concept of what men now call science." Paraphrasing one of Mumford's most poignant metaphors, humans, in what would become a repeated curse in their history, forged their own shackles.

Through the use of their own minds, I might add.

Such a reliance on the divine as the guiding beacon of human existence carried much danger, as is always the case with far-fetched mental abstractions. The truth of this statement can be thoroughly appreciated by the fact that, as Campbell points out, many ancient human civilizations made these beliefs a matter of life or death, no matter how abstract and unreal they were. In some instances, these intangible beliefs led to the complete demise of entire human cultures. Campbell points out the example of the "ancient Aztec civilization, where it was supposed that unless human sacrifices were continually immolated on the numerous altars, the sun itself would cease to move, time stop, and the universe fall apart. And it was simply to procure sacrifices by hundreds and by thousands that the Aztecs waged on their neighbors continuous war."

In corroboration with this thesis, according to Bertrand Russell, the obsession Egyptians developed with the cult of death and the afterlife led to such a degree of conservative religious thinking that Egyptian society simply ceased to invest any effort in evolving and innovating. As a result, Egypt was invaded and easily subjugated by the Hyksos, a Semite people, during the sixteenth and seventeenth centuries BC.

As it had happened before in Egypt and other major civilizations whose culture was dominated and unified by overwhelmingly powerful mental abstractions, during the medieval period the Catholic Church employed architecture as one of the most efficient and effective ways to spread its theology and exert its domination over its main followers: the European masses. That meant that Christian mythology—many of whose main tenets, by the way, such as the relationship between the Son, the Father, and the Holy Spirit, had to be settled by the vote of no more than a couple hundred bishops who met in a sporadic series of church councils—was now committed onto the rock walls, towers, naves, and altars of churches, including immense stone cathedrals. Such buildings were incommensurate with the small medieval communities in which they were erected, as pointed out by the art historian E. H. Gombrich: "The Church was often the only stone building anywhere in the neighborhood: it was the only considerable structure for miles around, and its steeple was a landmark to all that approached from afar. On Sundays and during services all the inhabitants of the town might meet there, and the contrast between the lofty building and the primitive and humble dwellings in which these people spent their lives must have been overwhelming. Small wonder that the whole community was interested in the building of these churches and took pride in their decoration."

But there was another side to this experience. Viewing these early gigantic medieval buildings, like Tournai Cathedral in Belgium or Durham Cathedral in England, or later Gothic monuments such as Notre-Dame de Reims and the Cologne Cathedral, it is not difficult to imagine the sense of overwhelming despondency and insignificance poor European peasants must have felt when they found themselves within these temples. Quite on purpose, these sumptuous medieval and Gothic cathedrals likely played a key role in imprinting the dominant mental abstraction of the Middle Ages—the worthlessness of humanity before God—on whole communities, assuring their subjugation and convincing them of humankind's negligible role in the universe when compared to the infinite power—not to mention the lavish real estate possessions—of God. As we will see in chapter 13, not much has changed in this perennial strategy to diminish humanity's role.

In the end, it was all about control. And the installed tyranny had no other culprit to blame than humans' own minds.

The devastation imposed by the medieval cosmological view, which debased rather than uplifted the human condition, was not unique to Christianity. The tragic end of the Muslim Renaissance—a period between the eighth and eleventh centuries when Muslim scholars, astronomers, and mathematicians, living and working in many cities of Central Asia, including Merv (in current Turkmenistan), Nishapur (in current Iran), Bukhara (in current Uzbekistan), and later in Baghdad, but also in the cities of Cordoba and Toledo, part of the Arab caliphate of Andalusia in Spain, were responsible for key advances at the frontiers of medicine, astronomy, mathematics, and philosophy, thanks to their reliance on and expansion of the classic Greek traditions—could be tracked to a single dogmatic Persian theologian, Abu Muhammad Al-Ghazali. Among other antiscience views, Al-Ghazali preached that the only book worth reading by a faithful Muslim was the holy Koran. Endowed with a talent for vehement rhetoric and backed by powerful friends in Baghdad, Al-Ghazali may have single-handedly succeeded in obliterating the intense brightness of the Muslims' scientific achievements and humanism for the next ten centuries.

It took another renaissance, born out of a unique collection of Italian geniuses, including Dante, Plutarch, Donatello, Brunelleschi, Leonardo da Vinci, and Michelangelo, just to name a few, to rescue humanity from the almost terminal black mental hole of the Middle Ages. With the flourishing of the Italian Renaissance, everything changed. Instead of uncountable portraits of angels, Madonnas, and saints, the new generation of drawings, paintings, and sculptures revealed the most minute details of the human body: muscles and veins; facial expressions of love, ecstasy, pain, and sorrow; the blazing stare of mortal, penetrating human eyes.

And thus, when Michelangelo's brain had the insight to dedicate the central fresco of the Sistine Chapel's ceiling to represent the moment in which the divine touch of God endowed man with his vital essence, in total contrast to his medieval predecessors, he painted the bodies of God and Adam with identical levels of biological splendor and exquisite detail.

More than certainly, that did not escape the astute retinas of Julius II, the penny-pinching pope who had commissioned Michelangelo—a sculptor by training—for the virtually impossible job of painting the ceiling almost as a punishment for his stubbornness in insisting to be commissioned to work on the pope's tomb instead. Deep inside, however, when he finally contemplated the ceiling's frescos, still wet, sometime on a summer day in 1512, Julius II likely knew at once that any further resistance on his part was indeed futile. For yet again, in no more than a fleeting tiny instant in humankind's millenary sorrowful history, through the work of his own hands, a man had exposed the intimate depths of his primate brain on a rock wall: liberating, by this single act of audacity and genius, the mental spark proclaiming its undisputed position as the only True Creator of the human universe.

9 • Building a Universe with Space, Time, and Mathematics

One winter morning sometime in the early 1300s, as the first dim sunlight tinged the frigid Swiss skies, revealing a rose-fingered dawn worthy of a Greek poem, the village of Saint Gallen, built around the stone walls of the iconic Saint Gall monastery erected by the Benedictine order in the eighth century, was about to find its last collective sleep cycle of the night abruptly interrupted yet again. As had happened for a while, every morning the inhabitants of that typical medieval community were taken away from their sweet dreams and warm beds by a sound that had changed their lives forever. As the immense iron bells in the monastery's towers began to toll, announcing the arrival of the first of the seven canonical hours of the day—known as matins, or daybreak—every brain touched by that holy sound wave once again became entrained as part of a well-synchronized brainet.

In accordance with the seventh-century bull of Pope Sabinianus, who instituted the tradition of canonical hours, during the next twenty-four hours the bells of the monastery would produce the same intimidating sound six more times (prime, at 6 a.m.; terce, at around 9 a.m.; sext, around noon; nones, around 3 p.m.; evensong or vespers, early evening; and compline, just before bed), reinforcing the irreversible grip the bells had acquired over those fourteenth-century human brains by dictating their daily routine. Since the practice had become law, one woke at 6 a.m. (at the prime hour), ate lunch at noon (at the sext), dined, and went to bed, all following the monastery bells' commands. And as this practice took hold of life around the Benedictines' stone walls, at each bell tolling, the villagers' brains were sure to remind their owners that time had ceased to be experienced as a continuous phenomenon, stretching fluidly from dawn to dusk without punctuation or meaning—simply flowing at the will of the rhythms of the natural world, its seasons and humors.

With the advent of the canonical hours—and later of the mechanical clock, which soon found its way to the same medieval monastery's towers—a new ruler—man-made discrete time—took over the daily living schedule, subjugating even the innate natural cadence of the human biological circadian rhythms. Although it wasn't until 1345 that a consensus emerged to divide an hour into sixty minutes and a minute into sixty seconds, the impact of time-dispensing on the way people thought, behaved, and lived was immense. Time-dispensing by medieval monasteries was a life-changing event, creating a new sense of order and further enhancing the regimental aspects of human life by imposing, quite artificially, what Lewis Mumford named "the regular collective beat and rhythm of the machine." Resonating with the central theme that informs my own view of how human brainets are established and account for powerful social behaviors since the emergence of our species, Mumford justifies his statement by adding: "For the clock is not merely a means of keeping track of the hours, but also of synchronizing the actions of men."

So powerful was this new reality of temporal entrainment of human affairs that one can credit the Benedictine monasteries of western Europe with introducing one of the key mental frameworks needed for the successful establishment of the industrial revolution—several centuries later—and the concurrent emergence and widespread acceptance of another powerful mental abstraction: capitalism. That is why Mumford assigns to the mechanical clock, not the steam engine, the glory of being the key invention that announced the coming of the industrial age as well as the birth of another "man-made religion" that he "baptized" as the Cult of the Machine (see chapter 13).

In a parallel development, time-keeping would also become essential for the flourishing of another very influential human mental abstraction: science. Again in Lewis Mumford's words: "The clock, moreover, is a piece of power-machinery whose 'product' is seconds and minutes: by its essential nature it dissociated time from human events and helped create the belief in an independent world of mathematically measurable sequences: the special world of science."

The impact of an official source of time-dispensing on human behavior can be better appreciated when one realizes that the monasteries of western Europe were not alone in their mission to dictate the living rhythm of large human populations. Since the days of the prophet Muhammad in the seventh century, the Muslim world has adopted the Salah times, the five instances of the day in which all believers should stop whatever they are doing to pray. The exact time of the five prayers depends on the position of the sun in the sky, so they vary according to the geographical location of each individual. They are the Fajr (morning twilight to sunrise), the Dhuhr (at the sun's zenith, midday), the Asr (afternoon), Maghrib (sunset), and Isha'a (halfway between sunset and sunrise, before midnight). Announced publicly by the singing of the muezzin from the minarets of mosques all over the world, the five daily calls for prayers have been a source of brain synchronization for Muslims and nonbelievers alike for the past fourteen hundred years since, legend has it, the Prophet Muhammad himself learned them directly from Allah. The Jewish Zmanim also represent specific moments of the day in which certain obligations, according to the Talmud, have to be performed.

The point of this brief historical background is that, essentially, from the Middle Ages on, Catholics, Muslims, Jews, and pretty much everybody else could not escape the new human mantra of time-keeping. In fact, to some degree, we can easily say that in the last seven hundred years since the first clocks were introduced in Europe, most human beings have become totally enslaved by the unrelenting ticking of man-made time. Despite changes in form and style, clocks and watches have remained practically the same over the centuries. And proof of the overwhelming success of this monopoly of the business of time-dispensing can be verified by the fact that, to this day, clocks still control our routine. If you have any doubt, just take a peek at your smartphone's clock app—and please remind yourself that you are staring at a trace of medieval technology that has been with us for more than seven centuries.

Today, one can only speculate how thoroughly different the world in which we live would be if time-keeping and time-dispensing had not been invented and achieved such a dominating penetration of the routines of human life. One can get a glimpse of this alternative way of life by observing how the few remaining societies and cultures that did not succumb to the artificial rhythm of time-keeping devices go about their living. Or we can use our imagination to return in time to an era in which there was no concept of time. For instance, a few million years ago, before language emerged, allowing humans to establish an oral tradition of storytelling that passed information from generation to generation, the longest temporal record an individual hominid could keep was the one she maintained inside her own brain, in the form of her own long-term memories. Imprinted on each individual's cortical mantle, these memories contained traces of the experiences each one of them lived or observed throughout a lifetime. But since only a fraction of someone's lifetime record could be consciously recalled—or declared, as our jargon goes—any attempt to reconstruct even an individual life history was destined to be incomplete, fractured, and biased. Yet the emergence of neurophysiological mechanisms that allowed long-term memories to be embedded in neuronal tissue and remain stored and accessible for future recall during a lifetime marks a fundamental departure in the natural processes of time-keeping by organic matter.

The tremendous impact that enduring long-term memories have had on our hominid ancestors and in our own modern lives can be clearly illustrated when this exceptional ability is lost, due either to a neurological disease or traumatic brain damage. In this domain, the case of Henry G. Molaison, immortalized in the neuroscience literature as patient H.M., offers by far one of the most emblematic accounts. Suffering from minor epilepsy since he was ten years old, by the time he was a young adult, H.M. had become completely incapacitated by the worsening of his seizures, which had stopped responding to the anticonvulsant medication available at the time. As a last-resort attempt to improve his condition, in 1953 H.M. was subjected to an extensive and radical neurosurgical procedure to remove a large amount of cortical tissue located in the medial temporal lobe, the origin of his seizures. As a result of this procedure, a considerable volume of H.M.'s hippocampus in addition to other key structures located in the medial temporal lobes were ablated bilaterally.

Upon recovering from this neurosurgical procedure, H.M. began to experience serious memory impairments. In his immediate postoperative period H.M. could not remember the staff that took care of him daily, nor could he recall any of the events that took place during his hospital stay. Although his attention, intellectual ability, and personality were unaffected, it soon became obvious to everyone that H.M. could not commit any new information he acquired or rehearsed in his mind into long-term memory. The most stunning impact of this memory deficit manifested itself when H.M. engaged in a conversation with a person he had just met. Despite being able to establish a dialogue and interact with a new acquaintance, a few minutes later, H.M. would recall neither the conversation nor the person he had had it with.

H.M.'s peculiar condition became known as anterograde amnesia. Basically, he could not create new long-lasting memories and then recall them. Even though he was able to learn to perform some new motor and perceptual tasks, he simply did not recall having carried out the repetitive actions involved in this type of learning, nor could he verbally describe these experiences or his interactions with the experimenters.

But that was not all.

Although most of his past memories acquired years before the surgical procedure were preserved, H.M. became incapable of recalling episodic autobiographical events from his prior life, suggesting that he had also developed some level of retrograde amnesia. As a result of these neurological impairments, from the moment he awoke from the general anesthesia to the end of his life, H.M.'s brain ceased to create a permanent record of the present, almost as if it had frozen the flow of time.

Following the development of neurophysiological mechanisms for the creation and maintenance of long-term memories, the next crucial step in biological time-keeping was taken when language became widely employed among our ancestors. For a variety of reasons, the emergence of oral language can be ranked as a watershed event, another true big bang of the human mind. In the context of human time-keeping, though, the ability to express one's thoughts through language meant that instead of being limited to a private and personal historical record of one's existence, *Homo sapiens* communities could now develop a collective and comprehensive account of their traditions, achievements, emotions, hopes, and desires. Despite the fact that this new collective historical account still required imprinting into the neuronal tissue of individuals for long-term preservation, it certainly contributed to a huge expansion of the notion of time among our early kin. The emergence of oral communication and speech among *Homo sapiens* tribes can, therefore, be considered the primordial mental record-keeping mechanism that gave rise to the never-ending process of building the human universe while reconstructing the cosmos that exists out there. Thus, once upon a time, the discipline of history was born around a fire as elderly men and women repeated to their children and grandchildren the legends and myths they had heard from their parents and grandparents, in a continuous process that echoed for millennia. That is why I like to say that history and time-keeping are twins born of a single mother: language.

Even though we often ignore it, the oral tradition of storytelling, through speech or singing, has dominated most of our species' communication strategy throughout its history, all over the world. For instance, before they became committed to a medium in the eighth century BC, most of the poetic passages that form *The Iliad* and *The Odyssey* were likely memorized—and sung—by innumerable generations of Greeks as a way to become initiated into the main tenets of Greek culture and tradition. So important was this oral ritual that, legend has it, both Plato and Socrates were adamantly against committing the great Greek poems to a written record, since they believed that this new medium would contribute to the quick erosion of their pupils' mental skills; students, according to Socrates, would become lazy because, having a written record, they would gradually abandon the tradition of learning the verses by oral repetition and knowing them by heart. As we will see in chapter 12, the fierce discussion about the effects of new communication media on human cognition has remained with us, pretty much intact, since the fifth century BC.

As we saw, our Upper Paleolithic ancestors found a new way to register their worldview by creating ways to reproduce some of their mental abstractions in an artificial medium, painting the rock walls of underground caves. In doing so, they assured that time was not only expanded in their own minds, but that it would also secure a lasting visual representation of past historical events that could be appreciated by present and future generations. It took about 350 centuries for time-keeping to move from cave paintings to another type of artificial medium. The appearance of the first astronomical calendars, built using the observation of recurrent celestial events, such as the relative movement of the Earth around the sun or the rhythm of the moon phases, introduced a new time standard. First created by the Sumerians, Egyptians, and a bit later by the Chinese, these astronomical calendars coincided with the appearance of the first records of written language, about 4000 BC. The Babylonian, Persian, Zoroastrian, and Hebrew calendars followed, showing that time-keeping became a serious business for all major human cultures during a short period of time.

The introduction of written language, calendars, and later mass printing (the latter thanks to the invention of methods first by the Chinese and then by Gutenberg in 1440) provided powerful new mechanisms for human brainet synchronization. In each of these examples, multiple individual brains could now be synchronized despite not being within sight or sound of one another. By committing their thoughts, insights, ideas, doubts, and theories to paper, and thanks to the mass dissemination of printed material, individuals could now communicate with greatly expanded human audiences, widely distributed in terms of both geographical space and historical time. Printed books, in particular, revolutionized the way brainets could be established, maintained, and expanded over time because they allowed one-way communication between people within and between generations. The neurophysiological mechanism for the formation for such brainets was similar to the one we discussed in chapter 7; the main difference was that the dopamine-dependent reinforcement of expectations and mental abstractions generated by the readers' brain in a given epoch happened when these readers got in contact with the written intellectual legacy left by previous generations. For example, as I write this paragraph, I can feel the clear influence that the printed words left by Lewis Mumford, among others, are bringing to the process of shaping my own ideas and writings. By the same token, every scientist worth his salt knows the sensation of having established an intellectual bond with minds that lived decades or centuries ago but whose ideas, immortalized in print, continue to influence, guide, and set one's philosophical views, ideas, and experimental agendas long after their physical deaths. This exquisite property, the ability to synchronize across vast expanses of time and space, is unique to the human brain, and as such plays a pivotal role in the process of dissipating energy to produce knowledge generation by the True Creator of Everything.

We can now move on from this brief discussion of time-keeping to the way the concept of space has evolved over the history of humankind. To justify this subject jump, I ask permission to recruit another powerful mind to the Mumford-brainet I belong to at this very moment. I am referring to Joseph Campbell, who presciently wrote: "Space and time, as Kant already recognized, are the 'a priori forms of sensibility,' the antecedent preconditions of all experience and action whatsoever, implicitly known to our body and senses even before birth, as the field in which we are to function. They are not simply 'out there,' as the planets are, to be learned about analytically, through separate observations. We carry their laws within us, and so have already wrapped our minds around the universe."

When we discussed the Passenger-Observer experiments in chapter 7, I briefly introduced a couple of brain-based mechanisms for keeping track of one's absolute position in space as well as other neurophysiological ways of computing relative spatial coordinates, like the distance to a reward or between members of a social group. Although such basic neural mechanisms have been identified in mammals and primates over the past decades of brain research, it is clear that they were also available to our hominid ancestors millions of years ago. But the notion of space, from the human brain's point of view, has also seen profound expansions since the emergence of *Homo sapiens*. One of the first ways in which *Homo sapiens'* notion of space was enlarged beyond the natural surrounding environment was likely through adventurous migrations, which took our ancestors first out of Africa to Europe, the Levant, and Asia, and then to the entire planet. Yet the historic records of these first epic expeditions of our kin remained committed only to the long-lost biological memories of our ancestors, since no artificial medium had been invented at the time to log the diaries of these primordial journeys.

The use of underground caves during the Upper Paleolithic as the sites chosen to express humanity's newly acquired artistic skills may also have contributed to a considerable expansion of the mental representation of space since, to some experts, it represented our ancestors' belief that the underground represented a completely new spatial realm, one built in the depths of their own minds to accommodate the territory of the "afterlife."

Later, when the sky became our main inspiration, humanity's concept of space expanded beyond the Earth's surface to the celestial heavens even though at the time nobody had any idea what was the shape of our planet and where its borders, if there were any, could be found.

By the time the first permanent human settlements were established in the Neolithic period, space, in the shape of land, began to be accrued as a way to establish social divisions within communities, and later on as a way to expand the reach of kingdoms and kings. Territorial expansion, by conquest and warfare, and intense building, exploiting the newly acquired knowledge of royal engineers and architects, became ancient civilizations' modus operandi to cement their domination over their own people and neighboring estates. Space had become a commodity, a mental currency that yielded social, economic, and state power to those who conquered, occupied, and reshaped it.

Thousands of years later, a major shift took place when space began to be codified in mathematical terms. That extraordinary mental feat came with the introduction of geometry ("earth measurement" in ancient Greek) by Euclid, a Greek mathematician who lived in the harbor city of Alexandria, Egypt, around the end of the fourth and beginning of the third century BC. Likely influenced by ancient Babylonian texts, Euclid's *Elements*, his classic multiple-volume geometry textbook, would remain as the only quantified mathematical formulation of space for the next twenty centuries, until the German mathematician Georg Friedrich Bernhard Riemann at the University of Göttingen, the mecca of German mathematics, proposed in the mid-1800s his version of non-Euclidean geometry. Riemannian geometry, which deals with multidimensional smooth manifolds, was rescued from academic anonymity about half a century after it appeared in print by none other than Albert Einstein when he employed this new view of multidimensional space in the formulation of his general theory of relativity.

But before Einstein's brain gave birth to a universe where space and time were fused into a space-time continuum, other revolutions reshaped the human mind's view of space, stretching its range and reach from the very tiny to the immensely huge.

Yet again, the transition in mental abstractions that drove the shift from the Middle Ages to the Renaissance in Europe played a pivotal role, this time in the process of space expansion and redefinition. This profound transition, which brought humanity back to the center of the dominant cosmological view, also altered the way space was perceived by regular people and, as has happened so often in history, how it was represented by artists, particularly painters. Once again, I resort to Lewis Mumford's words to highlight how fundamental a shift in space representation occurred during this transition. "During the Middle Ages spatial relations tended to be organized as symbols and values. The highest object in the city was the church spire, which pointed towards heaven and dominated all the lesser buildings, as the church dominated their hopes and fears. Space was divided arbitrarily to represent the seven virtues or the twelve apostles or the ten commandments or the trinity. Without constant symbolic reference to the fables and myths of Christianity the rationale of medieval space would collapse."

That explains why in medieval paintings the size of the characters was used to imply different levels of social importance in a group. Seen today, some of these paintings create a strange feeling: equivalent human bodies, which should be painted as having equal size because they share the same visual plane, are depicted with a great level of size disparity—if, for instance, one of them is a saint or a man of the church. By mixing in their paintings scenes related to Christ's life, which occurred hundreds of years in the past, with contemporary images, medieval artists had no problem in fusing multiple time epochs within the same spatial domain. As an example of this trend, Mumford cites Botticelli's *The Three Miracles of Saint Zenobius,* which merges three distinct moments in time in a single urban stage. In summarizing this medieval view of space, in which objects could appear or disappear, without any logic, or be placed in awkward or even impossible physical positions in a scene, Mumford concludes: "In this symbolic world of space and time everything was either a mystery or a miracle. The connecting link between the events was the cosmic and religious order; the true order of space was Heaven, even as the true order of time was Eternity."

The mighty blow that shook this artistic tradition of space depiction, as well as other millenary medieval institutions, was the immediate consequence of the triumph of a new mental abstraction, one that can easily be listed as yet another example of a major revolution in the history of the human mind. Following its original brief span in Greece's fifth century BC, a second major ascension of the common person to the center of the human universe took place in Europe between the fourteenth and seventeenth centuries AD. Among other things, this epic rebirth of the human being, no longer the unrepentant sinner but now cast as the new protagonist center of the universe, meant that the representation of the natural world had to be reframed spatially. From now on, space would cease to be considered and represented as a mere appendix to the divine order. Instead, the world had to be depicted as seen from the human eye's point of view. Within this new context, the discovery of the principles of perspective, and their application to the creation of an entirely new school of painting in Italy, gave rise to the visual projection, on colored canvas, of a new world order: the one seen and filled in by the human brain. The brain's own point of view now directed the painter's hand as it used contrasting colors and shadows to create an analog rendition of the surrounding world. After brewing for almost a thousand years inside humans' own minds, eventually, this new insight freed itself, spreading to hundreds or even thousands of human minds, synchronizing them into the brainet that, through its coherent and collective creative work and courage, gave birth to what became known as the Italian Renaissance. In Mumford's assessment: "Between the fourteenth and seventeenth century a revolutionary change in the conception of space took place in Western Europe. Space as a hierarchy of values was replaced by space as a system of magnitudes. . . . Bodies did not exist separately as absolute magnitudes: they were coordinated with other bodies within the same frame of vision and created in scale. To achieve this scale, there must be an accurate representation of the object itself, a point for point correspondence between the picture and the image. . . . The new interest in perspective brought depth into the picture and distance into the mind."

As the new center of the universe, humans reshuffled the world around themselves and painted it, first inside their minds and later on canvases that we worship to this day. By viewing some of the masterpieces of this period you can spend some time simply enjoying what the brain and hands of the Renaissance geniuses were capable of accomplishing.

To further confirm this freedom from the heavens, we can move beyond Renaissance art and focus on a totally different genre: cartography. Building on the Greek and Muslim mapmakers of the past, by 1436 the new Renaissance view of space had affected the way cartographers drew their maps. With the advent of latitude and longitude lines, all known space on Earth was endowed with a precise two-dimensional location. A new generation of maps and new technologies for open ocean navigation—astrolabes, ephemerides, the compass, and Jacob's staff, the predecessor of the eighteenth-century sextant—propelled the pioneering Portuguese and Spaniard navigators to launch the great age of sea explorations of the fifteenth and sixteenth century, generating yet another major impetus for the expansion of the Renaissance's view of space. Suddenly, after centuries of land-locked pious penitence in western Europe, exploring the vast—and at that time, totally unknown spatial extension and borders of Earth's oceans and the riches they hid—became the central obsession of European courts and adventurers, whose names, given the epic magnitude of their voyages—and, some would say, of their crimes—have remained household fixtures to this day. In several cases, on behalf of God and fortune, these European expeditions into Earth's unknown space led to horrible genocides against indigenous populations all over the world. Without forgetting this terrible and tragic stain, men like Columbus, Vasco da Gama, Pedro Álvarez Cabral, Amerigo Vespucci, Hernán Cortés, Francisco Pizarro, and Ferdinand Magellan produced by their deeds a major revolution in the collective medieval mind's notion of what Earth's space really contained. No wonder the new territories in which Vespucci, Columbus, and Pedro Álvares Cabral landed became known as the New World; as far as the medieval European concept of Earth's space went, the discoveries of the Americas amounted, metaphorically, to what the identification of a new exoplanet in a distant solar system would be in the twenty-first century.

So foreign were these new territories that European courts were deeply shocked by the discovery of the tremendous diversity of the New World's animals, vegetation, and food stocks, not to mention its native inhabitants and their culture. But their shock was easily appeased by the stunningly large amounts of gold, silver, and precious stones their envoys extracted from their new possessions and brought back to their kings and queens.

For the sixteenth-century royals, the New World's space, not time, meant money.

As far as the human notion of space is concerned, the two hundred years from the mid-fifteenth to the mid-seventeenth centuries were rather turbulent ones. If the discovery of the New World was not enough of a stunning event, the brainet created by the synchronization of the thoughts and discoveries generated by the unique minds of Nicolaus Copernicus, Johannes Kepler, Galileo Galilei, Isaac Newton, Robert Hooke, and Antonie van Leeuwenhoek, among many others, most certainly accounted for the unleashing of arguably one of the greatest supernova explosions in the concept of space in the entire history of humanity. Indeed, only the space expansion that took place between the late nineteenth to the mid-twentieth centuries, thanks to the introduction of Einstein's theory of general relativity and quantum mechanics, would be able to rival it.

The impact of this fifteenth–seventeenth century brainet began to materialize when Nicolaus Copernicus (1473–1543) moved from his native Poland to Italy to enroll at the University of Bologna where, in one of the most ironic developments in history, he was later awarded a doctorate in canon law, of all fields. During the first decades of his life, Copernicus carried out his own astronomical observations. Through the analysis of his own measurements and extensive readings of the works of Greek and Muslim astronomers, Copernicus began to identify profound flaws in the classic Ptolemaic model of the solar system, proposed around 100 AD, which granted an immobile Earth the center position, not only in the solar system but in the entire universe. Although Ptolemy is usually credited as the creator of this geocentric view, his model actually embodies a refined version of similar models developed in Greece by multiple astronomers who lived a few centuries before him. Despite this apparent consensus, other Greek astronomers—like the great Aristarchus of Alexandria—doubted the idea of a universe centered on Earth. These doubts were recorded and, likely, survived to the time of Copernicus' life.

In the geocentric Ptolemaic model, all the stars, the planets of the solar system, the moon, and the sun itself orbit an immovable Earth. Five centuries ago, the discussion on the true position of Earth in the universe carried profound political and religious overtones, particularly for the institution whose survival depended on the continuous and unquestionable acceptance of the central mental abstractions that gave rise to the Middle Ages. I say that because for the medieval societies of western Europe, the singular spatial position occupied by Earth as the epicenter of the entire universe was more than an abstract astronomical or scientific issue; it constituted unambiguous proof that validated two of the most cherished beliefs of those times: the uniqueness of humanity as God's privileged progeny, and the undisputable claims of the Catholic Church—and its institutional spokespeople, the army of cardinals, bishops, nuns, and priests—as God's only truthful representative on Earth. In this context, the geocentric model of the universe proposed by Ptolemy represented a very powerful tool of domination by the Catholic Church, one that was stoically and brutally defended to the very last millisecond of credibility, no matter how much human suffering it had to produce or how many lives it had to eliminate.

Although today it is easy to disdain this provincial geocentric view of the universe, one has to realize that for the better part of fifteen centuries, the Ptolemaic model was routinely employed to generate uncountable astronomical predictions of planet trajectories with a surprising level of accuracy. As the physicist Lee Smolin discusses in *Time Reborn: From the Crisis in Physics to the Future of the Universe*, by adopting the concept of epicycles and following a few refinements introduced by Islamic astronomers, the Ptolemaic model could predict the position of planets, the sun, and the moon with a minuscule error of 0.1 percent, or one part in one thousand!

In about forty pages of a monograph entitled *Commentariolus* (Little Commentary), which was never officially published but circulated widely among scholars of the early sixteenth century, Copernicus wrote what amounts to the preamble of what would become his defining and enduring contribution to science, the treaty entitled *De revolutionibus orbium coelestium* (On the Revolutions of the Celestial Spheres), published just before his death in 1543. In that work, in a single stroke of genius backed by over half a century of study, Copernicus expelled Earth, with all its human inhabitants, animals, mountains, oceans, deserts, and Old and New Worlds, along with the entire Catholic Church and its bureaucracy, from the center of the universe. In its place, Copernicus settled the sun, near to which he allocated the new center of the entire cosmos. In this new configuration, it took about a year of the Gregorian calendar for Earth to complete an orbit around the sun. The daily rotation of Earth accounted for the day and night cycle we all experience. Copernicus also deduced that, when compared to Earth's distance from the stars, its distance from the sun was negligible.

Copernicus did not live to witness how profound and widespread was the impact of his heliocentric model, nor how brutally the Catholic Church reacted to it. Summarizing the shock Copernicus and his disciples caused, Joseph Campbell wrote: "What Copernicus proposed was a universe no eye could see but only the mind imagine: a mathematical, totally invisible construction, of interest only to astronomers, unbeheld, unfelt by any others of this human race, whose sight and feelings were locked still to Earth."

Yet the heliocentric model prevailed, despite the ultimate sacrifice paid by many of those who defended it against the geocentric dogma preferred by the Church. The fate of the Italian Dominican friar Giordano Bruno, a Copernican disciple, who dared to propose that stars were just distant suns around which planets like Earth orbited, offers the most well-known example of the

Church's reaction to the new cosmological model of Copernicus. For his collective "heresies," Bruno was tried and convicted by the Holy Inquisition. In 1600, in the middle of the Italian Renaissance, he was burned to death at the stake as punishment for his "crimes."

Picking up the baton from Copernicus, the German astronomer Johannes Kepler was the next major contributor to the expansion of the human perception of space. Using the methodical observational data painstakingly collected by the last of the great naked-eye astronomers of human history, the Dane Tycho Brahe, Kepler focused all his energy on trying to account for a small discrepancy produced when the Ptolemaic model was used to predict the orbit of Mars. From this small error in the Ptolemaic prediction, Kepler was able to derive a completely new mental abstraction, shaped in the form of mathematical language, of how planets orbit the sun. With his laws of planetary motion, he demonstrated that all planets of the solar system followed an elliptic—not circular—orbit around the sun.

Kepler's impact was much deeper than one may think. I say that because with his work Kepler extended the most successful mental abstraction of space of his time—Euclidean geometry—to the heavens. This, in turn, endowed the Copernican heliocentric model with a much finer degree of mathematical accuracy—since until then even Copernicus had used epicycles to account for the lack of circularity in Mars's orbit. Kepler's elegant solution also set the stage for the works of two other geniuses: Galileo Galilei and Isaac Newton.

Galileo Galilei is credited as being the creator of multiple fields of experimental physics, including tool-based observational astronomy. He also printed in his mind the birth certificate of the very method of inquiry that dictates the procedure of scientific investigation to this day, the so-called scientific method. His pioneering observations of the Milky Way, the natural satellites of Jupiter, the phases of Venus, and the sunspots, craters, and mountains of the moon resulted from the routine utilization of one of the two most powerful new instruments of space expansion produced by the Renaissance: the telescope. Like its counterpart, the microscope (1595), the telescope was introduced (1608) thanks to the perfection of the process of lens production. Like many other examples in the history of technology, the lens industry benefited from developments that took place centuries before: the significant increase in the production of glass during the twelfth and thirteenth centuries due to the never-ending demand for stained glass panels to decorate the windows of churches all over Europe. With the foundation in the thirteenth century of the glass works at Murano, a community near Venice, the Italian Renaissance facilitated the emergence of a gift that forever changed the way we explore different spatial ranges: from the very big and distant to the very small and near realms, spatial domains never explored before became accessible to humans' observation, reflection, and wonderment.

The introduction of the microscope produced a sudden expansion of the visible limits of space to the range of the micrometer level (1 µm = 1 × 10⁻⁶ meters). In this domain of the microscopic world, Robert Hooke was able to observe, identify, and name the key functional unit of both animal and plant tissues: the cell. In 1665, Robert Hooke described this and other discoveries in his *Micrographia*. After reading Hooke's book, Antonie van Leeuwenhoek, a Dutch tradesman with no schooling or formal scientific education, decided to learn how to produce lenses and build his own microscopes. As a product of this effort, driven purely by his intellectual curiosity, van Leeuwenhoek used his own microscopes to discover the existence of bacteria—using a sample of his own saliva—and a large variety of microscopic parasites and other life-forms.

To people's astonishment, the work of Hooke, van Leeuwenhoek, and other microscopists soon indicated that there was a vast microscopic world as rich and diverse as the one we could see with our naked eye. The very brain of humans, it was soon found, was formed by a mesh of billions of microscopic cells, which were named neurons.

Looking in the opposite direction, to the heavens, Galileo used the telescope to make astronomical observations of planets, the sun, and distant stars, and his upholding of Giovanni Bruno's idea that these stars were basically similar to our own sun—all examples of celestial furnaces—further expanded the human concept of celestial space to the limits to which the telescope-aided human eye could see. Kepler—who was Galileo's contemporary—and Galileo spoke of the possibility that comprehending the universe could be within humanity's reach, particularly through the use of the emergent—by the seventeenth century—new mental abstraction that Kepler had relied upon: mathematics, the ciphered symbolic language used since then to describe all that exists around and within us.

By demonstrating that all objects, no matter how heavy or light, fall toward the ground with a constant acceleration, following a similar curve, a parabola, that can be described by a simple mathematical equation, Galileo originated the premise that laws derived on the surface of Earth by abstract mathematical thinking and ingenuity could also apply to much larger territories of the universe. Most people didn't know it yet, but space had exploded many orders of magnitude in range, at least in Galileo's mind.

The individual who would take the decisive leap into fulfilling one significant aspect of Galileo's original research program—that is, transforming mathematical abstractions and objects, derived solely by the inner electromagnetic dynamics of the human mind, into laws that apply to the entire vast cosmos—was born on the very day of Galileo's death. As another distinguished member of the brainet that changed humanity's sense of space forever, Isaac Newton projected the human mind to never before visited spatial territories, the vast realms of the known and unknown universe whose limits remain mysterious even today, with his introduction of the concept of gravity.

It is difficult to describe the true magnitude of Newton's mental conquest. For two centuries after its formulation, Newton's theory of gravity remained the first and only description of a fundamental force of nature—one capable of acting at a distance, following the same principle, anywhere and everywhere in the universe. That such an earth-shattering discovery could be described by a simple formula became, for generations, the premier example of the epic triumph of humanity's rational thinking over mysticism. In due time, Newtonian physics became the self-propelled rocket that catapulted materialism to the dominant philosophical position it still occupies in science today.

One of the great insights of Newton's discovery, and the way he dramatically expanded on Galileo's view, was the realization that "orbiting is a form of falling." In understanding this, Newton had succeeded in unifying Galileo's findings dealing with the fall of objects on Earth with Kepler's laws of planetary motion into a single elegant idea: gravity.

Newton's model yielded many more predictions, not to say impositions, about the way the universe should behave. For starters, in Newton's universe, space was a given, an absolute entity that did not require any explanation regarding its origins, nature, or behavior; it was simply there, an endowment to the cosmos and all it contained, including us. That view also implied that space was no lesser blessing to mathematicians, although they, according to Newton, should not worry about it at all. Space was there to support the beautiful show of forces acting on objects to create precise motion. As such, we should simply leave it alone to do its job quietly, anonymously, and without creating any unnecessary and irksome mathematical difficulties for us.

Perhaps even more stunning than making space an afterthought, in the Newtonian account of the universe, time had no ticket to the celestial show.

All the events that took place inside the cosmic theater of Newton's universe were entirely deterministic. This means that, given the initial conditions of the system and the force(s) that impinged on a given object, by applying Newton's laws of motion one could directly predict the entirety of the object's future movement by deriving things like the object's acceleration, its movement direction, and overall trajectory. Put in other words, if one knows the initial conditions of the system, the forces, and then uses Newton's laws of motion, one can calculate the future position of an object in a straightforward way, even before the object gets there. That is why there are no surprises of any sort in a Newtonian universe; nothing is left to chance; every step into the future is well predicted, way ahead of time, before the future arrives. Using the computation analogy I made in chapter 6, the Newtonian universe is like a Turing machine, a digital computer; given an input and a program, one always gets the same outcome, and time has no bearing on that outcome because its flow changes neither the computer program nor the way the computer reads the original input. Moreover, like a digital computer, in the Newtonian universe, one can reverse time as easily as turning around; given a certain motion outcome, by reversing the direction we apply to the laws of motion, one can recover the initial conditions that led to that particular movement.

The Newtonian view of nature became known as determinism—the belief that all natural phenomena, including our own human intentions, can be determined by a well-defined cause. Nobody defined better the consequences of the adoption of a deterministic philosophical mind-frame based on the central axioms of a Newtonian universe than the French mathematical genius Pierre-Simon Laplace, who maintained that "if he were given the precise position and motions of all the atoms in the universe, together with a precise description of the forces they were subject to, he could predict the future of the universe with total accuracy."

Newton was not alone in his views: Copernicus's, Kepler's, and Galileo's universe models essentially shared the same attributes of absolute space and timelessness.

Newton's universe had no role for an observer either. Things simply happened, independently of whether we—or anyone or anything else, for that matter—were present to observe the show.

By the end of the nineteenth century and during the first two decades of the twentieth century, humankind experienced a renewed expansion and redefinition of the concept of space. As was the case in the seventeenth century, the notion of space exploded again in two main directions: toward the very,

very big—the billions of light-years that define the whole universe—and, conversely, to the very, very tiny—the nanometers (1×10^{-9} meters) and angstroms (1×10^{-10} meters) that define the atomic world. First, let's focus briefly on the explosion toward the very big.

During the first two decades of the twentieth century, Albert Einstein's revolutionary mental abstractions single-handedly changed the dominant view of relative movement, space, and gravity and, in the process, created a very distinct universe from the one imagined by Isaac Newton. With the publication of the special theory of relativity in 1905, Einstein brought the observer's reference point to center stage. He did that by examining the question of whether two observers, who are located far apart and moving with distinct velocities in relation to each other, can agree that two events separated by great distances are occurring simultaneously. In raising this question, Einstein was greatly influenced by the ideas of the eminent Austrian physicist Ernst Mach, who believed that all movement that takes place in the universe is relative. In other words, things move in relation to other things, not by themselves. Einstein's genius was to realize that if one takes Mach's relative view of motion and adds to it one more fundamental assumption—that the speed of light is a universal constant, meaning that any pair of observers, no matter how far apart they are, will obtain the same value (186,282 miles per second or 299,792 kilometers per second), if they measure it—neither time nor space can be considered absolute entities anymore. Faced with this dilemma, Einstein did not hesitate: he simply unleashed a total rupture with the Newtonian view of space and time by proposing what Paul Halpern properly calls in his book, *Einstein's Dice and Schrödinger's Cat*, "more malleable notions" of these primitives. In doing so, Einstein discovered that time and the very judgment of the simultaneity of events taking place far from each other are relative and ambiguous.

The classic example used to illustrate Einstein's special relativity theory is based on the interaction of two observers, represented, for instance, by two twin brothers. One is on board a spaceship traveling at a velocity close to light speed, far from Earth, where his brother has stayed to wait for his return. Next to each twin there is a clock, by which they can measure the elapsed time. Under these conditions, if the Earth-bound brother could look at his sibling's clock, located inside the faraway, fast-moving spaceship, he would verify that time was flowing slower there than according to his own clock, located on Earth's surface. This time dilation, as the effect is classically known, would mean that, upon his return to Earth, the astronaut would discover that his Earth-bound brother had aged much more than he. Interestingly, from their own individual brain's perspective, time would have elapsed as it always did, no matter if one remained on Earth while the other flew on a spaceship.

By the same token, if the Earth-bound brother managed to use a very powerful telescope to evaluate the length of his brother's spaceship during the trip, he would notice that it had shortened a little as it flew close to the speed of light. This length contraction basically means that, as movement velocity approaches light speed, space itself is compressed!

Put in other terms, Einstein's special relativity shows that judging the simultaneity of two events is not a trivial business, since two far-apart observers, traveling at different velocities, will disagree in their assessments. Much more than creating a bit of confusion for the twin brothers' clock synchronization, this conundrum further shattered the existence of an absolute concept of time in the universe. Even more disturbing, Einstein's special relativity called into question one's ability to objectively discern whether two events, occurring far apart from each other, share any causal relationship, meaning that one led to the occurrence of the other. In Lee Smolin's words: "So there is no right answer to questions that observers disagree about, such as whether two events distant from each other happen simultaneously. Thus, there can be nothing objectively real about simultaneity, nothing real about 'now.' The relativity of simultaneity was a big blow to the notion that time is real."

Smolin continues: "Hence, to the extent that special relativity is based on true principles, the universe [proposed by Einstein] is timeless. It is timeless in two senses: There is nothing corresponding to the experience of the present moment, and the deepest description is of the whole [universe] history of causal relationships at once. The picture of the history of the universe given by causal relations realizes Leibniz's dream of a universe in which time is defined completely by relations between events. Relationships are the only reality that corresponds to time—relationships of a causal sort."

By proposing this timeless universe, Einstein completed the "coup d'état" engendered by his brainet companions, Galileo and then Newton, to establish the so-called block universe, where time is basically considered as another spatial dimension. This transformation became even more evident when, in 1909, barely four years after Einstein published his theory, one of his former professors in Zurich, the mathematician Hermann Minkowski, introduced a purely geometrical description of Einstein's special relativity. Minkowski accomplished that by fusing the traditional three dimensions of space with time, creating a four-dimensional space-time continuum that could account for all movements in the universe in geometric terms.

Suddenly, in the blink of a mathematician's eye, a Swiss mental abstraction, the space-time continuum, made time vanish altogether from the entire universe.

Once again, Lee Smolin provides a perfect metaphor to what this meant, in the big scheme of things, by citing the great mathematician Hermann Weyl who, in reflecting about the magnitude of Einstein's accomplishment, had this to say: "The objective world simply is, it does not happen. Only to the gaze of my consciousness, crawling upward along the world line of my body, does a section of the world come to life as a fleeting image in space which continuously changes in time."

By now you may have guessed why, as a neuroscientist, I am taking you to a journey deep into the mind-frame that moved Einstein to command his revolution. Keep Weyl's words in your long-term memory for a while, because I will come back to them in a few paragraphs.

If there was any barrier still hindering Einstein's determination to pursue an even deeper mathematical depiction of the universe, particularly one that included a new view of gravity, the widespread impact and thorough acceptance of Minkowski's mathematical treatment of special relativity likely tipped Einstein over the edge.

For the next decade, Einstein would seek obsessively for a new geometric description of the universe. The end result of this epic search became known as his general relativity theory. By adopting the mathematics that describes the behavior of multidimensional curves, or manifolds, also known as Riemannian geometry, Einstein innovated many times over, yet again. The first major revolution triggered as a result of his mental abstraction was to introduce the concept that the universe scaffolding, the Minkowski space-time continuum, is not rigid and fixed but rather dynamic. That meant that it can bend and fold, allowing the propagation of waves.

But what was the source of the waves that travel through the universe's space-time continuum? The answer, which caused the instantaneous implosion of the Newtonian universe, could not be more shocking: gravity!

By continuing the tradition of generalizing the concept of objects falling, Einstein proposed that gravity was not manifested all over the universe as a force acting at a distance—Newton's classic view—but rather as a bending of the space-time continuum caused by the mass of planets and stars. According to a very nice description by Lee Smolin, "Planets orbit the sun not because the sun exerts a force on them but because its enormous mass curves the geometry of space-time so that the geodesics, the shortest pathway between two points in a sphere or curved surface, curve around it."

In the Einsteinian universe, gravitational waves are generated by the motion of massive celestial bodies all over the cosmos, and carry in them information about the minimal details of this heavenly dance. Even more spectacular, because gravitational waves have been generated since our universe exploded into existence as a result of the big bang, finding new ways to detect them may provide us with unique records of the cosmological events that occurred prior to the photon decoupling time, during the so-called recombinant epoch, a period in which photons could be emitted and radiate away in the form of light, before being rapidly recaptured by other particles. In this context, the space-time continuum could be compared to an immense vibrating string ensemble, whose never-ending oscillation carries in it the waxing and waning of the entire historical record of the cosmos. It was this vibration of the space-time continuum, in the form of tiny gravitational waves, that has been recently measured, for the first time, by the Laser Interferometer Gravitational-Wave Observatory (or LIGO) project, a discovery that again confirmed Einstein's general relativity theory and resulted in the 2017 Nobel Prize in physics being awarded to the three pioneer investigators who led the project.

The radicalization of Einstein's intellectual move can be appreciated in yet another perfect Lee Smolin quote: "Matter influences the changes in geometry just as geometry influences the motion of matter. Geometry becomes fully an aspect of physics, just like the electromagnetic field. . . . That geometry is dynamical and influenced by the distribution of matter realizes Leibniz's idea that space and time are purely relational."

As had been the case before, by applying Einstein's general relativity, physicists improved their predictions of the orbits of planets around the sun, particularly in the case of Mercury. But some other profound predictions embedded in Einstein's new model took physicists by surprise. For example, when reversed and solved backward in time, eventually the equations of general relativity converged to a point in which neither space nor time existed anymore; at this point, the equation yielded only infinites, and could not be solved analytically. This hypothetical limit is known as a singularity. Using the same comparison to a Turing machine I used to describe the Newtonian universe, this means that the "Einstein computer" would never stop. In this particular case, this hypothetical singularity marked what many believe is the beginning of our universe: the primordial big bang.

Einstein's projection of his mental abstractions to describe the entire cosmos transformed the status of mathematics and mathematical objects, elevating them to the zenith of the official scientific language and the very fabric of the creation.

At long last, science had touched the divine and seen the face of its own God and his commandments, all of which were written using the grammar of elegant mathematics applied to a background of space and time.

But where did time and space come from?

Given the long historical debate that surrounds this question, the answer I am about to give may be considered by some as one of the most contentious in the entire book. Yet, as I anticipated in the previous chapter when I introduced figure 8.1, the relativistic brain theory offers a very straightforward answer to the mystery of the origins of time and space: they are both creations of our human brains.

Shocking as it may sound to some initially, I am now ready to reveal why, from the point of view of the relativistic brain theory, time is like pain and space is like the sense of self. What I mean by this statement is basically that the very primitive concepts of space and time are also mental abstractions created by the human mind in order to reduce the dimensionality of complex potential information obtained from the outside world. Furthermore, I propose that as basic mental abstractions, time and space emerge as a result of the process of natural selection—that is, through interactions with the natural world—as a way of enhancing our evolutionary fitness. Put in other words, by filling the human universe with a continuous scaffold made of time and space, our brains enhance our chances to survive the contingencies imposed by the environment in which we have been immersed since the origins of our species.

My argument in defense of a brain-based origin of both time and space is pretty straightforward. There is no physical manifestation of either time or space we can speak of in the external world. Indeed, as we saw above, for most cosmological models proposed throughout history, time and space were considered either to be absolute quantities (as in the Newtonian universe), or reduced to a geometrical description (in the case of relativity). No one has ever proposed the existence of either "a time or a space fundamental particle"—no time or space boson—that would serve as the physical entity responsible for the existence and property of these two primitives. This is the first argument

I use to assert that this happens because neither time nor space exists per se in the outside world. Instead, both represent brain-built mental abstractions that allow us to make sense of the continuous changes in physical states and objects that occur in the outside world—which we perceive as the passage of time—or of the stuff that exists between objects that we individualize—which we call space. Coherent with this view is the fact that we normally do not measure time at all, only the passage of time, or "delta time."

Given this brief introduction, I can now explain why time is like pain. The short explanation is because neither of them exists in the outside world per se. Neither time nor pain can be directly measured or detected by any peripheral sensory apparatus. Instead, both time and pain result from the brain's coalescing of a variety of potential information provided by the outside world. Once this information is integrated and matched against the brain's own point of view, it is experienced by each of us as the very primordial sensations of time and pain. Essentially, according to the relativistic brain theory, time is the manifestation of a brain-built emergent property.

You may recall that I wrote in the beginning of this chapter that prior to the introduction of artificial ways of keeping time, things like the monastery bells and mechanical clocks, time was perceived as a more continuous entity, defined by the gradual and continuous transition from daylight to night defining a day, and the progressive transition of the seasons in a year. The impact of such environmental phenomena on organisms has been the focus of many decades of research on the origin of our circadian rhythms—that is, the intrinsic biological process that oscillates with a cycle close to twenty-four hours. Observed in all forms of living organisms, from bacteria to plants, animals, and humans, biological circadian rhythms likely emerged early on during the process of natural evolution as a way to maximize synchronization of key biological processes with the twenty-four-hour variation in life-supporting variables, like environmental oxygen levels. Thus, to maximize their chances of survival, organisms had to embed in their biological routine a twenty-four-hour organic clock. Because of their key role in entraining biological processes to a twenty-four-hour cycle, external environmental signals that vary on a circadian rhythm are known by the German word *Zeitgebers* (or "time givers" in English). The fundamental importance of circadian rhythms in controlling biological processes was recently recognized when Jeffrey Hall, Michael Rosbash, and Michael Young received the 2017 Nobel Prize in medicine and physiology for having elucidated the neuronal circuits and genes involved in the generation of circadian rhythms in the fruit fly *Drosophila melanogaster*.

In mammals like us, many key physiological processes follow a circadian rhythm. Those include the sleep cycle and the production of hormones, just to mention two. The maintenance of such a circadian beat is dictated by a brain-based clock: a tiny cluster of neurons located in the hypothalamus known as the suprachiasmatic nucleus that generates and distributes a circadian cadence that eventually reaches the entire brain and body. The suprachiasmatic neurons are capable of performing this task because they receive a direct projection from cells in the animal's retina that signal the presence of light in the outside world. Moreover, some suprachiasmatic neurons exhibit an endogenous twenty-four-hour cycle that can persist during complete darkness. As such, the suprachiasmatic nucleus, and the neuronal circuits that receive projections from it, likely played an essential role in the original emergence of time in our ancestors. At that point, however, time was perceived as a continuous entity, varying gradually according to the level of light in the outside world.

The existence of such a primordial brain-based circadian clock provides me with a clear example to illustrate how outside environmental signals—in this case, the variation in the intensity of sunlight—could have been used by the human brain to generate the experience of time elapsing that is so common to all of us. Indeed, the passing of time can be generated by our brains from any process that is continuously changing, either in the outside world or inside our minds. In the latter case, the passing of time is naturally perceived as a continuous phenomenon because it is primarily associated with mental phenomena that require the expression of Gödelian information. This includes our emotions and feelings, which can be embedded into the rhythm in which we sing a particular song or recite a poem. This latter mixture is guided by another mental abstraction: our sense of aesthetics. It is no wonder, therefore, that time always occupies a key, albeit mysterious, place in most of our scientific theories created to explain what takes place in the external cosmos.

The origins of circadian rhythms also help me make sense of all the historical data described at the beginning of this chapter by saying that clocks and all other artificial means of "keeping time" can influence neuronal tissue to generate the experience of time elapsing in a discrete way. Indeed, having been exposed as a species to the artificial concept of a second, a minute, or an hour for centuries, each of us is capable of experiencing what each of these time measurements feels like, albeit not realizing that they are artificial impositions created by man-made technologies and mental abstractions, as we saw above.

As in the case of brain-made time, the notion of space can also be attributed to the True Creator of Everything. Simply put, our sensation of space is nothing but a brain-made inference for what can be found between objects that we identify and detach from the background out there in the outside world. In that sense, the neurophysiological mechanisms behind the genesis of space are very similar to those that endowed our brains with the capacity to compile a variety of sensory inputs (tactile, proprioceptive, visual, and so on) to generate our sense of self and the vivid experience of occupying a finite body, separated from the outside world.

Let me present an example that, albeit simple, allows me to introduce the hypothesis on how the human brain builds our common notion of space. As I write this paragraph, I can use my peripheral vision to look at a glass of water placed on my desk. Since the pioneering experiments carried out by Hans Geiger and Ernest Marsden between 1908 and 1913 in the lab of the New Zealander physicist and Nobel laureate Ernest Rutherford at the University of Manchester, we know that the atoms that form the glass and the water that I am now perceiving as distinct continuous entities are basically formed by a very, very tiny and massive nucleus, a cloud of electrons, and a humongous amount of nothing but empty space. That means that most of the volume occupied by each atom is basically devoid of anything. The classic Geiger-Marsden experiments revealed this basic atomic structure by showing that when a very thin layer of gold foil was bombarded with a beam of alpha particles (a helium nucleus formed by two protons and two neutrons), most of the particles went through the metal. Yet you and I, and all our fellow human beings—as well as other animals on Earth—experience the glass and the water as occupying a continuous three-dimensional space, out there in the world. There is no sign of any empty space when we look at them, just as I am doing right now; we see only a continuous structure, despite the vast emptiness that exists in each of them at the atomic level.

For the relativistic brain theory, what we call space is basically a product of our brains, a mental abstraction created by neuronal circuits as a way to allow us to make sense of the scene presented in front of us, in particular how individual objects are positioned in relation to each other. This emergent notion of space may not feel so strange if we can accept that the particular macro properties of the glass and the water that we perceive—that is, the liquidity of the water or the smoothness of the glass surface—cannot be anticipated if one analyzes the properties of individual atoms or even small groups of atoms that constitute either the water or the glass. Put in other words, when structures made of atoms are projected from their natural nanometric scale to the macroscopic world in which we live and perceive things, we experience object properties—things like "liquidity" of the water or "smoothness" of glass—that could not be derived by any thorough description of their individual atoms. A system that produces this effect is called complex, and the structure that emerges as a result of the interaction of its elements is known as an emergent property. Because our brains tend to generate abstractions, they continuously produce emergent properties like the liquidity of water and the smoothness of glass. That means that what we live and experience in our daily lives results from or depends on emergent properties produced when our brains interpret the potential information provided to us by such complex systems.

Up to this point we are talking about well-accepted concepts—complex systems, emergent properties—that, by 2019, do not cause major controversy anymore, although they certainly did in the not so distant past. After thinking deep and hard about this issue, I came to realize that our brains are constantly busy generating emergent properties to build a continuous representation of the external world that can make sense to us. Reflecting about this, I have concluded that, without an observer like ours, meaning one in which a brain is actively attempting to make sense of the external world to enhance our chances of survival, how could emergent properties be experienced in the first place? Borrowing from a metaphor originally proposed by the physicist Julian Barbour in *The End of Time* to illustrate the core of his timeless cosmology theory, let's consider another concrete example—a cat. At the quantum level, a cat is nothing but a gigantic collection of atoms disposed in a particular and rather complex molecular arrangement. From moment to moment, this enormous pile of atoms assumes different states or configurations, which when seen from the atomic scale would not mean much. Yet at our level of observation—and likely the one of a poor mouse—a cat is a whole different beast: a living and breathing being that we experience as a continuous entity that can jump, run, scratch us and, sometimes, once in a while, sit calmly on our laps and concede to us the unique privilege of being able to pet it. If we behave properly, that is.

With this metaphor in mind, the first idea I had was that to explain why we do not experience any empty space when we look at the heap of atoms that form a glass of water or a cat, but rather perceive continuous objects at this macroscopic level, one has to consider the way the human brain, in particular our visual system, reacts when it is confronted with unexpected discontinui-

Figure 9.1. The visual filling in phenomenon. (Image credit to Custódio Rosa.)

ties in the outside world. The general neurophysiological phenomenon associated with this contingency is called visual filling in. To understand that you
need to remember my favorite aphorism—we see before we watch. I am using
this aphorism to emphasize that our relativistic brains are always relying on
their own internal model of the world to decide what it is that they are about to
see in the near future. The visual filling in phenomenon (figure 9.1) clearly illustrates this fundamental property. Although there is no white triangle drawn
in figure 9.1, your brain simply generates an image that corresponds to it by
combining the empty space left by the particular placement of the interrupted
circles and black triangle in the image. Patients with retinal lesions experience
the same filling in phenomenon. That explains why often they remain totally
unaware of even serious visual deficits, until they are tested by an ophthal-
mologist or begin bumping themselves on door frames or hitting the edges of
their garages with their cars. That is the essence of the filling in phenomenon;
the brain basically fills the blind spot—or a scotoma caused by a retinal lesion—with the surrounding elements of the scene.

The phenomenon of filling in is also manifested in other sensory channels, which suggests that it defines a general brain strategy employed to make
sense of scenarios in which some information is missing. Thus, for the same
reason that, for our brains, the world should not have a "hole" in the middle
created by the blind spot, we normally "fill in" words as we listen to partial or
interrupted sentences during a conversation. Through the same mechanism,
sequences of discrete tactile skin stimulation, when delivered in a particular

frequency, can be perceived as a continuous touch on our arms. From the point of view of the relativistic brain theory, the phenomenon of filling in, which seems to take part at the cortical level, represents yet another example of the power of neuronal electromagnetic fields to generate widespread neuronal synchronization that yields a continuous analog description of the outside world.

Altogether, this suggests that the generalization of the phenomenon of filling in by the brain is advantageous from an evolutionary point of view, insofar as it provides an optimal way to render the external world in the presence of a localized "hole" of information provided by sensory receptors located at the body's periphery. As I was writing this paragraph, coincidently, a clear demonstration of the potential role of visual filling in manifested itself next to me. Since my eyesight was focused on my laptop screen, I barely noticed my iPad sitting on its stand in the peripheral visual field of my left eye. Suddenly, I jumped from my chair because I had the vivid sensation that a cockroach was rushing toward me on my desk. As it turned out, the supposed cockroach was nothing but a brown sphere moving horizontally in an advertisement bar that had popped up on the iPad screen. Thanks to the filling in phenomenon, my brain converted an innocuous brown sphere into a potentially harmful threat—a North Carolina cockroach—and made me jump away from its trajectory.

I propose that visual filling in has a lot to do with the way we perceive continuities at the macroscopic level of what are totally discontinuous entities at the quantum level. Put in another way, without the kind of brain we have, the projection of the quantum world into our realm would not be perceived as continuous objects. To fully explain how this may take place, however, one has to introduce a mechanism through which the visual system is trained to expect object continuity—and use it as its main benchmark or normal standard—and then strive to produce it, no matter what level of discontinuity it may observe in a scene or in objects, for the rest of our lives. I believe that our brains have been shaped to do so, first during the long evolutionary process that got us here, but also during our long postnatal development. In the latter case, I believe that other sensory modalities, particularly the sense of touch, are used to calibrate our sense of vision (and vice versa) so that, by trial and error, our brains converge through a final solution that indicates that objects should be perceived as continuous entities. In this context, it is relevant to remind you that, in reality, we never touch anything. Because of the Pauli exclusion principle and the fact that the electrons on the surface of any object tend to repel the electrons on the surface of our body (negative charges repel each other), our fingertips get very microscopically close but never quite touch any object's surface. In what may be one of the greatest ironies in sensory neurobiology, what we all experience as a touch is nothing but the product of electromagnetic repulsion.

In addition to multimodal calibration, our brains are likely influenced, during our early postnatal life, by a multitude of social interactions that help each of us to learn a consensual model of what to expect from the external world. When mothers talk to their babies and instruct them about all aspects of life— "Be careful with the hot water; Do not touch the knife's sharp edge"—they are likely helping their children's brains to consolidate a particular model of how objects should be perceived. Combined, these mechanisms—evolution, multimodal calibration, and postnatal social consensus—likely account for how our brains generate the type of emergent properties that allow us to experience elaborate solid and continuous objects made primarily of empty space at the atomic level.

If we take this hypothesis a bit further, it is not too difficult to imagine that the very primitive notions of space and time as we experience them could also be considered as emergent properties produced by our brains, using a similar but somewhat expanded version of the phenomenon of brain filling in. So far, the best evidence in favor of this hypothesis derives from reports from subjects under the influence of hallucinogens. For instance, it is well known that some people under the influence of LSD report that the space around them has suddenly become rather liquid. My classic example is the report I heard many years ago, while I was in medical school, of an individual who, a few minutes after ingesting LSD, suddenly decided to take a dive on a rather solid concrete sidewalk because he believed it had become a swimming pool. In *The Doors of Perception*, Aldous Huxley describes in great detail what he felt half an hour after taking a small dose of mescaline. When he was asked how he was experiencing the space around him, Huxley reported, "It was difficult to answer. True, the perspective looked rather odd, and the walls of the rooms no longer seemed to meet at right angles. The really important facts were that spatial relationships had ceased to matter very much and that my mind was perceiving the world in terms of other than spatial categories. At ordinary times the eye concerns itself with such problems as Where?—How far?— How situated in relation to what? In the mescaline experience the implied questions to which the eye responds are of another order. Place and distance cease to be of much interest."

When asked to describe the furniture in the room, he had this to say about what happened to the spatial relationship between a typing table, a wicker chair, and a desk placed behind the chair: "The three pieces formed an intricate pattern of horizontals, uprights, and diagonals—a pattern all the more interesting for not being interpreted in terms of spatial relationships. Table, chair and desk came together in a composition that was like something by Georges Braque [together with Pablo Picasso, one of the founders of cubism] or Juan Gris, a still life still recognizably related to the objective world but rendered without depth, without any attempt at photographic realism."

When asked about his perception of time, Huxley was even more categorical: "There seems to be plenty of it. Plenty of it, but exactly how much was entirely irrelevant. I could, of course, have looked at my watch; but my watch, I knew, was in another universe. My actual experience had been, was still, of an indefinite duration or alternatively of a perpetual present made up of a continuously changing apocalypse."

Later on, reflecting about his very unusual experience, Huxley concluded: "But in so far as we are animals, our business is at all costs to survive. To make biological survival possible, [our] mind at large has to be funneled through the reducing valve of the brain and nervous system. What comes out at the other end is a measly trickle of the kind of consciousness which will help us to stay alive on the surface of this planet."

Most people use accounts like that of Huxley to claim that by messing with the brain we only alter the way we perceive time and space, implying that time and space are still entities that exist by themselves in the outside world. That is the current mainstream viewpoint. I beg to differ radically with this interpretation. Essentially, my hypothesis proposes that space and time are instead true mental abstractions, created by our brains through neurophysiological mechanisms that include the phenomenon of filling in. This is very reminiscent of an idea originally proposed by the German polymath Gottfried Wilhelm Leibniz—Isaac Newton's bitter rival—who in the seventeenth century argued that space cannot be seen as an entity by itself but instead should be considered as an emergent property derived from the relationship established between objects. The same kind of relational view has been suggested to apply to time by some philosophers, as Lee Smolin describes in his book.

I believe that our peculiar senses of space and time are included in the "package" our brains have to create in order to optimize our chances of survival. But as Huxley and a large number of other people can testify, the fine structure of this brain-sculptured, space-time continuum can be easily disturbed.

I suspect that by now you may be asking yourself, but what about Prigogine's arrow of time, or the notion that by strictly enforcing the second law of thermodynamics, nature may be providing a guiding signal from which time can emerge? Well, it is one thing to have a potential natural clock and another thing, completely different, to extract time out of it. My contention is that time needs an observer—and more specifically, an observer's brain—to materialize and be perceived. Furthermore, according to Henri Poincaré's famous recurrence theorem, after a very long but finite time, a dynamic system that has evolved into a particular configuration may eventually return to its original state. In this context, Prigogine's potential arrow of time, in a very, very long scale, may simply vanish as a system returns to its original state.

10 • The True Origins of the Mathematical Description of the Universe

Having described my brain-based hypothesis for the generation for time and space, I can now move on to discuss another key mental abstraction employed by the True Creator of Everything to build a tangible account of reality and the external world. To begin this discussion, I need to pose a very basic question: Where does mathematics come from?

Essentially, this question is at the core of another famous inquiry, one that puzzled not only Albert Einstein himself but also several leading mathematicians of the twentieth century. For example, in his Richard Courant's lecture in mathematical sciences delivered at New York University, the mathematician and physics Nobel laureate Eugene Wigner referred to the "unreasonable effectiveness" of mathematics in explaining the outside world. At the root of this puzzle is the repeated demonstration, over the past four centuries and change, that mathematical objects and formulations, as we saw above, seem to describe with great accuracy the behavior of natural phenomena in the universe that surrounds us. The astonishment that recurring verifications of this claim caused in many of the most brilliant minds that contributed to the quantum revolution is exemplified by another wonderful statement by Wigner, as quoted by Mario Livio in *Is God a Mathematician?* "The miracle of the appropriateness of the language of mathematics to the formulation of the laws of physics is a wonderful gift which we neither understand nor deserve. We should be grateful for it and hope that it will remain valid in future research and that it will extend, for better or worse, to our pleasure, even though perhaps also to our bafflement, to wide branches of learning."

According to the braincentric cosmology, to solve this mystery one has to begin by identifying the true creator of mathematics, the "language" multiple human brainets have created, groomed, and promoted as the best grammar to generate a comprehensive and accurate description of the cosmos.

It is not a secret to anyone that the majority of professional mathematicians believe that mathematics has an existence of its own in the universe, which means that it is totally independent of the human brain and mind. Mathematicians hold that theory mainly as a matter of professional expediency, because that allows them to have a better grip on the domain in which they work. Yet taken to the limit, this view would basically imply that all mathematics we know emerged as the product of pure discovery by its practitioners. Members of this intellectual camp are usually named Platonists because they defend the existence of Platonic mathematics. For the Platonists, there is no doubt that God—if he exists—is a member of their brotherhood. Ironic as it may sound, Kurt Gödel, the man who demonstrated the inherent incompleteness of axiomatic formal systems, was a devoted Platonist himself.

On the other extreme of this discussion, cognitive neuroscientists and psychologists, such as George Lakoff and Rafael Núñez, almost consensually refute the Platonistic view of mathematics. Instead, they argue forcibly, and with lots of experimental evidence to back their claims, that mathematics is another pure creation of the human brain. Consequently, they believe that all mathematics is invented in our minds and then used to generate a description of natural phenomena occurring in the outside world, or even to predict the occurrence of events not yet observed. In the introduction to their *Where Mathematics Comes From: How the Embodied Mind Brings Mathematics into Being,* Lakoff and Núñez state, "All that is possible for human beings is an understanding of mathematics in terms of what the human brain and mind afford. The only conceptualization that we can have of mathematics is a human conceptualization. Therefore, mathematics as we know it and teach it can only be humanly created and humanly conceptualized mathematics." They continue: "If you view the nature of mathematics as a scientific question, then mathematics is mathematics as conceptualized by human beings using the brain's cognitive mechanisms."

Thus, in addressing the essential question of why mathematicians and physicists have been able to use mathematics, time after time, to formulate comprehensive and precise theories about the universe, Lakoff and Núñez do not hesitate in replying: "Whatever fit there is between mathematics and the world occurs in the minds of scientists who have observed the world closely, learned the appropriate mathematics well (or invented it), and fit them together (often effectively), using their all-too-human minds and brains." According to this view, there is no doubt what the origin of mathematics is: mathematics comes from us or, more precisely, from the type of brain and mind we have.

As discussed by Mario Livio in *Is God a Mathematician?*, over the years, many distinguished mathematicians have broken ranks with their brotherhood to publicly defend the notion that mathematics is a human creation, brewed and packed inside our brains. For example, the distinguished British-Egyptian mathematician Michael Atiyah, a Fields and Copley Medal winner, believed: "If one views the brain in its evolutionary context then the mysterious success of mathematics in the physical sciences is at least partially explained. The brain evolved in order to deal with the physical world, so it should not be too surprising that it has developed a language, mathematics, that is well suited for the purpose." Atiyah had no qualms about openly admitting that "even a concept as basic as that of the natural numbers was created by humans, by abstracting elements of the physical world."

Interestingly, the view that defends brain-built mathematics frontally challenges Albert Einstein's famous aphorism: "The most stunning thing about the universe is that it can be understood." When examined from the point of view of an evolution-built, human brain–based mathematics, Einstein's astonishment is unwarranted. Indeed, as the computer scientist Jef Raskin points out, "The groundwork for mathematics had been laid down long before in our ancestors, probably over millions of generations."

As cited in Livio's book, Raskin argues that mathematics had to be consistent with the physical world and, as such, it is a human-created tool that serves to describe the universe that exists outside our heads. Therefore, there is no big mystery why mathematics offers a good fit for the surrounding world, simply because it was this world and all its peculiar features that led to the embedding of the primitives inside our brains that resulted in the emergence of logic and mathematics in the first place.

The evolutionary nature of mathematics is strongly supported by the demonstration that other animals, including other vertebrates, mammals, and our close ancestors, monkeys and apes, also express rudimentary mathematical skills, particularly numerical abilities. Lakoff and Núñez list a series of compelling examples collected over the past six decades. For example, rats can be trained to press a lever a specific number of times to obtain a food reward. Rodents also learn to estimate a finite number through their perception of a sequence of tones or light flashes, demonstrating that they have some sort of general brain-generated number estimation capability that is sensory-modality independent.

The experimental evidence also indicates that nonhuman primates are better "mathematicians" than rats. For example, wild rhesus monkeys seem to exhibit a level of arithmetic proficiency that rivals that of human infants. Other studies have shown that chimpanzees can perform sum operations that involve the use of fractions, such as one-quarter, one half, and three-quarters; when presented with one-quarter of a fruit (apple) and a glass half filled with a colored liquid, a chimp would invariably select three-quarters as the answer for this mathematical puzzle.

In summary, there is a consensus that, unlike those of humans, the brains of rodents and primates are not equipped to express mathematic skills that go beyond some elementary primitives. As such, they cannot create an abstract description of the natural world as we do.

For over half a century now, neuroscientists have realized that individual neurons in the primary visual cortex of mammals and primates exhibit the exquisite property of firing maximally when lines of light presented at different orientations, or even moving bars, are placed inside the neuron's visual receptive field. That suggests to me that the primitives of geometry, like straight lines, were imprinted in the brains of animals during the evolutionary process as a result of their interactions with the external environment. And since this imprinting yielded a considerable evolutionary advantage, it has been passed from generation to generation and from species to species until it found itself hosted deep in the visual cortex of our own human brains.

So far I have talked about mammals and primates. However, a couple of years ago, Ronald Cicurel brought to my attention a video he watched during a scientific conference. The video describes the mating ritual the Japanese puffer fish performs at the bottom of the ocean to attract females. This tiny fish, which is naturally almost invisible in the bluish ocean waters, expends a full week to get a single date—working twenty-four hours a day, seven days a week, without a break—to complete his geometrical masterpiece. Using a blueprint embedded by evolution in his diminutive brain, this fish is able to use his fins to plow the seabed to sculpt a magnificent three-dimensional "calling to mate sign" made of pristine sand and pure mathematical instinct. Paraphrasing the British naturalist and broadcaster David Attenborough, if this puffer fish does not convince you that the primitives of both mathematics and geometry have been embedded in animal brains, including ours, long ago during the evolutionary process as a result of interactions with the surrounding world, nothing will. Indeed, in commenting on the puffer fish video, Ronald touched directly on a crucial point: "We have not been selected by evolution to see or experience reality as it is but to maximize our ability to survive most circumstances posed by the world that surrounds us. These are two distinct things. Experiencing reality as it is does not guarantee fitness at all; it may even be a handicap. So, there is no reason for our brains to be 'realistic' in their account of the world. Instead, their function is to anticipate and mitigate potential risks that we may incur while immersed in this world, even if we never experience it as it really is, but through the point of view created and provided by our primate brains."

Lakoff and Núñez support this notion by providing a comprehensive list of studies showing that some of our mathematical skills are innate, being expressed by babies at a very early postnatal age. In this domain, these authors emphasize that all humans, no matter their cultural or educational background, are capable of instantaneously telling whether they are facing one, two, or three objects. All experimental evidence points to the fact that this ability, known as "subitizing," is innate. Some basic aspects of arithmetic operations, such as grouping, addition and subtraction, and some primitive geometric concepts may also be inborn as far as humans are concerned.

Over the past years, neurophysiological and imaging methods have been utilized to show which parts of the brain are involved in the process of "doing mathematics." In one of the most unusual findings of this line of research, neuroscientists have been able to identify a small number of patients in whom epileptic seizures are triggered the moment the subject begins doing arithmetic calculations. Properly named *Epilepsia arithmetices*, these epileptic attacks have been shown to originate in a region of the inferior parietal cortex. Further imaging studies have also implicated the prefrontal cortex during the performance of more complex arithmetical operations. Interestingly, rote memory, the one used to remember multiplication tables, requires the involvement of subcortical structures, such as the basal ganglia. Likewise, algebra seems to involve different brain circuits than those utilized for arithmetic calculations.

Lakoff and Núñez put forward the idea that a key reason humans have been able to expand their innate mathematical skills resides in our ability to build what they call "conceptual metaphors." This is very similar to my idea of mathematics as another type of elaborate human mental abstraction. Lakoff and Núñez define this as our exquisite mental ability as a species to translate what may originally be only an abstract concept into a much more tangible projection of it. In supporting this contention, these authors suggest that arithmetic, which has become a very concrete tool in human life, may have its mental origins rooted as a metaphor of object collection. By the same token, they propose that the more abstract algebra that characterizes Boolean logic may have arisen from a metaphor that links classes to numbers.

In closing this part of the debate, I find it only fair to give to Lakoff and Núñez the final word on the view that proclaims the True Creator of Everything as the copyright owner of mathematics and all mathematical objects ever created to account for the natural phenomena of the human universe. "Mathematics is a natural part of being human. It arises from our bodies, our brains, and our everyday experiences in the world. Mathematics is a system of human concepts that makes extraordinary use of the ordinary tools of human cognition. . . . Human beings have been responsible for the creation of mathematics, and we remain responsible for maintaining and extending it. The portrait of mathematics has a human face."

Having exhausted the main arguments, there is one thing that not even the Platonists of mathematics can deny. If one day they could find proof for their point of view, that proof would come from a human brain, like all the other proofs they have produced in the history of the field.

Somehow, there is no escaping from the True Creator of Everything.

At this point I can say that there are very far-reaching consequences to accepting that mathematics is brain-made. If one accepts the evolutionary origins of mathematics as a given, neither human logic nor mathematics can be seen as universal. That means that theories built using human mathematics cannot be accepted as the only truthful description of the cosmos. Logically, it follows that, assuming that other intelligent life-forms exist in the universe, and that one day we may be able to establish contact and communicate with them, particularly those that evolved in a different part of the universe under very different natural conditions than ours—let's say, on a planet revolving around binary stars—our logic and mathematics may not make any sense to these aliens. Instead, they may offer an alternative explanation for the universe that will be totally foreign to us. What this means, basically, is that all cosmological views of the universe can be seen only as "relativistic" since different intelligent life-forms, because they likely evolved different biological substrates for their intelligence, will be prone to come up with a distinct view of the cosmos. Essentially, that means that Ernst Mach's concept of relative motion, which so deeply inspired Einstein to come up with his special theory of relativity, should be expanded from the restricted domain of analyzing motion to be applied as a new framework to account for a completely new cosmological view of the universe. That is precisely what my braincentric cosmology intends to do.

This idea can be illustrated by a very simplistic mathematical analogy, borrowed from the mathematician Edward Frenkel. In this analogy, the description of the same simple vector is defined from the viewpoint of two different frames of reference, or coordinates. Depending upon which frame of reference one chooses, the same vector will be defined by a different pair of numbers. This is precisely what I mean when I say that cosmological descriptions can be seen only as relativistic: as in the case of the vector, depending on the frame of reference applied by different intelligent life-forms living in distinct parts of the universe, the same cosmos will be described in very different ways.

According to the relativistic brain theory, the nonlinear nature of the neuronal electromagnetic interactions that characterize the hybrid analog-digital engine of the human brain would allow higher-order mental abstractions, like elaborate mathematics, to be produced within a single brain. Subsequently, through social exchanges with other mathematicians over multiple generations, mathematical concepts and objects could evolve naturally. Essentially, I see the internal nonlinear dynamics of both an individual human brain and large human brainets generating the same type of unpredictable behavior the French mathematical genius Henri Poincaré observed in his nonlinear equations when he slightly changed the initial conditions, or when Ilya Prigogine observed complex spatiotemporal structures emerge from certain chemical reactions (see chapter 3). Because of its propensity to generate rich dynamic interactions and mixing, the long-term operation of a human mathematics brainet over hundreds of generations could certainly account for the emergence of all flavors and layers of mathematical complexity, moving from the humble beginnings, seeded by the imprinting of mathematical and geometric primitive kernels in the brains of our animal and hominid ancestors. Therefore, the whole body of accumulated mathematical knowledge could be seen as another type of emergent property produced by a human brainet, dispersed in time and space, over the entire history of humankind.

But why does this debate matter at all? At stake here are two concepts that most scientists, in particular physicists, have been holding for quite some time because, as Schrödinger put very well in *What Is Life?*, they serve as the key foundations of the type of science we have chosen to do since the time of Galileo. Without them, a lot would have to change in our approach to exploring the world, or, at the very least, in the way we interpret our findings. The two bedrock concepts are the existence of an objective reality that is independent of the human mind and causality. As you may have noticed already, the braincentric cosmology that I propose here challenges the notion that one can speak of such an objective reality without taking into account the interference provided by our brains in describing what is out there in the universe. Although this debate has been raging for quite some time now, fortunately for me, in the past several years very distinguished thinkers have supported the braincentric view espoused here, even though they have never used this term to describe their viewpoints. In the last part of this chapter, my goal is to bring to the foreground some of these physicists, scientists, writers, and philosophers who laid the ground-work that supports today the braincentric cosmology proposed in this book.

Although one could argue that the intense debate between the distin-guished Austrian physicists Ernst Mach and Ludwig Boltzmann, which took place in the late decades of nineteenth-century Vienna, offered the first salvo of the modern battle for the true nature of reality, I would like to illustrate the chasm between the two contrasting views with a different encounter. I am re-ferring to a conversation that can easily be ranked as one of the greatest intel-lectual duels of the twentieth century. This epic collision of worldviews began on July 14, 1930, when the Nobel laureate Bengali poet and Brahmo philoso-pher Rabindranath Tagore paid an afternoon visit to Albert Einstein's home in Berlin. During this first meeting, the following dialogue was recorded:

EINSTEIN: There are two different conceptions about the nature of the universe: (1) The World as a unity dependent on humanity. (2) The world as a reality independent of the human factor.

TAGORE: When our universe is in harmony with Man. The eternal, we know it as Truth, we feel it as beauty.

EINSTEIN: This is the purely human conception of the universe.

TAGORE: There can be no other conception. This world is a human world—the scientific view of it is also that of the scientific man. There is some standard of reason and enjoyment which gives it Truth, the stan-dard of the Eternal Man whose experiences are through our experiences.

EINSTEIN: This is a realization of the human entity.

TAGORE: Yes, one eternal entity. We have to realize it through our emotions and activities. We realized the Supreme Man who has no in-dividual limitations through our limitations. Science is concerned with that which is not confined to individuals; it is the impersonal human world of Truths. Religion realizes these Truths and links them up with our deeper needs; our individual consciousness of Truth gains universal significance. Religion applies values to Truth, and we know this Truth as good through our own harmony with it.

EINSTEIN: Truth, then, or Beauty is not independent of Man?

TAGORE: No.

EINSTEIN: If there would be no human beings any more, the Apollo of Belvedere would no longer be beautiful.

TAGORE: No.

EINSTEIN: I agree with regard to this conception of beauty, but not with regard to Truth.

TAGORE: Why not? Truth is realized through man.

EINSTEIN: I cannot prove that my conception is right, but that is my religion.

TAGORE: Beauty is in the ideal of perfect harmony which is in the Universal Being; Truth the perfect comprehension of the Universal Mind. We individuals approach it through our own mistakes and blunders, through our accumulated experiences, through our illuminated consciousness—how, otherwise, can we know Truth?

EINSTEIN: I cannot prove scientifically that Truth must be conceived as a Truth that is valid independent of humanity; but I believe it firmly. I believe, for instance, that the Pythagorean Theorem in geometry states it is something that is approximately true, independently of the existence of man. Anyway, if there is a reality independent of man, there is also a Truth relative to this reality; and in the same way the negation of the first engenders a negation of the existence of the latter.

TAGORE: Truth, which is one with the Universal Being must essentially be human, otherwise whatever we individuals realize as true can never be called truth—at the least the Truth which is described as scientific and which can only be reached through the process of logic, in other words, by an organ of thoughts [the brain] which is human. According to Indian Philosophy there is Brahman, the absolute Truth, which cannot be conceived by the isolation of the individual mind or described by word but can only be realized by completely merging the individual in its infinity. But such a Truth cannot belong to Science. The nature of Truth which we are discussing is an appearance—that is to say, what appears to be true to the human mind and therefore is human, and may be called maya or illusion.

EINSTEIN: So according to your conception, which may be the Indian conception, it is not the illusion of the individual, but of humanity as a whole.

TAGORE: The species also belongs to a unity, to humanity. Therefore the entire human mind realizes Truth; the Indian or the European mind meet in a common realization.

EINSTEIN: The word species is used in German for all human beings, as a matter of fact, even the apes and the frogs would belong to it.

TAGORE: In science we go through the discipline of eliminating the personal limitations of our individual minds and thus reach that comprehension of truth which is the mind of the Universal Man.

EINSTEIN: The problem begins whether Truth is independent of our consciousness.

TAGORE: What we call truth lies in the rational harmony between the subjective and objective aspects of reality, both of which belong to the super-personal man.

EINSTEIN: Even in our everyday life we feel compelled to ascribe a reality independent of man to the objects we use. We do this to connect the experiences of our senses in a reasonable way. For instance, if nobody is in this house, yet the table remains where it is.

TAGORE: Yes, it remains outside the individual mind, but not the universal mind. The table which I perceive is perceptible by the same kind of consciousness which I possess.

EINSTEIN: If nobody would be in the house the table would exist all the same—but this is already illegitimate from your point of view—because we cannot explain what it means that the table is there, independent of us. Our natural point of view in regard to the existence of truth apart from humanity cannot be explained or proved, but it is a belief which nobody can lack—no primate beings even. We attribute to truth a super-human objectivity; it is indispensable for us, this reality which is independent of our existence and our experience and our mind—though we cannot say what it means.

TAGORE: Science has proved that the table as a solid object is an appearance and therefore that which the human mind perceives as a table would not exist if that mind were naught. At the same time it must be admitted that the fact, that the ultimate physical reality is nothing but a multitude of separate revolving centers of electric force, also belongs to the human mind. In the apprehension of Truth there is an eternal conflict between the universal mind and the same mind confined in the individual. The perpetual process of reconciliation is being carried out in our science, philosophy, in our ethics. In any case, if there be any Truth absolutely unrelated to humanity then for us it is absolutely non-existing. It is not difficult to imagine a mind to which the sequence of things happens not in space but only time like the sequence of notes in music. For such a mind such a conception of reality is akin to the musical reality in which Pythagorean geometry can have no meaning. There is the reality of paper, infinitely different from the reality of literature. For the kind of mind possessed by the moth which eats that paper literature is absolutely non-existent, yet for Man's mind literature has a greater value of Truth than the paper itself. In a similar manner if there be some Truth which has no sensuous or rational relation to the human mind, it will ever remain as nothing so long as we remain human beings.

EINSTEIN: Then I am more religious than you are!

TAGORE: My religion is the reconciliation of the Super-personal Man, the universal human spirit, in my own individual being.

In a second meeting, on August 19, 1930, the extraordinary dialogue continued.

TAGORE: I was discussing . . . today the new mathematical discoveries, which tell us that in the realm of the infinitesimal atoms chance has its play; the drama of existence is not absolutely predestined in character.

EINSTEIN: The facts that make science tend towards this view do not say goodbye to causality.

TAGORE: Maybe not; but it appears that the idea of causality is not in the elements, that some other force builds up with them an organized universe.

EINSTEIN: One tries to understand how the order is on the higher plane. The order is there, where the big elements combine and guide existence; but in the minute elements this order is not perceptible.

TAGORE: This duality is in the depths of existence—the contradiction of free impulse and directive will which works upon it and evolves an orderly scheme of things.

EINSTEIN: Modern physics would not say they are contradictory. Clouds look one [way] from a distance, but if you see them near, they show themselves in disorderly drops of water.

TAGORE: I find a parallel in human psychology. Our passions and desires are unruly, but our character subdues these elements into a harmonious whole. Are the elements rebellious, dynamic with individual impulse? And is there a principle in the physical world which dominates them and puts them into an orderly organization?

EINSTEIN: Even the elements are not without statistical order: elements of radium will always maintain their specific order now and ever onwards, just as they have done all along. There is, then, a statistical order in the elements.

TAGORE: Otherwise the drama of existence would be too desultory. It is the constant harmony of chance and determination, which makes it eternally new and living.

EINSTEIN: I believe that whatever we do or live for has its causality; it is good, however that we cannot look through it.

[*SCIENCE AND THE INDIAN TRADITION: When Einstein Met Tagore* by Gosling, David L. Reproduced with permission of Routledge in the format Book via Copyright Clearance Center.]

If you had asked me five years ago who had won this debate, I would have immediately answered that Einstein had come out on top. Today, I have no problem in admitting that the poet Tagore won this exchange fair and square by forcing Einstein to admit, finally, that his lifetime obsessive defense of the existence of a human-independent objective reality was nothing but the product of his own personal religious belief, not to say private intellectual bias. Therefore, Tagore, in a way that only poets of divine magnitude can do, offers the most appropriate summary of the argument I have tried to synthesize in the last two chapters. Indeed, if one reads this dialogue a few times to get used to Tagore's jargon and oratory style, one can easily identify in his discourse the presence of key concepts of the braincentric cosmology proposed here: things like Gödelian information, Gödelian operators (like belief), the use of humans' mental abstractions as an attempt to explain the outside world, and the inexorable realization that whatever scientific description of the universe we come up with, no matter how well it is validated experimentally, it will always be limited by the neurobiological properties of the human brain because, in the end, the only reality we have access to is the one sculptured by our brains. That simply means that our very human condition works both as a gift and a constraint in our obsessive desire to give meaning to the cosmos.

Tagore's philosophical position also helps me to briefly touch on another major discussion: causality. According to the relativistic brain theory, our brains build internally a vast map of causal-effect relationships, which they

single impact, not an interferometry pattern. Put in other words, when a measurement is made prior to crossing the slits, light behaves like a particle. This particle-wave duality continues to be a major mystery in the interpretation of quantum mechanics.

To account for this wavelike interferometry pattern, three main interpretations have been proposed. According to the so-called Copenhagen consensus, originally formulated by the collaboration of the renowned physicists Niels Bohr and Werner Heisenberg, the interferometry pattern emerges because what really crosses the slits is not light per se but probability wave functions that describe the potential different states that light can assume when it is eventually measured. Once these functions reach the screen behind the slit, and the observer looks at it (and this is the crucial point for us), these functions are said to collapse, producing, in this case, the interferometry pattern observed by Young. Conversely, if a detector is placed before the slit, the wave function collapses in a different way, producing instead a particle-like impact.

But why does that happen? Classically, this is known as the measurement problem of quantum mechanics. Essentially, the Copenhagen consensus proposes that the observation act performed by an external observer, directly or via an instrument, is needed to reduce a set of probabilities describing the potential properties of a physical system—the wave function—into a single one (particle or wave). Before this measurement is done, quantum mechanics can describe the physical system only through a mathematical (or mental) construct, the wave function.

A second explanation, known as the many worlds hypothesis, formulated by the American physicist Hugh Everett in the late 1950s, denies any role to the observer of the experiment triggering the collapse of the probability wave function, as proposed by the Copenhagen consensus. Instead, it proposes that the interferometry patterns emerge because, even though we are performing the experiment in our own universe, the photons—or electrons—we generate to run the experiment, by the time they reach the slits, interfere with identical particles that exist in a variety of other universes. Essentially, according to this theory, the interferometry pattern we observe simply reflects the product of the complex interaction that occurs among an infinite number of other worlds.

Finally, the third interpretation is known as the pilot-wave theory, or De Broglie–Bohm, in honor of the French Nobel laureate physicist Louis De Broglie and the American physicist David Bohm. In a very simplified description of this view, the interferometry pattern would emerge because each photon—or electron—rides on a pilot wave, which crosses both slits at the same time. The interferometry pattern we observe, therefore, would emerge from the interference of the pilot wave to which each particle is attached. Again, as in the many worlds theory, no role for the observer is postulated by the pilot wave explanation.

Although most mainstream physicists would likely disagree with me, the braincentric cosmology proposed in this book is consistent with the Copenhagen interpretation of the double-slit experiment. First, the probability wave function proposed by the Copenhagen interpretation is essentially identical to my definition of potential information as the raw input provided by the external world to an observer. Second, both views recognize an active role for the observer in determining an outcome at the quantum level. This is made clear in the Copenhagen interpretation's requirement for an observer to produce the "collapse of the wave function." The convergence of the braincentric view and the Copenhagen interpretation of quantum mechanics can be further illustrated by a quote from Neils Bohr, who had this to say about the very scientific field he fathered: "There is no quantum world. There is only an abstract quantum physical description. It is wrong to think that the task of physics is to find out how *nature* is. Physics concerns what we can *say* about nature."

Neils Bohr was not alone in his view. Here is what the great British astronomer and physicist Sir Arthur Eddington had to say at the dawn of quantum mechanics: "Our knowledge of the nature of the objects treated in physics consists solely of readings of pointers [on instrument dials] and other indicators."

Bertrand Russell espoused the same view when he said: "Physics is mathematical not because we know so much about the physical world, but because we know so little; it is only its mathematical properties that we can discover."

In *An Inquiry into Meaning and Truth*, Bertrand Russell wrote, "We all start from 'naive realism,' i.e., the doctrine that things are what they seem. We think that grass is green, that stones are hard, and that snow is cold. But physics assures us that the greenness of grass, the hardness of stones, and the coldness of snow, are not the greenness, hardness, and coldness that we know in our own experience, but something very different. The observer, when he seems to himself to be observing a stone, is really, if physics is to be believed, observing the effects of the stone upon himself. Thus science seems to be at war with itself: when it means to be most objective, it finds itself plunged into subjectivity against its will. Naive realism leads to physics, and physics, if true, shows that naive realism is false. And therefore the behaviourist, when he thinks he is recording observations about the outer world, is really recording observations about what is happening to him."

As Philip Goff rightly wrote in the *Guardian*, what both Bertrand Russell and Arthur Eddington wanted to say is that "while physics may be great at telling us what matter does, it doesn't really tell us what it is." Goff continues: "What do we know of what matter *intrinsically* is beyond how it affects our instruments? Only that some of it—i.e. the stuff in brains—involves consciousness. Consciousness then should be our starting point in trying to work out what matter is, rather than something we try to squeeze in as an afterthought."

Goff's point of view is clearly illustrated when we try to follow the usual infinite regress, known as reductionism, employed by physicists to describe reality. Initially, we are told that the entire universe is made of atoms. Okay, that is fine. Moving deeper, we learn that atoms are made of elementary particles like electrons, protons, and neutrons. So far, so good. Going deeper yet, we are told that protons and neutrons are made of strange entities named quarks. Okay. So strange are quarks that, to this date, no scientist has been able to see one of them with their eyes or fine instruments. That is because they exist only as mathematical objects that are very useful in predicting the behavior of matter. But what are quarks made of? If you believe in the latest mathematical abstraction brought to the table by reductionism, you have to accept that quarks are made of infinitesimally small vibrating strings (10^{-35} meters) coiled in many more dimensions than the usual four we recognize in daily life (three-dimensional space and time). Despite these strings being considered one of the hottest topics in modern theoretical physics, there is no experiment that can test their existence. They can exist only as highly elaborated mathematical objects created by the mental abstractions of very gifted mathematicians' brains. And, as such, they seem to be extremely useful to mathematicians.

Eugene Wigner concurred with Russell's, Eddington's, and Goff's thinking when he wrote in *Remarks on the Mind-Body Question: Symmetries and Reflections*:

When the province of physical theory was extended to encompass microscopic phenomena, through the creation of quantum mechanics, the concept of consciousness came to the fore again: it was not possible to formulate the laws of quantum mechanics in a fully consistent way without reference to the consciousness. All that quantum mechanics purports to provide are probability connections between subsequent impressions (also called "apperceptions") of the consciousness, and even though the dividing line between the observer, whose consciousness is being affected, and the observed physical object can be shifted towards the one or the other to a considerable degree, it cannot be eliminated. It may be premature to believe that the present philosophy of quantum mechanics will remain a permanent feature of future physical theories; it will remain remarkable, in whatever way our future concepts may develop, that the very study of the external world led to the conclusion that the content of the consciousness is an ultimate reality.

Since the onset of the quantum mechanics revolution, several renowned physicists have explicitly supported the principles of the braincentric cosmology I am defending here. In his essay "The Mental Universe" published by *Nature* in 2005, a distinguished astronomer at Johns Hopkins University, Richard Conn Henry, cites some of these physicists to staunchly argue the case for the adoption of a braincentric view of the universe. Among other distinguished luminaries of the quantum revolution and physics world of the twentieth century, Henry cites the English physicist Sir James Hopwood Jeans, who had this to say: "The stream of knowledge is heading towards a non-mechanical reality: the universe begins to look more like a great thought than like a great machine. Mind no longer appears to be an accidental intruder into the realm of matter . . . we ought rather hail it as the creator and governor of the realm of matter."

Further support for a braincentric cosmology can be found in new interpretations of quantum mechanics. For example, in 1994, the Italian physicist Carlo Rovelli, from the Centre de Physique Théorique de Luminy at Aix-Marseille University, introduced a new theory that he called relational quantum mechanics. In it, Rovelli provides an argument in favor of the notion that there are no absolute physical quantities. Instead, he proposes that the state of any quantum system is relational, meaning that it depends entirely on the correlation or interaction established between the system and the observer. Essentially, Rovelli's approach argues in favor of using the observer's frame of reference to define any physical system, pretty much like my braincentric cosmology proposal.

To Ronald Cicurel and me, the key to obtaining a better understanding of what happens during the wave function collapse may be related to the well-known phenomenon of quantum entanglement, which has become a major area of research in modern physics. Briefly, particles are entangled when their quantum states cannot be described independently of each other. As such, if you perform a measurement of a given physical property of one particle—

let's say its spin—you affect the same property of its entangled twin instantaneously. So, if the initial measurement returns a spin value of –1/2 for the first particle, its entangled twin assumes the spin value of 1/2. By definition, therefore, entangled particles are correlated. Ronald and I believe that when we make an observation—like looking at light crossing a double slit—particles located in our retinas become entangled with the photons of the light beam and produce the wave function collapse predicted by the Copenhagen interpretation. Further exploration of this hypothesis in collaboration with physicists may provide us in the future with a quantum justification for the adoption of a braincentric cosmological view.

Coming back to our main thread, I have one final example that supports my thesis that mental abstractions are behind all our scientific theories. According to the latest canon in particle physics, one of the main properties of elementary particles, their mass, is endowed to them by their interaction with an abstract mathematical entity, the Higgs field, an operation intermediated by the now famous Higgs boson. Again, a vital component of the mainstream explanation of reality—the mass of particles—can be defined by physicists only as a mathematical object. Again, these facts suggest that, as far as physics is concerned, the entire cosmos that exists out there consists of nothing but a humongous soup made of potential information that can be described by us only through the employment of highly elaborated mental constructs—or mathematical objects—built inside some of our most brilliant human brains. That is why I say, to the chagrin of some (but not all) of my physicist friends, that the human universe is the only possible description of the cosmos available to us. Again, Lewis Mumford has it all figured out: "It is only through illumination by the mind of man that either the cosmic or the human drama makes sense."

And before anyone claims that I am exempting my own brain theories from the limitations I discussed above, there is no doubt in my mind that any man-made theory suffers from the same neurobiological constraints. Thus, as neuroscientists try to find explanations for how our own brains work, we suffer from the same limitations physicists have in explaining material reality. The only minor advantage we have over our colleagues in physics is that more of us are ready to admit that it is about time to move the human brain to the center of the human universe and take the observer's brain into account when we propose our scientific theories.

In summary, one cannot ignore any longer that all the mathematical abstractions proposed to account for the existence of such a human-independent objective reality are true by-products of the human brain, not of any independent process that exists in the universe. Physicists usually answer this by saying that since the universe existed much before our species emerged on Earth, neither human existence nor its subjective experiences and perceptions can account for the reality that existed before us. Yet, if we use the same reasoning, one can say that it does not make any sense for a universe that existed for billions of years before we came about to be explained by logic and mathematics that derive from the intrinsic neurobiological properties of the human brain. The chances that would happen are less than miniscule; they are equal to zero. Therefore, physical laws derived by us can be considered universal only as far as the human mind and its most stunning creation, the human universe, is concerned since, ultimately, both their conception and validation depend on theoretical formulations, experimental testing, and tools carried out and created by the same entity: the human brain. Indeed, physics suffers from a problem well known in biomedical research: the lack of a control group! To really prove that the physics laws derived by human brains are universal, we would have to verify that other equivalent intelligent life-forms who evolved in different parts of the universe have derived and accepted the same laws we humans have come up with to explain the cosmos. Unfortunately, this is not feasible. So far, at least.

What does this all mean? Simple. For humans, there is no escaping from Plato's cave. As Tagore so poetically but forcibly explained to the great Einstein, what we loosely call the universe can be experienced, described, and comprehended only through the shadows of an elusive reality that is continuously built and finely sculpted in the depths of humans' own minds, as a product of the brain's own point of view. And, as the views of the philosopher Ludwig Wittgenstein and the conclusions of Kurt Gödel seem to indicate, a purely mathematical description of reality may not suffice to describe the full complexity and richness of the human universe. Shocking as it may still sound in some quarters of the academic world, this means that we scientists have to be humble enough to admit that our traditional way of doing science may not be broad enough to describe the entirety of the human universe. As a scientist myself, I do not see this as any sort of tragedy or defeat at all, but rather an enormous opportunity for reflection and to change old habits. And by that I do not mean at all that we scientists need to resort to any mystical, religious, or metaphysical approaches—rather, we must simply be aware of the limitations of our trade.

After centuries of intellectual struggles in which a series of elaborated mathematical abstractions fought brutal battles for the domination of the privilege of defining our species' best account for the universe out there, the plateau we have reached binds us to a few harsh conclusions. Not only is nature, at its very quantum core, noncomputable and therefore nonpredictable in the Newtonian-Laplacian sense but, contrary to Einstein's deepest religious belief, there is no objective reality to speak of without the filter imposed by the observer's brain. In the case of the human universe, that means us. And that is just fine because, as Tagore teaches, contrary to abstract notions, there is only one universe that really counts for us: the human one.

In Bohr's words: "In physics we deal with states of affairs much simpler than those of psychology and yet we again and again learn that our task is not to investigate the essence of things—we do not at all know what this would mean; but to develop those concepts that allow us to speak with each other about the events of nature in a fruitful manner."

That brings us to the final two points I would like to make. Another surprising and humbling outcome of the proposed braincentric cosmology is the prediction that, if there are things out there in the cosmos that are more complex than our brains, they will remain forever beyond the comprehension of human beings. In this context, what we call random processes may only represent phenomena that lie outside the boundaries of logic that can be properly reached by the human brain before it feels utterly lost and guideless. Seen from this viewpoint, as my friend Marcelo Gleiser likes to say, the human universe could be metaphorically compared to a small island of knowledge surrounded by vast oceans of entropy which, given the limitations of the human brain, will never ever be sailed by any human mind.

Far from being disappointing, I see this analogy as providing the true dimension of the preciousness and uniqueness of what we, as a species, have been able to collectively achieve by carefully building such an island of knowledge over millions of years. I say that because until concrete evidence for the existence of extraterrestrial intelligent beings is found, the human universe constitutes the greatest mental accomplishment ever attained by any intelligent life-form that dared to emerge, rise, and still have the extra amount of courage and determination to establish its own mental footprint in this mostly bare, cold, and forever incomprehensible cosmos.

To finish, it is important to emphasize that the braincentric approach suggests that to reach the most accurate definition of what is out there in the cosmos that surrounds us, the ultimate description and interpretation of one's reality has to include the viewpoint of the observer's own brain. When extrapolated to include the entirety of the human universe, this view proposes that to thoroughly describe this human cosmos one would have to take into account the viewpoints of the brains of all observers who ever lived to survey, even if for a single millisecond, the wonders around them. Indeed, in a universe made of potential information, nothing really happens, nothing really leads to anything, and nothing acquires any relevance at all until an observer's brain or a brainet stamps meaning on raw observations and, by doing so, one collects another tiny pebble of knowledge to add to the shores of that island, the one first laid down on the open sea, by the observations made by the first of our human ancestors who dared to raise her head to contemplate the night sky and, taken by sheer awe, wonder for the first time where it all came from.

According to this view, at any moment, the human universe is defined by the collective amalgamation, into a single ever-growing entity, of every single act of living, observing, thinking, reflecting, creating, remembering, wondering, loving, worshiping, hating, understanding, describing, mathematizing, composing, painting, writing, singing, talking, perceiving, and experiencing that emanated from every human brain that has ever lived.

You do not need to take my word for that. Just listen to what the great American physicist John Wheeler had to say at the end of his career. Taking the conflicting results obtained by the double-slit experiment, Wheeler proposed a thought experiment in which he predicted that by observing today the light generated billions of years ago by a faraway star, a human observer could change the very manifestation of a split beam of that light, generated by the same star, that spread to other regions of space. Essentially, Wheeler proposed that an observation made today by a human observer could change the nature of light emitted in the past by a star located in a part of the universe billions of light years away from us. Based on this theoretical speculation, Wheeler introduced his theory that the universe can be described only as a participative cosmos, since all that happens in it depends on the cumulative observations performed by all intelligent life-forms that inhabit its confines.

Years after Wheeler published his little idea, experiments demonstrated the validity of his concept by using what became known as the delayed choice quantum eraser apparatus. Briefly, this experiment showed that if a beam of light is split in order to generate a continuous stream of entangled photon pairs and each of the photons in a pair is rerouted to different locations in the setup, one can show that the act of observing one of these photons, which has been deviated to a longer path inside the experimental apparatus, is capable of altering the properties of the other member of the pair, which was routed to a shorter path, even if the latter had been detected eight nanoseconds in the past. Thus, if as an observer I looked at the photon that followed a longer route and determined that it behaved like a particle, this simple act of observation would induce the other photon that defined the entangled pair to behave like a particle too—and not as a wave—even if the latter had already reached a detector a few nanoseconds prior. Since multiple labs have now reproduced this finding, the results generated yet another unsolved puzzle in quantum physics. That is, of course, unless you accept Wheeler's braincentric interpretation of it: that the most comprehensive account of the cosmos is the one created by the sum of all observations made by all intelligent life-forms that inhabit it.

As you can see, the more we search for the origins of all things that seem to be used to define our human notion of reality, things like space, time, mathematics, and science, the more all roads seem to lead to the same culprit: the True Creator of Everything.

11 • How Mental Abstractions, Information Viruses, and Hyperconnectivity Create Deadly Brainets, Schools of Thought, and the Zeitgeist

Precisely at 7:30 a.m. on July 1, 1916, the silence was suddenly broken around the placid waters of the Somme River in the northern French countryside. In terrifying unison, hundreds of military whistles blew all over the muddy serpentine ditch shared by the Allied armies. Upon hearing the fateful sound they had eagerly waited, and dreaded, for the past few days, more than one hundred thousand heavily armed men, consisting of all social echelons of British and French society, left behind everything that had truly mattered in their less than half-lived lives to rise together as a single organic wave from the relative shelter of their deep trenches, without any hint of hesitation, and plunged into a future no one could imagine, let alone predict.

In what looked from a distance like a perfectly rehearsed ballet, that immense human horde reenacted a tragic choreography witnessed many times over in our species' history: each man, in synchrony with the whole around him, climbed up the ladder to deliver his fate into the harsh hands of what soldiers prophetically referred to as "no-man's-land," four hundred yards of open space, totally devoid of any sense of humanity, that separated the Allied soldiers from the Germans' first line of defense.

Once at the gate of the life and death bifurcation that was no-man's-land, chance alone would dictate from which side each man would exit.

Deeply imbued by their devotion to their homelands and their private and collective sense of honor and duty, these brave men came out into the clear daylight, hoping for a quick dash that could carry them either to the safety of an intermediary crater, generated by one of the almost 2 million shells deployed during the previous seven days of intense bombardment that preceded that infantry charge or, better yet, over the first line of the German trenches, supposedly emptied by the casualties produced by the rain of iron and fire that had befallen the enemy for that whole week.

Instead, many of the British soldiers did not even have time to feel the texture of the soil of no-man's-land under their boots. The only thing they experienced was that ground's deadly bitter taste. Contrary to the conventional wisdom that became prevalent, both among the troops and the high command of the Allied forces, before that fateful infantry charge, the Germans had stoically endured the bombing by hiding deep in their fortified dugouts for the entire seven days. Then, having noticed the change in the British artillery pattern on the morning of July 1, they emerged from their hideouts to populate their first line of trenches and aim their machine guns at the advancing British and French infantry. By the time their targets rose out of the ground, they were more than ready for them.

The first wave of the British charge, at the north left flank and at the center, was received with a dense and sustained wall of machine-gun fire and artillery shelling, indicating at once that the advance would come at a very high human cost. It was at that juncture that, having finally received the answer from a fate they most feared, scores of men began to fall, badly wounded or simply dead, all over the emptiness of no-man's-land.

And yet for the greater part of that morning, wave after wave of British infantry continued to pour out from the trenches and dive into an inferno that, by all accounts, looked and felt like the closest definition of certain death one could have ever experienced.

To this day, the dimension of the human tragedy that took place at the margins of the Somme as a result of the callous decision of the British high command to continue to send men to their sacrificial pyre remains an open wound on Britain's conscience, particularly after it became well known that this tragic outcome was born out of a military strategy and tactics full of egregious misconceptions and illusions. By the end of this first day of that World War I battle, British casualties alone had reached the staggering number of 57,470 people, of which 19,240 were fatalities. That meant that almost six out of ten men who participated in that initial attack were either wounded or killed by day's end.

As a result of the stalemate trench warfare, no weapon became off limits, no matter how ghastly its effects, as long as it could bring devastation to the other side. Such a philosophy of engagement meant that troops on both sides were to be mauled, day after day, until they were either pulverized or reduced to unrecognizable bits of human flesh, by the latest technology for mass killing available to the military on both sides of the conflict. As a matter of fact, the well-timed convergence of multiple new technologies orchestrated by the British and German military-industrial complexes played a central role in determining not only the unprecedented levels of casualties and fatalities but also the exceptional degree of the severity of the wounds sustained on the Somme battlefield. According to Peter Hart in *The Somme*, this almost fetishistic reliance on new mass-killing technologies included, from the British side alone: about 1,537 modern pieces of artillery—each placed roughly 20 yards apart throughout the more than 25,000-yard front line—capable of annihilating, in the case of the 60-pounder guns, targets as far as 10,500 yards away; machine guns that could spit death at a rate of 500 rounds per minute over a range of up to 4,500 yards; and powerful hand grenades, mines, and the latest generation of repetition rifles. Poisoned gas was used in profuse quantities to mercilessly torment the enemy trenches on both sides of no-man's land. In the Somme, the British army also introduced tanks for the first time in a battlefield, presaging what would become the norm in World War II.

In *The Age of Extremes*, the distinguished British historian Eric Hobsbawm addresses the central question that still flabbergasts anyone who confronts the level of devastation that was brought upon Europe by the 1914 decision of its main powers to settle their differences not by diplomacy but through a total war of annihilation. In Hobsbawm's view, the main question to be asked to explain this multidimensional human tragedy is: "Why, then, was the First World War waged by the leading powers on both sides as a zero-sum game, i.e. a war which could only be totally won or lost?"

To which he offers the following answer: "The reason was that this war, unlike earlier wars, which were typically waged for limited and specific objectives, was waged for unlimited ends. In the age of Empires, politics and economics had fused. International political rivalry was modeled on economic growth and competition, but the characteristic feature of this was precisely that it had no limits. . . . More concretely, for the two main contestants, Germany and Britain, the sky had to be the limit, since Germany wanted a global political and maritime position like that now occupied by Britain, and which therefore would automatically relegate an already declining Britain to inferior status. It [the war] was either/or."

In the braincentric view I propose, nations, empires, and multinational corporations are, like elaborate mathematics, emergent properties of the primitive principles embedded in our brains. While high-order mathematics emerged from basic principles of logic, geometry, and arithmetic, as we saw in chapter 10, I believe that large-scale political and economic suprastructures have their origin, as surprising as it may sound, in the primordial social interactions honed during our humble tribal origins hundreds of thousands of years ago. At the limit, these political or economic suprastructures try to overcome the fact that, as we saw in chapter 2, once human social groups exceed 150 individuals, there is a dire need to impose supervising systems—like the management levels of a company, the constitution and laws of a country, or rules and regulations for the economy—if one has any hope to make such large human societies viable. In that sense, the fact that large numbers of people are willing to die on behalf of their allegiance to symbolic entities, like the motherland, a given political ideology, or an economic system, constitutes another demonstration of how powerful—and deadly—mental abstractions and beliefs can be in determining human group behavior and fate. That can be perfectly illustrated by the numbing dimension of the human tragedy witnessed during the battle of the Somme. By the end of hostilities, on November 18, 1916, close to 3 million soldiers had been brought into the fight. Of those, more than a million—or one out of three—left the battlefield as a casualty. These numbers become even more horrendous when one realizes that, in exchange for their 623,917 casualties, the Allied forces advanced no further than five miles into German-controlled territory during the entire duration of the battle. As such, at the end of the human grinding that cost so many lives, no side could claim anything close to a decisive victory. From the battle of the Somme, as in so many other battles, the only things left were historical records, painful memories, medals made of cheap metal, an army of orphans and widows, and vast cemeteries.

I choose the battle of the Somme to illustrate my viewpoint not only for its profound symbolism in terms of demonstrating the futility and horror of warfare, but also because it depicts in a very powerful and tragic way how particular mental abstractions have been employed throughout humankind's history to lock hundreds of thousands, even millions, of minds into a brainet so cohesive, so synchronized that its participants, usually normal regular people when seen in the isolation of their ordinary daily lives, become willing to risk everything, including their own mortal lives, for a cause that, most of the time, they can barely define or fully comprehend. Without diminishing in any way the tremendous levels of heroism and bravery demonstrated by all those men who fought and died on both sides of the graveyard they called no-man's-land, the battle of the Somme offers a perfect example of how large groups of human beings can be driven to the limits of what they can endure physically and mentally when a mental abstraction is employed, exploited, or manipulated enough, to synchronize these individuals' brains into a single collective entity. Just for a moment, put yourself in the place of an eighteen- or nineteen-year-old foot soldier crouched inside a trench, dazed by the relentless shelling and the incessant cries of the wounded, witnessing the ongoing massacre, seeing with your very own eyes thousands of dead bodies spread over no-man's-land, and imagine how you would react when your number was called by that fateful whistle blow. What would it feel like to climb up that ladder, which has already taken so many of your buddies into a deadly fiery rain of bullets, knowing that the outcome was a fait accompli? What reason could possibly make you actually take that final step, expecting nothing but death's embrace or, at best, a lifetime of suffering and disfigurement? I can hear you answering already: courage, patriotism, bravery, a sense of moral duty to your country and family. Most certainly these and other lofty feelings and emotions would be part of the driving forces that would make you face such irrational odds and defeat your own fear and terror, instead of either refusing to come out of the trenches, as a small fraction of soldiers did, or run away as fast as you could from that hell. But where did these feelings come from? And what allowed them to overwhelm the rationality of each and every one of those soldiers to the point that they ignored any kind of logical judgment that would have intuitively propelled them to ensure their own self-preservation and safety?

Surprising as it may sound, I propose that such counterintuitive behaviors and attitudes take place because our human brains are extremely susceptible to fall prey to mental abstractions that appeal to our basic instincts or primitive archetypes that, embedded in the neuronal circuits of our primate ancestors by the process of natural selection millions of years ago, were transmitted through the human evolutionary tree and now remain buried, deep in our own *Homo sapiens* brains, as part of a silent inheritance granted to us.

By elucidating the potential brain-based mechanism that accounts for the formation of such brainets, I believe that I can provide a neurophysiological hypothesis to explain how, over the four thousand years of human recorded history, there have been innumerable instances in which large cohorts of human beings synchronized their brains into powerful and deadly brainets.

Once exposed to an overwhelming call to defend or uphold a dominant mental abstraction—a nation, a religion, an ethnic group, an economic system, or a political ideology, to name just a few—that appeals to deeply entrenched primitive human beliefs, these tightly locked human brainets become capable of waging total war against their own kin or engaging in the decimation of another human group, as happens in genocides. From the immense Greek armada that between the thirteenth and twelfth centuries BC sailed across the Mediterranean Sea all the way to the shores of Troy in order to avenge the bruised marital honor of Menelaus, king of Sparta, and rescue his wife Helen from Paris's arms, at the cost of destroying the whole Trojan civilization to the Syrian war waged, more than three thousand years later, in a nearby Mediterranean battlefield, which by now has killed, wounded, and dislodged millions of civilians, the pattern seems to repeat itself with very little nuance, examined from a braincentric point of view. Although the specifics vary, the central motif accounting for this phenomenon seems to be always the same: first a mental abstraction, without any tangible link to reality, is selected as the rationale for waging war or genocide, then a message supporting that call for action is widely disseminated through a human social group, using the most efficient medium for communication available at the time, so that large human contingents can synchronize their brains into a brainet that will sustain the intended goal of victory at any cost and total eradication of the enemy. For the most part, that is what my theory proposes is beneath the process of formation, synchronization, and engagement of highly cohesive human brainets that have fought and committed despicable atrocities on the grounds of, to name just a few reasons, religious disputes, ethnic, social, and racial prejudices, imperial conflicts, national economic interests and territorial borders, trade monopolies, sheer financial gain, political ideological disputes, geopolitical maneuvering, and many other equally abstract concepts that nonetheless gave the appearance of offering a more than solid justification to demonize, ostracize, injure, segregate, maim, kill, and exterminate the appointed enemy of each occasion. Once engaged in such brainets, as in the battle of the Somme, hordes of men are ready to march toward their set common goal even if that means, in extreme cases, facing self-annihilation or, conversely, joining forces to commit horrible atrocities against other human beings—actions that, individually, and before they were recruited to be part of such a brainet, they would never have conceived, let alone executed. As François-Marie Arouet, also known as Voltaire, once said, "If they can make you believe absurdities, they can make you commit atrocities."

Voltaire's eighteenth-century wisdom captures the essence of my thesis since it helps us begin to understand how the highly combustive mixing, during the past 150 years, of enduring and powerful human mental abstractions—things like religious zealotry and the modern incarnations of tribalism, patriotism, and nationalism, ethnic and racial superiority, and material greed—with a continuous and accelerating process of refinement of technologies for mass communication and killing has contributed to define a period that includes the so-called ages of modernity and the postmodern, from the mid-nineteenth century to the present.

As had happened many times before, it was through art that the dominant mental abstractions of the time were vividly expressed and dissected. In the case of the first four decades of the twentieth century, a period in which the entire planet was engulfed by two cataclysmic wars, two paintings seemed to capture the sentiment of horror and despair experienced by tens of millions of people worldwide. I am referring to *The Scream*, by the Norwegian expressionist painter Edvard Munch, and one of Picasso's most recognized masterpieces, *Guernica*, the larger-than-life oil on canvas mural in which the Andalusian painter depicted, through a sinister pallet containing only black, white, and gray tones, the tragedy and outrage caused by the bombing of a Basque village, the true opening salvo of World War II, by German and Italian plane squadrons in support of the forces of the future dictator Francisco Franco during the Spanish Civil War.

To appreciate how devastating have been the consequences of mental abstractions transmitted by "information viral infections" that are capable of synchronizing millions of otherwise benign and ordinary human beings into brainets that unleashed indescribable levels of human destruction, just consider three more emblematic examples of anthropogenic tragedies: the civil war that ravaged China between 1851 and 1864, the Taipan Rebellion, triggered by a religious dispute between the reigning Manchu Qing dynasty and a Christian movement (the Taiping Heavenly Kingdom), led to the loss of 40 to 100 million lives, depending on different accounts; the Spanish conquest of Mexico and Peru, motivated by an insatiable search for gold and silver, killed about 33 million Aztecs and Incas; and World War II, in which fatalities reached the staggering number of 60 million people, or 3 percent of the world's population at the time.

Only twenty-five years ago, one of the most perplexing examples of human mass killing unfolded under the eyes of the entire world in a period of less than one hundred days in Rwanda, a tiny central African country once nicknamed the Switzerland of Africa because of the splendor of its marvelous mountain chains laced with dense tropical forests. The Rwanda genocide of 1994 offers a gruesome reminder that once a widely accepted mental abstraction runs amok, anthropogenic catastrophes of epic proportions are very likely to ensue.

In Rwanda's case, 1 million people lost their lives due to a totally artificial ethnic conflict that can be traced to the decision of European colonial governments to arbitrarily divide in half a native population that from an anthropological point of view was absolutely homogenous, sharing the same language, culture, and religion, into two competing groups. Despite having lived together, the native people of Rwanda began to be formally and forcibly assigned by colonial powers that occupied the country, first the Germans and then the Belgians, as belonging to one of two ethnic groups, Tutsi or Hutu. This happened despite the fact that intermarriage across the two groups was widespread, making any imposed division totally arbitrary and meaningless. Yet in addition to promoting such a division, the European authority instituted a policy that benefited the Tutsi with better opportunities for education, economic gain, and social ascension, since this group was arbitrarily selected to fill the best public servant jobs and serve as the local puppet regime that governed the country according to the interests of the European colonizers. The maintenance of such a social and economic segregation over many decades contributed to a growing tension between Hutus and Tutsis, eventually leading to the outbreak of major violent conflicts. Those initial incidents foreshadowed that the occurrence of an out-of-control massacre was just a question of time. And all of this because the Belgian authorities had decided to decree, out of nowhere, that Rwandans should be ethnically divided, belonging to the Tutsi group if they were taller, lighter skinned, and thinner, with finer bone structure and features, and to the Hutu group if they were shorter, darker skinned, and stockier, with softer features.

Ironically, one of the major factors guiding the classification of a Rwandan as a Hutu or a Tutsi—a difference in height—turned out to reflect the same gap in average value (twelve centimeters) that separated the rich and poor people living in different neighborhoods of major European cities, like London, in the nineteenth century.

It all started on April 6, 1994, when the plane carrying the president of Rwanda, Juvénal Habyarimana, the leader of a Hutu-based government, and the president of Burundi, Cyprien Ntaryamira, was shot down on its final approach to the airport in Kigali, Rwanda's capital. Next morning, inflamed by radio broadcasts that called upon the Hutus to avenge the death of their leader by unleashing their wrath upon the Tutsi enemy, the country's armed forces, the police and Hutu militias, began to hunt down and execute, in cold blood, thousands of unarmed Tutsi civilians. During the buildup to this tragedy, the radio was instrumental in the process of demonizing the Tutsis. Months before the genocide, radio stations incited the Hutu population against Tutsis and broadcast messages indicating that orders for the final attack would be issued at the right moment and that everybody must be prepared to act upon them.

And when that moment finally came, Hutus, armed with machetes, knives, sickles, and any tool that could inflict a deadly blow, went about the business of systematically exterminating their Tutsi neighbors, classmates, work colleagues, and peers of all sorts. Nobody was spared in the tsunami of blood that engulfed the country; no mercy was shown to women, children, or the elderly. Any one captured whose identification revealed him or her to be a Tutsi was at once slaughtered on the spot.

The Rwanda genocide offers a somber example and a grim reminder of both the range and magnitude of the type of lethal human forces that can be unleashed when a large group of people synchronize their brains in response to a widely disseminated message, particularly one that elevates a distorted mental abstraction to the status of ultimate truth: something whose veracity is perceived to be so unquestionable and irrefutable within each of the minds that have become synchronized that no rational intervention is capable of derailing or dislodging it from the pool of co-opted brains. By the time such a brainet synchronizes, it does not matter whether one is talking about Hutus and Tutsis in Rwanda or the generals of the conflicting armies in World War I; once lost inside the disorienting fog of a perceived life-or-death conflict between different human groups, no one is immune to becoming synchronized into a brainet capable of producing lethal outcomes never imagined or even condoned by its individual members.

After reflecting on the nature of these catastrophic collective human behaviors in the context of the results I have obtained by studying brainets in my lab, I came to the conclusion that such examples of total collective blindness to rational thinking, exemplified by the battle of the Somme, the Rwandan genocide, and so many other equally appalling human-caused calamities, can be at least partially explained by the mechanism discussed in previous chapters in this book. In a nutshell, I propose that catastrophic outcomes generated by large human groups basically result from the capacity of our social primate brains to establish highly synchronous brainets involving large numbers of individuals. In the extreme cases illustrated in this chapter, the interbrain synchronization involves cortical and subcortical structures other than the motor cortex, which mediated the Passenger-Observer brainet discussed in chapter 7. For this type of destructive large-scale interbrain synchronization to happen, though, a few conditions have to be satisfied. First, a powerful mental abstraction has to emerge and become so widely disseminated across a human social group that it reaches a point in which it becomes accepted by the vast majority of the individual members of that group as representing a consensual worldview or truth. Invariably, to achieve that threshold, the mental construct has to appeal in a very basic way to one of the most primitive instincts and archetypes embedded in our social brains: the almost obsessive desire to belong to a cohesive and selective tribal group that shares common values, beliefs, prejudices, and worldviews, and, even more important, the willingness to fight and repel the "enemy," the archetypal representation of "evil," the symbol of the perfect antithesis that must be neutralized and destroyed at any cost, since it is the reason and cause of all ills that threaten the tribe's way of living. Put in another way, because, like all animals, humans contain in the depths of their brains traces and remnants of fixed patterns of behavior and reasoning that proved to be of great adaptive value in the past, they are very susceptible to becoming entrained by a mental abstraction that appeals to these innate primitive beliefs and stereotyped mental patterns. Thus, being recruited into such brainets can easily happen despite the fact that our overgrown cortex, as well as our ability to learn new social norms and ethical and moral values through the process of education, imposes some deterrent on the expression of these primitive instincts.

Once a mental abstraction becomes dominant within a human social group, like the hatred created by an imposed ethnic division between Hutus and Tutsis, all that a brainet needs to be put in motion is a triggering message, what I like to call an information virus, and a medium to disseminate it widely. And then, once the size of such a human brainet crosses a certain critical threshold, it may begin to operate pretty much like Poincaré's nonlinear dynamic system whose global behavior becomes utterly unpredictable. And as they enter into such an uncontrollable dynamic regime, human brainets can perpetrate the kind of unbounded violence one sees in wars, revolutions, genocides, and other anthropogenic atrocities.

My notion of a shared repertoire of innate modes of reasoning and acting pervading all human brains as a precondition for the creation of large-scale human brainets is somewhat similar to the classical concept of the collective unconscious originally proposed by the Swiss psychiatrist Carl Jung. This can be verified by one of his descriptions of the human unconscious realm, which Jung proposed to divide as follows: "A more or less superficial layer of the unconscious is undoubtedly personal. I call it the personal unconscious. But this personal unconscious rests upon a deeper layer, which does not derive from personal experience and is not a personal acquisition but is inborn. This deeper layer I call the collective unconscious." In a subsequent passage, Jung further elaborates on the definition of the collective unconscious: "I have chosen the term 'collective' because this part of the unconscious is not individual but universal; in contrast to the personal psyche, it has contents and modes of behavior that are more or less the same everywhere and in all individuals. It is, in other words, identical in all men and thus constitutes a common psychic substrate of a suprapersonal nature which is present in every one of us."

Although Jung seemed to apply some mystic overtones to his theory, something that I do not subscribe to, when applied to the context in which I am framing the present discussion, Jung's notion of the "collective unconscious," aligned with the accumulated neuroscientific knowledge available today, may allow us to describe how large-scale brainets form once a number of brains are invaded by a message that works inside their minds like a type of virus, providing the final link to allow life-or-death relevance to be assigned to highly abstract concepts like the motherland, ethnic or racial superiority, religious principles, or distinct political ideologies or economic views.

In Jung's view, there are four levels of mental processes that modulate the way we behave. First, there is the level determined by our social relationships with other people: family, friends, and acquaintances. Altogether, this interpersonal social realm sets certain boundaries for which kind of behaviors are acceptable and which are not, by imposing some kind of "social filter" or restraining force that circumscribes our daily actions. Then, there is the conscious mode of action, the one that confers to each of us an identity, an ego, a sense of being and thinking by ourselves. The next two levels are reserved for the unconscious, which Jung further divides into personal and universal components. The personal unconscious is determined primarily by the multitude of individual life experiences that become gradually stored in our brains, away from our conscious access. Underneath this personal unconscious, according to Jung, reside the innate set of instincts and fixed patterns of behaviors and thinking that constitute the collective unconscious that we all share, more or less, as members of the same human species. Jung was aware of the tragic consequences that may ensue when the potential energy of collective human forces stored in large-scale human brainets is released in the form of collective behaviors and actions:

The unconscious no sooner touches us than we are it—we become unconscious of ourselves. That is the age-old danger, instinctively known and feared by primitive man, who himself stands so very close to this pleroma. His consciousness is still uncertain, wobbling on its feet. It is still childish, having just emerged from the primal waters. A wave of the unconscious may easily roll over it, and then he forgets who he was and does things that are strange to him. Hence primitives are afraid of uncontrolled emotions, because consciousness breaks down under them and gives way to possession. All man's strivings have therefore been directed towards the consolidation of consciousness. This was the purpose of rite and dogma; they were dams and walls to keep back the dangers of the unconscious, the "perils of the soul." Primitive rites consist accordingly in the exorcizing of spirits, the lifting of spells, the averting of the evil omen, propitiation, purification, and the production by sympathetic magic of helpful occurrences.

Jung continues: "My thesis, then, is as follows: In addition to our immediate consciousness, which is of a thoroughly personal nature and which we believe to be the only empirical psyche (even if we tack on the personal unconscious as an appendix), there exists a second psychic system of a collective, universal, and impersonal nature which is identical in all individuals. This collective unconscious does not develop individually but is inherited. It consists of pre-existent forms, the archetypes, which can only become conscious secondarily and which give definite form to certain psychic contents."

Coincidently, the cardinal example selected by Jung to illustrate the power of the collective unconscious pretty much reflects the climate that became pervasive over Europe during the months that preceded the outbreak of World War I, leading to the carnage of the battle of the Somme and many others that followed during the four years of that conflict. As Peter Hart writes in *The Somme*: "Popular jingoism was certainly stirred then as now, by cynical politicians and morally opaque newspaper proprietors; however, it had its wellspring deep within the dark corners of the popular consciousness. The political imperatives of defending the bloated empire, the endemic racism and all-embracing casual assumption of moral superiority of the age, the overwhelming reliance on blunt threats to achieve what might have been better achieved by subtle diplomacy—these were all part of the British heritage in 1914."

The thesis that deeply entrenched neuronal routines embedded in the human brain provided a great portion of the drive needed for millions of people to embrace war in the early twentieth century, instead of pressuring their politicians to find a peaceful settlement that could satisfy the geopolitical greed of the dominant European powers, is further reinforced by the systematic occurrence of innumerable other large-scale human conflicts that were triggered under the same circumstances throughout our history.

In Jung's description of the collective unconscious, "when a situation occurs which corresponds to a given archetype, that archetype becomes activated and a compulsiveness appears, which, like an instinctual drive, gains its way against all reason and will, or else produces a conflict of pathological dimensions, that is to say, a neurosis." The consequences, then, become obvious, "when in a state of violent affect one says or does things which exceed the ordinary. Not much is needed: love and hate, joy and grief, are often enough to make the ego and the unconscious change places. Very strange ideas indeed can take possession of otherwise healthy people on such occasions. Groups, communities, and even whole nations can be seized in this way by psychic epidemics."

Even without mentioning the role of natural evolution in seeding deep into our brains the mental programs that eventually release, in each of us, a facsimile of the thinking patterns, instincts, and behaviors that define the collective unconscious, Jung emphasizes this "historical" component and how it helps in shaping what I have called throughout this book the brain's own point of view: "Whereas we think in periods of years, the unconscious thinks and lives in terms of millennia." Jung here clearly proposes that conscious thinking is a more recent evolutionary by-product of the human mind and, as such, one that can be kidnapped at any time by the older and more dominant unconscious repertoire of mental programs. "Consciousness grows out of an unconscious psyche which is older than it, and which goes on functioning together with it or even in spite of it. . . . Also, it frequently happens that unconscious motives overrule our conscious decisions, especially in matters of vital importance."

The convergence of Jung's ideas with the more operational and neurophysiologically grounded proposal I am making in this book suggests a very interesting approach to dissect the mechanisms that have allowed, over millennia, the creation of large-scale human brainets. In this scheme, one must separate the role of mental abstractions that over eons of history have become dominant among human social groups, and hence acquired the power to appeal to primitive human archetypes, from the means of communication that have allowed human brains to tightly synchronize, once an "information virus" becomes widely disseminated and capable of infecting large numbers of human brains.

At this point, it is important to emphasize that my definition of an information virus is distinct from a similar neologism used by the British evolutionary biologist Richard Dawkins to explain how memes spread across populations. In my definition, the information virus is basically a mental abstraction capable of working as a powerful synchronizing signal that allows the formation of large human brainets. Dawkins coined the term *meme* in *The Selfish Gene* to describe how an idea, a behavior, or a new cultural manifestation can spread across a population in a process that resembles a viral infection. Following his original definition, some authors have proposed that a meme is equivalent to "a unit of culture" whose transmission—or infection—across human populations is governed by natural selection, like other biological traits. Although very interesting, this latter view is not the one that I am using when I talk about information viruses in the context of brainet synchronization.

Let's now turn our attention to another essential component in the process of mental amalgamation that leads to the establishment of large-scale brainets capable of going to war or committing despicable atrocities against members of their own species. For that, we need to discuss how different natural communication strategies and man-made technologies impact collective human behaviors. Such a discussion, to be minimally satisfactory, requires that we introduce some key concepts in media theory, originally proposed by the Canadian professor and philosopher Marshall McLuhan in *Understanding Media: The Extensions of Man*.

McLuhan's major insight was that the different means of communication, or media, utilized by humans—whether natural, like oral language and music, or resulting from the introduction of new man-made technologies, like written language, printed books, telephones, or the radio—have the common effect of enhancing the reach of our species while collapsing the dimensions of time and space to the point at which all humanity would be reduced, from a communication point of view, to nothing more than "a global village."

Consider music, for instance. It is no coincidence that countries have national anthems, armies rely on brass bands to play martial songs, most religions utilize sacred songs and communal singing, and movies employ soundtracks to entice audiences worldwide. In all these examples, music likely plays the pivotal role of the synchronization signal that allows large numbers of people to become bound to a common set of mental abstractions, like being part of a nation, fighting for an army, or sharing a religious belief. Thus, after listening to a large crowd singing at full lung capacity "La Marseillaise," what French citizen would not be ready to go fight for country, flag, or a political ideology? By the same token, who would not become a pious believer if exposed earlier in life to singing as in Wagner's Pilgrims' Chorus in *Tannhaüser* or Handel's *Messiah*?

McLuhan appeals to one of my favorite examples to describe the work of a brainet—a sports stadium full of fans—as a way to introduce the power and reach of a medium to synchronize people and allow them to jointly express primitive social behaviors and emotions that remain deeply entrenched in our collective unconscious. According to McLuhan, "The wide appeal of the games of recent times, the popular sports of baseball and football and ice hockey seen as outer models of inner psychological life, become understandable. As models, they are collective rather than private dramatizations of inner life. Like our vernacular tongues, all games are media of interpersonal communication, and they could have neither existence nor meaning except as extensions of our immediate inner lives. . . . Sport, as a popular art form, is not just self-expression but is deeply and necessarily a means of interplay within an entire culture."

Over the past fifty years, I have witnessed McLuhan's thesis demonstrated over and over again during my excursions to soccer stadiums in cities all over the world. No matter the country or culture, the behavior patterns and emergent outcomes I observed firsthand are always the same all over the globe: once inside the stadium, people of all social backgrounds—blue-collar workers, doctors, judges, engineers, and yes, scientists—tend to abdicate all of their routine social rules of conduct, strictly followed in the outside world, so that they can seamlessly merge into the crowd and become one single voice and body rooting for their favorite team. Once synchronized, these fans may sing songs they would never sing in the outside world, say things they would never afterward admit they said, and act in ways that they themselves may never condone in their routine lives, once they are detached from the stadium brainet.

According to McLuhan's views, sport events "are extensions, not of our private but of our social selves," basically another manifestation of a mass medium capable of synchronizing large groups of people during a ritual that resembles aspects of our primordial tribal origins. As such, such events seem to be needed to maintain social cohesion in any culture, no matter how sophisticated it is. That may explain why the Roman Empire devoted so much effort and attention to keeping crowds entertained by a variety of deadly games played in the Coliseum and many other arenas, or why the Greeks valued so much the heroes who excelled in their Olympic games. The fact that in our times professional sports athletes command the largest salaries and are granted celebrity status in most societies seems to simply confirm what the Romans and Greeks knew all along.

Half a century before people would be talking about globalization in economic terms, McLuhan had already predicted that this would be one of the likely outcomes of the process triggered by the introduction of successive waves of mass communication technologies in the twentieth century. As such, McLuhan's ideas, proposed in the late 1950s and early 1960s, predicted in many ways a future that we are living in now, at the end of the second decade of the twenty-first century. As a matter of fact, using a terminology different from mine, he reached the same conclusions about how different synchronizing signals provided by different types of communication technologies would lead to the generation of large-scale human brainets. In McLuhan's terms, all media communication technologies share a common property: the extension of humans' reach, first including our bodies and senses and, ultimately, with the introduction of "electrical media," through the expansion of our own central nervous systems. As a consequence, the widespread employment of such mass media technologies would have a profound impact on human society. In McLuhan's words: "The use of any kind of medium or extension of man alters the patterns of interdependence among people, as it alters the ratios among our senses."

From oral language and music, the most ancient synchronizing signals of human social groups, to written language, art, printed books, telegraphs, telephones, radios, the movies, all the way to television, McLuhan exposed the essential role played by mass communication in shaping our beliefs, worldviews, ways of life, and collective behaviors. So prescient was this work that McLuhan was able to predict even the impact that digital computers and other "electrical media," like the internet, would have on human social, economic, and political interactions. Accordingly, some passages of *Understanding Media*, even though they were conceived in the remote 1960s, sound as if they were just written a few days ago and take into account the potential impacts of the twenty-first-century digital age. Take this one, for example: "Our new electric technology that extends our senses and nerves in a global embrace has large implications for the future of language. Electric technology does not need words any more than the digital computer needs numbers. Electricity points the way to an extension of the process of consciousness itself, on a world scale, and without any verbalization whatever."

In further exploring the same theme, McLuhan basically anticipated the concept of brainets that I have discussed, even though he had no way of knowing—and clearly not the remotest intention of discussing—the potential neurophysiological mechanisms that may be involved in establishing such a neural construct. Nevertheless, it is truly amazing to read what he had to say more than half a century ago: "The tendency of electric media is to create a kind of organic interdependence among all the institutions of society, emphasizing de Chardin's view that the discovery of electro-magnetism is to be regarded as 'a prodigious biological event.' . . .

If political and commercial institutions take on a biological character by means of electric communications, it is also common now for biologists like Hans Selye to think of the physical organism as a communication network." According to McLuhan, this "organic network" will materialize itself because "this peculiarity about the electric form, that it ends the mechanical age of individual steps and specialist functions, has a direct explanation. Whereas all previous technology (save speech, itself) had, in effect, extended some part of our bodies, electricity may be said to have 'outered' the central nervous system itself, including the brain." Consequently, he asserts, "We live today in the Age of Information and of Communication because electric media instantly and constantly create a total field of interacting events in which all men participate. . . . The simultaneity of electric communication, also characteristic of our nervous system, makes each of us present and accessible to every other person in the world. To a large degree our co-presence everywhere at once in the electric age is a fact of passive, rather than active, experience."

McLuhan also tried to explain how the introduction of artificial media, like the radio, to a global audience led, and likely would continue to lead, to major changes in human behavior. Having known the central role radio broadcasts played in the unfolding of the Rwandan genocide before I got to know McLuhan's work, the first time I read his predictions, made exactly three decades before the occurrence of that central African tragedy, a wave of goose bumps spread all over me. The reason for such a shivering reaction can be easily identified when we read some of McLuhan's ideas about the impact the introduction of radio broadcasts had on human societies, particularly in cultures that had not been exposed to other forms of mass media communication before. "Radio affects most people intimately, person-to-person, offering a world of unspoken communication between writer-speaker and listener. That is the immediate aspect of radio. A private experience. The subliminal depths of radio are charged with the resonating echoes of tribal horns and antique drums. This is inherent in the very nature of this medium, with its power to turn the psyche and society into a single echo chamber." That statement had Rwanda written all over it. He continued, "Radio provided the first massive experience of electronic implosion, that reversal of the entire direction and meaning of literate Western civilizations. For tribal peoples, for those whose entire social existence is an extension of family life, radio will continue to be a violent experience."

According to McLuhan, unsavory scenarios involving violence could become a real concern if radio broadcasting were to become a tool of domination by authoritarian regimes willing to impose a single world viewpoint on a largely uninformed or uneducated population with no independent means to verify or critically analyze this information. In that sense, McLuhan again gets very close to my definition of an information virus as the potential trigger of a human brainet that becomes widely synchronized and, upon being "infected," decides, in a collective way, to perform all sorts of acts people would not enact when alone or not incited. What would McLuhan say if he were alive today facing the tsunami of so-called fake news disseminated through social media?

McLuhan believed that mass communication technologies provide the means through which Jung's collective unconscious can be synchronized and released from the depths of the brains of a human community and overcome rational conscious thinking or any social pressures that would normally block its full expression in the form of unsavory collective behaviors. "Radio provides a speedup of information that also causes acceleration in other media. It certainly contracts the world to village size, and creates insatiable village tastes for gossip, rumor, and personal malice."

McLuhan's prediction that the widespread dissemination of electric media around the world would lead to the emergence of his global village, this state of full human connectivity, in which both communication space and time collapse has become one of the most iconic metaphors associated with his work. The other is his famous aphorism, "The medium is the message."

In one of his more prophetic pieces of writing, McLuhan states without hesitation the kind of life the electric communication technologies of the future would bring to us: "In the electric age, when our central nervous system is technologically extended to involve us in the whole of mankind and to incorporate the whole of mankind in us, we necessarily participate, in depth, in the consequences of our every action. It is no longer possible to adopt the aloof and dissociated role of the literate Westerner."

For the most part, I tend to believe that McLuhan got it right: the inception of wave after wave of new electric media, and in particular the wide dissemination of the internet to all corners of Earth, has indeed created a global village. However, there is growing evidence that McLuhan's utopia of a global village seems to more properly reflect the level of potential connectivity, or penetration, attained by our latest and most advanced communication technology, the internet, rather than offer an appropriate description of what kind of dominant social effects it has induced—or, should I say, liberated. Ironically, a growing consensus seems to indicate that the more connected we become, the more fragmented and confrontational our social interactions are. Anecdotally, one only needs to look at the effects produced by hugely popular social media to find support for the notion that, if anything, people living in our hyperconnected modern societies increasingly tend to limit their regular social interactions in favor of primarily virtual encounters. Invariably, the latter tend to occur within the very constrained borders of carefully built or chosen virtual social groups, which usually restrain their discussions to a very narrow, focused set of themes, values, and worldviews. Dissent in this new era of social media seems neither to be tolerated very well nor desired as a form of intellectual or social interaction. Rather, being surrounded and interacting regularly with people who think like you and share your political, religious, ethical, moral, or cultural views seems to be a much more desirable and popular option.

Furthermore, in the virtual world of social media, the removal of some of the more traditional restrictions that characterize real-life social interactions— the first level in Jung's scheme—may account for the widespread prevalence of aggressive, prejudicial, and even violent language in cyberspace, as well as the existence of infamous "virtual gang lynching or bullying attacks" that have become almost routine in these environments. In this context, I often wonder how McLuhan would react if he had lived to study our present communication routine, when all the ease of communication and the hyperconnectivity status afforded to the global village have induced, as their main side effect, an apparent return to a more tribal mode of life that characterized our hunter-gatherer ancestors tens of thousands of years ago. How would McLuhan respond to the observation that at the present juncture in the history of human social interactions, the bias introduced by our avid desire to develop technologies that enhance our communication skills, to the limit of actually fusing our own brains into one distributed virtual nervous system, has basically made us become yet another demonstration of the ubiquity of Poincaré's recurrence theorem?

I can almost hear McLuhan's reply: "Once a member of a tribe, forever a tribesman; no matter the medium!"

To avoid leaving you with the impression that the combination of mental abstractions, viral information infections, and mass connectivity can foster only brainets that spread devastation out of our tribal mental heritage, I would like to close this chapter by emphasizing the existence of another and much brighter side of the same neurobiological coin. For the relativistic brain theory, the very same neurophysiological mechanisms discussed above may explain how human groups throughout history have constantly engaged in massive collaborative efforts that produced the greatest technological and intellectual achievements of humankind.

As we saw before, the original power of the human way of collaborating comes from the possibility of interacting intellectually with large groups of people who share the same methods, ideas, and concepts. Such an enduring human tradition led to the creation, nurturing, and maintenance, over long periods of time, of innumerous schools of thoughts, countless cultural traditions, and a huge number of artistic and scientific movements. The advent of new ways of mass communication (written language, print, and other media), in addition to oral language, allowed human collaborations to extend to the point of involving people who are located far from one another (spatial resonance) or, even more extraordinarily, to people who have lived in different times (temporal resonance).

The ease with which human brainets can form as a result of an information virus infection may also explain how large social groups shared and assimilated a novel mental abstraction, a mood, an idea, a sense of aesthetics, or a new worldview that, once created, was widely disseminated throughout a human community in a given period. This pervasive mental trend that resonates across large ranges of space and time in human societies is commonly known as the zeitgeist. For the relativistic brain theory, the zeitgeist can also be seen as the product of a widely disseminated information virus infection, one that synchronizes large numbers of individual brains into a brainet because it appeals to deeply entrenched and primitive human instincts or archetypes.

Essentially, by relying on information viruses, beliefs, mental abstractions, and different types of mass communication to form highly cohesive brainets, human social groups have acquired the means to generate extremely benefi-cial outcomes from a cooperative neurobiological process that promotes and optimizes collective thinking.

Call it the ultimate in human brainstorming!

A few examples may help clarify what I mean. Think of Athens in the fifth century BC, where large numbers of Greek brains became deeply synchro-nized (and inebriated) into a brainet driven by a series of revolutionary mental abstractions, things like mathematics, science, philosophy, and democracy. Another good example would be the zeitgeist that "contaminated" the immor-tal artists of Florence during the Italian Renaissance and led to the rediscov-ery of the beauty of the human body as well as the depiction of the world seen through the perspective of the painter's eye. The so-called Vienna Circle, congregated for a couple of decades in the early twentieth century by world-renowned philosophers, mathematicians (like Kurt Gödel), scientists, histori-ans, economists, social scientists, and other Austrian intellectuals who gravi-tated around the University of Vienna's core group, could be used as a third typical example of a highly influential brainet that dictated the zeitgeist of a human group during a historical period.

These three examples reveal some common properties. For example, once a zeitgeist takes hold in a human group, it spreads like a wave from all strata of the community, influencing habits, moods, the culture, and aesthetic tastes. As such, it often manifests itself across many human endeavors. For instance, it is well known that the enormity of the technological and social impact caused by the English industrial revolution was documented by the colors, composition technique, and images created by the legendary J. M. W. Turner, the greatest romantic landscape painter of this most transformative period in British history.

As one of the leading painters of Victorian England, Turner not only partici-pated in frequent meetings and social gatherings held at the Royal Academy, he also attended some key scientific discussions and lectures delivered at the Royal Society, which at the time conveniently shared the same building with the Royal Academy. In one of his visits, Turner may have attended a lecture by the famous astronomer William Herschel on the dynamic nature of the sur-face of the sun and its emission of infrared light. He could have also learned more about Goethe's theory of color, which is said to have influenced some of his paintings. What we know for sure is that during one of these mixed artistic and scientific gatherings, Turner befriended a certain Michael Faraday, one of the greatest experimentalists of all time, who was destined to become the legitimate successor of Thomas Young and the protector of Young's wave theory of light at the Royal Institution. Coincidently, Turner lived and worked in an atelier located at 47 Queen Anne Street, half a block from the polymath Thomas Young's house at 48 Welbeck Street.

More than anyone else, Turner was responsible for creating the most enduring visual records of what it meant for England and its inhabitants to be swept by the many simultaneous economic, technological, scientific, and social earthquakes that shook first Britain, then Europe, and later on the whole world. In a series of unrivalled paintings created in the early decades of the nineteenth century, Turner documented the English countryside, as well as its coast and sea, as nobody before—and arguably nobody after—had done. He accomplished this by including in his many paintings a new treatment of light, mixed with a variety of objects and scenes that represented the sweeping technological and scientific revolutions taking place around him. The elements of technological innovation that became Turner's objects of worship included the steam engines and mills that invaded the otherwise typical rural English landscapes (*Crossing the Brook*), great engineering feats (*Bell Rock Lighthouse*), a paddle wheel–steam tug towing a relic wooden ship of the Royal Navy, the ninety-eight-gun HMS *Temeraire,* to its final docking (*The Fighting Temeraire*), and steam-spewing locomotives crisscrossing the Great Western Railway at the surreal—for the time—speed of thirty to forty miles per hour (*Rain, Steam, and Speed—The Great Western Railway*). Through these and thousands of other canvases and drawings, Turner became the industrial revolution's artistic herald par excellence, the unofficial chronicler of a time of enormous change for humanity—for the better and for the worse.

At the final stage of his career—a moment in which, despite producing masterpiece after masterpiece, he was accused of losing his touch for his insistence on pursuing a totally revolutionary way of mixing light, sea, and skies while blurring the contours of the main concrete objects present in his composition—Turner's entanglement with the multiple technological and scientific revolutions taking place around him was so intense that some art historians suggest that one of his most revered paintings, the magnificent *Steamboat during the Snowstorm in the Ocean*, hid in its depiction of dynamically fused sky, ocean, fog, and snow a sketch of the magnetic fields observed by Michael Faraday in his experiments at the Royal Institution.

Another example of an enduring zeitgeist came during a period that became known as the Belle Époque, a time of great enthusiasm and optimism in Europe that lasted from the 1870s to the years prior to World War I. As we saw in chapter 5, during this period, the French impressionists covered their canvases with the "relativistic" mood championed by Ernst Mach. As in other moments in history, the jubilant tone of the Belle Époque's zeitgeist infected not only painters and scientists but also musicians and writers, once again demonstrating the tremendous spatial resonance this mental information infection commanded. The fact that the Belle Époque remains the object of much intense study illustrates the tremendous temporal resonance of its zeitgeist.

During the transition between the nineteenth and twentieth centuries, another zeitgeist began to synchronize the brains of both artists and scientists into a brainet that profoundly affected our definition of reality. Basically, this brainet sought to rely on pure geometrical forms to represent and explain everything in the natural world, from individual objects to the entire relativistic cosmos that embraces us all. In the arts, this geometric credo emerged from the post-impressionist brush strokes of the French master Paul Cézanne. He was soon followed by an equivalent scientific pursuit, represented first by Hermann Minkowski's geometrical depiction of Einstein's special theory of relativity, and later by Einstein's own general theory of relativity. Later on, this geometric fascination inspired Pablo Picasso and Georges Braque to launch cubism, the birth of modern art. As had happened before in Athens, Florence, Paris, and Vienna, once again, the dominant zeitgeist of an ebullient period in our history become the sculptor of parallel mental revolutions taking place in multiple areas of human endeavor. Indeed, in *Einstein, Picasso: Space, Time, and the Beauty That Causes Havoc*, Arthur I. Miller argues that one has to evoke concurrent scientific, mathematical, and technological developments of the time if one wants to gain a deeper insight into the factors that led Picasso to paint *Demoiselles d'Avignon*, the masterpiece that served as the first cubist salvo. In Miller's opinion, "Relativity and *Les Demoiselles* represent the responses of two people—Einstein and Picasso, although geographically and culturally separated—to the dramatic changes sweeping across Europe like a tidal wave."

In my own relativistic terms, as two individuals infected by the same information virus that defined the zeitgeist of their time, Picasso and Einstein expressed two distinct Gödelian-rich mental abstractions of reality that, once generated inside their brains, were projected onto the external world in the shape of two particular manifestations of geometric language: general relativity and cubism.

From that point on, it did not take much for both quantum physicists and avant-garde artists to synchronize into brainets capable of conceiving even more elaborate mental abstractions in order to build their Gödelian information-rich views of reality. When these two groups revealed these newly conceived mental constructs, they did so by getting rid of the traditional object forms we normally encounter in our daily lives. That is why Miller points out, "Just as it is pointless to stand in front of a Mondrian or Pollock, for instance, and ask what the painting is of, so it is pointless to ask what the electron under quantum mechanics looks like."

My final example to illustrate all the good that these synchronized human brains can produce concerns the way we humans practice the art of science. Thanks to the evolutionary gift of being capable of using ideas and abstractions produced by previous generations to synchronize our present thoughts, we scientists can build brainets that cross centuries of human history. For example, thanks to the contribution of a six-century-long brainet (figure 11.1), formed by the interlinked mental legacy of people like Petrus Peregrinus de

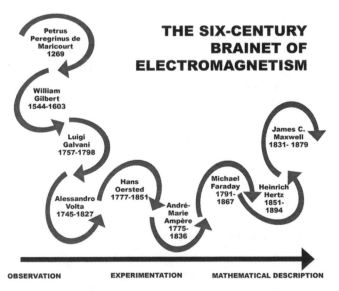

Figure 11.1. The six-century human brainet responsible for the identification and description of the phenomenon of electromagnetism. (Image credit to Custódio Rosa.)

Maricourt, William Gilbert, Luigi Galvani, Alessandro Volta, Hans Christian Oersted, André-Marie Ampère, Michael Faraday, Heinrich Hertz, and James Clerk Maxwell, among many others, the entire description of electromagnetism, one of the most pervasive phenomena in the entire cosmos, has been reduced to just a few lines of otherwise mundane mathematical symbols.

After realizing the enormity of what this and innumerable other human brainets have accomplished, I bet that, like Turner, you may also suddenly feel an urge to take some brushes and colors to express on canvas your own feeling of awe. And who could blame you for experiencing this itching?

After all, deep inside our brains, we all are ready to simply synchronize.

12 • How Our Addiction to Digital Logic Is Changing Our Brains

Back in the mid-2000s, the early signs were already very clear to anyone who cared to look around and connect the dots. It took me a while, but I finally noticed what was going on.

During a subway ride in the midst of Tokyo's rush hour in the fall of 2004, I was impressed by the profound silence that reigned inside the crowded coach. At first, I thought this was simply a reflection of the Japanese culture. A quick visual survey, however, revealed that the reason for the silence was very different than the one I had first imagined: all the commuters were staring at their smartphones, their silence an indication that, despite being physically inside the train, most of the passengers' minds were surfing somewhere else, in the distant and not yet fully delimited borders of the newly discovered cyberspace. As one of the pioneers in the process of mass market creation of cell phones and their more elaborate next generation, smartphones, Japan has become a social laboratory of sorts for a phenomenon that has gone viral worldwide. Today, of course, in any public place, whether an airport or a soccer stadium prior to the game, many of us are immersed in our cell phones—browsing, texting, posting on social media, taking selfies or other photographs—instead of relating to the people and circumstances around us.

Fast-forward to 2015. Standing on the sidewalk of the main avenue of Seoul's fashion district, I was waiting with my South Korean host for a cab to return to my hotel after giving a talk about the future of technology. To fill the time, I tried to engage the young undergraduate student with some small talk. "How many people live in South Korea right now?" I asked, trying to establish some line of communication.

"I am so sorry, but I do not know. Let me ask Google!"

Surprised by this matter-of-fact answer, which carried much more meaning to me than the student intended, I tried the next question in my routine list. "How is South Korea doing politically? What about the tension with North Korea these days?"

"I really do not know. I do not pay attention to these political issues. They really do not have anything to do with my life."

Having visited the Korean Demilitarized Zone myself back in 1995, I had witnessed firsthand the tremendous amount of tension that engulfs the border between the two Koreas and how the conflict between the two countries still dominates most of Korean life. Because of this knowledge, I was totally stunned by the young Korean student's complete lack of interest in the subject.

When a cab arrived, I listened to the student's instructions on how to communicate with the driver—who, I noticed, was totally enclosed in a hermetically sealed Plexiglas cabin constituted by both front seats of the modern Korean-manufactured black sedan. "After you sit in the back and fasten your seatbelt, just insert this card, where I wrote your hotel address, in the slot in front of you and the driver will take you back to the hotel. When you get there, just insert your credit card in the same slot and wait for your receipt."

After a formal good-bye, Korean style, I got into the cab and at once experienced the sensation of having boarded an alien spaceship by mistake. For starters, the driver, facing the windshield, did not even nod or emit any sound of greeting. Looking around, I noticed that I was totally isolated by the Plexiglas wall, which, on my side, contained only the thin slot that the student had mentioned and a television monitor showing some midafternoon program. A small video camera at the corner of the intersection between the Plexiglas and the car frame became evident only after a second scanning. Certainly, there was a microphone there too, and a loudspeaker next to it for allowing bidirectional communication with Korean-speaking customers, but I never had the chance to experience the quality of this potential vocal exchange. The moment I sat down and fastened the seat belt, an LED on top of the slot turned on and a computer-created female voice requested in English the card containing the directions. Having no other alternative at hand, I obliged by inserting the card with the address facing up. As I lost sight of the card, I noticed a light coming up on the driver's dashboard. It was then I realized that my analogy of the alien spaceship was not too far removed from the truth. Focusing my eyes to take in the overwhelming accumulation of electronic gear stuffed in that dashboard, I wondered how in heavens a human being would be able to survive and not go mad, having to spend every work day, which could amount to ten to twelve hours, driving around traffic-heavy Seoul surrounded by so many flashing lights, GPS systems, and all sorts of other digital paraphernalia. By my estimate, the cab had at least three distinct digital GPS systems, each with a different degree of resolution and complexity. The most elaborate one provided a clear three-dimensional rendition of the streets of Seoul that looked pretty realistic to me at first glance. Curiously, all systems spoke at the same time: different female voices, likely generating the same set of driving instructions, but in different tones and pitches.

Not being able to indulge in one of my favorite hobbies—engaging in small talk with taxi drivers in different parts of the world to learn what really goes on in the city—I resigned myself to watching Seoul flash by through the window.

When we arrived at the hotel's main entrance, sure enough, the slot's LED flashed again. Promptly, I inserted my credit card and waited for a minimal sign that I was in the company of another human being: a good-bye. What I got instead was my credit card, a receipt, and a computer-generated warning not to slam the door.

No human contact, no human voice, no social synchronization of any sort transpired during that Korean ride. I was treated very cordially, as I always am when I go to Korea, but efficiency was the name of the game, not social engagement. I was delivered to the right address, the fare was fair, and that should suffice.

Or should it?

In retrospect, even though I spent a great deal of time after that ride feeling sorry about the kind of life that Korean driver had to endure every day—the loneliness, the physical and mental stress that must come from being confined in a tight Plexiglas cockpit—later I realized that his fate, as miserable as it was in my view, could not be considered the worst possible scenario. After all, in 2015 he still had a job and could earn a wage performing a task that may soon be removed from the roll of menial motor occupations that humans do for a living. In the fast-evolving world of digital automation, self-driven cars are just around the corner—or so some manufacturers insist. And as has already happened to millions of jobs in the past, and certainly will happen to many millions more in the future, driving a car for pay may soon belong to the history books.

In *Rise of the Robots: Technology and the Threat of a Jobless Future*, futurist Martin Ford shows how the exponential increase in digital and robotic automation may carry us in the near future to a perfect storm of massive unemployment and economic collapse, due to a depression in the consumer market that will result from a world in which hordes of jobless people far outnumber those who can make a living through their own labor. In his book's introduction, Ford reminds us that "the mechanization of agriculture vaporized millions of jobs and drove crowds of unemployed farmhands into cities in search of factory jobs . . . [and] later, automation and globalization pushed workers out of the manufacturing sector into new service jobs."

Yet, if his predictions are correct, the unprecedented levels of unemployment that we may encounter in the next two decades of this century—something around 50 percent—thanks to exponential advances in robotics and digital technologies of all sorts, are going to surpass all waves of unemployment seen in the past due to the disruption caused by the introduction of new technologies. According to Ford, the current wave of human displacement from the job market poses a central risk to the entirety of the world economy and the survival of billions of people. Paradoxically, the first impact of this perfect storm is likely to be felt in the most advanced countries, like the United States, where both digital/robotic automation and the growth of the financial component of the GDP are likely to contribute to the most massive destruction of jobs in the shortest amount of time possible.

In his book, Ford indicates that during the first ten years of this century, instead of the 10 million jobs needed to keep pace with the natural growth in the country's workforce, the U.S. economy yielded an eye-opening *zero* net gain in new jobs created. On top of that alarming statistic, by plotting the U.S. economy's productivity and workers' rise in income from 1948 to 2017 (see figure 12.1 for the most recent data from the Economic Policy Institute), it becomes apparent that these two curves, which traditionally ran together at the same pace for twenty-five years, began to diverge significantly starting in 1973. As a result, by 2017, while worker compensation had grown 114.7 percent, productivity had soared to a 246.3 percent increase. That meant that instead of reaching a median household income of $100,800, as expected if the staggering gains in productivity in the period had been fairly transferred to worker compensation, American families had to deal with soaring costs in health care, education, and other basic living expenses while earning a median of approximately $61,300.

Ford asserts that the same phenomenon has been taking hold, albeit at different times, in thirty-eight out of fifty-six economies around the world, including in China, where the routine of massive layoffs as a consequence of industrial automation has already been computed as an integral part of the job market reality. In some countries, the share of worker compensation has

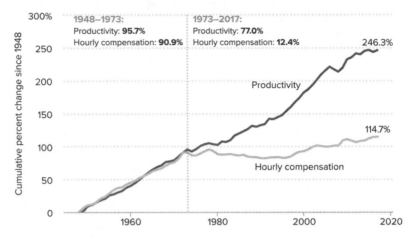

The gap between productivity and a typical worker's compensation has increased dramatically since 1973

Productivity growth and hourly compensation growth, 1948–2017

Notes: Data are for compensation (wages and benefits) of production/nonsupervisory workers in the private sector and net productivity of the total economy. "Net productivity" is the growth of output of goods and services less depreciation per hour worked.

Source: EPI analysis of unpublished Total Economy Productivity data from Bureau of Labor Statistics (BLS) Labor Productivity and Costs program, wage data from the BLS Current Employment Statistics, BLS Employment Cost Trends, BLS Consumer Price Index, and Bureau of Economic Analysis National Income and Product Accounts

Economic Policy Institute

Figure 12.1. Comparison between cumulative percentage change in U.S. productivity and workers' hourly compensation from 1948 to 2017. (Reproduced with permission from Economic Policy Institute, "The Pay-Productivity Gap, 1948–2017," August 2018.)

plunged even deeper than in the United States. As a result, during the first decade of the twenty-first century, there has been a significant increase in social and economic inequality and a worrisome trend toward a massive elimination of jobs. Quoting Martin Ford again: "According to [a] CIA analysis, income inequality in America is roughly on par with that of the Philippines and significantly exceeds that of Egypt, Yemen, and Tunisia."

To make things even worse, Americans born today are likely to experience much lower levels of economic mobility than their counterparts in most European nations, a finding that, as Ford properly remarks, makes a serious statistical dent in the widespread claim that the American dream of climbing the ladder by sheer effort, merit, and persistence is still alive and well. The picture gets even more alarming when one realizes that it is not only menial, manu-facturing, and other blue-collar jobs that are evaporating in the modern global economy: the jobless tsunami is already reaching the shores of the white-collar job paradises, including professions that most thought could never be vapor-ized by the digital revolution. Journalists, lawyers, architects, junior bankers, doctors, scientists—and, ironically, even highly qualified workers in the very sector that is driving this trend, the digital industry—are feeling the pinch. As Ford says, the traditional 1990s idea that a degree in computer science or engineering would assure young people entering the U.S. job market a good position has become nothing but a myth in the present climate.

Ford cites a couple of examples illustrating the mindset of those who forget to carefully imagine, for even a brief millisecond, the social consequences of a world in which 50 percent or more of the workforce is out of work. Take, for instance, the unbelievably callous prophecy offered by Alexandros Vardakos-tas, co-founder of Momentum Machines, who, speaking about his company's main product, said, "Our device isn't meant to make employees more efficient . . . it's meant to completely obviate them."

We will come back to this paradox in chapter 13 when we discuss the in-teresting "coincidence" that these economic ideas seem to originate from the same quarters that proclaim, as if it were an irrevocable law of nature, that the human brain is simply a digital machine and hence can be simulated by digital computers. But first let's address a concern that is even more terrifying to the future of humanity, if anything can be, than a jobless world.

One of the most troubling conclusions, at least for me, raised by some famous U.S. economists cited in Ford's book is the belief that workers should forget about the idea of competing with machines and instead lick their wounds, bury their chauvinistic organic pride deep inside, and face reality: according to these economists, in order to survive, the only viable strategy in the future will be to learn the best way to play second fiddle to machines. In other words, our only hope is to become machine and computer babysit-ters, their helpers and assistants—a nice euphemism for being downgraded to their servants or slaves rather than their masters. Indeed, without most of us knowing, something very similar to this scenario is already happening with pilots, radiologists, architects, and a large variety of highly skilled people. The call for surrender has been heard loud and clear and, as a response, some hu-man troops are already forfeiting their mental arms and accepting their defeat.

As disturbing as this scenario is, I believe that there is something that could be even more devastating for the future of humanity: the erasing from our brains of the very traits that have defined our human condition since the emergence of the modern human mind about one hundred thousand years ago. Far from being something out of a bad science fiction movie, I consider this as a very pertinent and worrisome possibility, one that has already been raised by many authors who conclude that our continuous and thorough immersion with and total surrender to digital technology in every minute of our conscious life—minus a few hours of sleep a day, for now—may be corrupting and quickly eroding the basic operation of our brains, their unique reach and range, not to mention their ability to generate all that defines the splendor and distinctiveness of the human condition. If the thought of a 50 percent unemployment rate did not shock you, how would you react to knowing that, by the time that prognosis becomes reality, a much larger percentage of us may have already morphed into nothing more than mere biological digital zombies rather than proud descendants and carriers of the genes and cultural traditions of those early members of the *Homo sapiens* clan, those who from humble primate origins, after enduring all sorts of life-threatening challenges, from glaciations to famine to pestilence, lived to prosper while creating their own private human universe out of a jelly-like lump of white and gray organic matter and one picotesla of magnetic power.

My assessment, based on a variety of evidence and findings from psychological and cognitive studies, is that the risk should be taken very seriously. The human brain—being the most competent neural chameleon ever created by nature—when exposed to new world statistical contingencies, particularly those associated with strong hedonic experiences, usually reshapes its own internal organic microstructure and then uses this newly embedded information as templates for guiding future actions and behaviors. Accordingly, in the particular context of our interaction with digital systems, there is a credible possibility that the establishment of a routine of constant positive reinforcement obtained by our continuous interaction with digital computers, algorithm logic, and digitally mediated social interactions, just to mention a few examples, may gradually reshape the way our brains acquire, store, process, and manipulate information.

Using the relativistic brain theory as a background, I believe that this continuous daily digital onslaught may simply corrode the normal process of storage and expression of Gödelian information and production of noncomputable behaviors by our brains while favoring an increase in the central nervous system's reliance on Shannon-like information and algorithmic-like actions to conduct its daily business. Essentially, this hypothesis predicts that the more we become surrounded by a digital world and the more the mundane and complex chores of our lives are planned, dictated, controlled, evaluated, and rewarded by the laws and standards of algorithmic logic that characterize digital systems, the more our brains will try to emulate this digital mode of operation in detriment to the more biologically relevant analog mental functions and behaviors engendered over millennia by the process of natural selection.

This digital chameleon hypothesis predicts that as our obsessive infatuation with digital computers takes a deeper grip on the way we perceive and respond to the world around us, unique humanlike attributes, such as empathy, compassion, creativity, ingenuity, insight, intuition, imagination, outside-the-box thinking, metaphoric speech and poetic discourse, and altruism—just to name a few typical manifestations of noncomputable Gödelian information—will simply succumb and vanish from the repertoire of the human mental capacities. Taking this reasoning one level deeper, I can easily see that in this potential future scenario, whoever controls the programming of the digital systems surrounding us will have a grip on dictating the future mode of operation of the human mind, both at the individual and collective levels. Worse yet, in the long run, I dare to say that this control could extend itself to become a crucial influence on the evolution of our whole species.

Essentially, the digital chameleon hypothesis provides a neurophysiological framework or backing for an idea that has been floating around since Sir Donald MacKay first argued against embracing Shannon information as a description of how the human brain processes information. In *How We Became Posthuman*, N. Katherine Hayles writes that at the end of World War II, "the time was ripe for theories that reified information into a free-floating, decontextualized, quantifiable entity that could serve as the master-key for unlocking secrets of life and death." Ironically, the particular political and economic context of the postwar U.S. cleared the many intellectual objections that could have stopped a context-free theory of information locomotive from derailing before it left the station.

In *The Closed World*, Paul Edwards describes how both the cybernetics movement and its spin-offs, computer science and artificial intelligence, were heavily influenced by the agenda—and funding—of the U.S. Department of Defense during the Cold War. As early as July 8, 1958, barely two years after a historic conference held at Dartmouth University launched artificial intelligence as a credible scientific field, the *New York Times* published an article whose headline—"New Navy Device Learns by Doing; Psychologist Shows Embryo of Computer Designed to Read and Grow Wiser"—forecasted the imminence of a time in which smart machines, funded by the Department of Defense, would replace humans in the process of decision making in matters of national security and defense and also in the marketplace. Even in the late 1950s the propaganda hype machine—the conjoined twin of artificial intelligence—was at full throttle; the text of the article with the following announcement: "The Navy revealed the embryo of an electronic computer today that it expects will be able to walk, talk, see, write, reproduce itself and be conscious of its existence."

Needless to say, the navy never got to play with the self-aware talking device for which it paid top money. Indeed, sixty years after that *New York Times* article was published, there is no sign that such a device will ever see the light of the day, in the U.S. or abroad. In fact, for the past six decades, artificial intelligence has lived through an endless sequence of boom-and-bust cycles, which my good friend the futurist Alexander Mankowsky, an executive at Daimler-Mercedes in Berlin, has described in the graph depicted in figure 12.2. According to Alexander's picture, invariably this cycle starts with the repackaging of the old idea that building smart machines is just around the corner. After a few years of growing enthusiasm—and sizeable public and private investments, notably by military agencies, like the Defense Advanced Research Projects Agency (DARPA)—the results prove to be rather disappointing and the whole field, as well as small companies that were created during the latest boom phase, experiences a Permian type of extinction process. Indeed, two such events almost did in artificial intelligence for good: the Lighthill report, created in response to a request by the British Science Research Council, all but devastated artificial intelligence in early 1973 by showing that the big promises made by the field had not materialized at all; and the utter failure of the so-called Japanese robots in a project that aimed at creating autonomous intelligent mechanical machines capable of performing tasks that only humans can do. A tragic example of the failure of this Japanese initiative was made explicit when no robot available in Japan was able to penetrate the damaged nuclear reactors of Fukushima to perform the repairs needed to stop the consequences of the worst nuclear accident in the country's history. Instead, human volunteers had to perform these tasks, and many of them perished while executing their heroic duty. Meanwhile, a series of the latest Japanese robots lay destroyed in the pathway that led to the now deadly reactors.

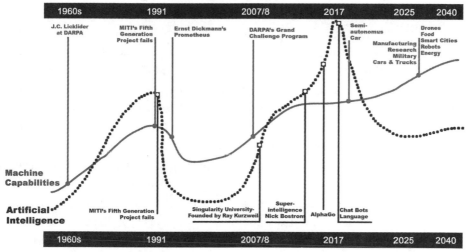

Figure 12.2. The boom-and-bust cycle of artificial intelligence during the past several decades. (Image credit to Custódio Rosa.)

But by the end of World War II, the increasing processing power acquired by digital computers—the perfect embodiment of machines that could take advantage of Shannon's concept of information—became utterly irresistible. That led many to predict that simulating human brain performance was just a matter of time. Reflecting on the zeitgeist of that period, Joseph Weizenbaum, the MIT computer scientist who in the 1960s pioneered one of the first interactive computer programs, ELIZA, had this to say:

> By the time the digital computer emerged from university laboratories and entered the American business, military, and industrial establishments, there were no doubts about its potential utility. To the contrary. American managers and technicians agreed that the computer had come along just in time to avert catastrophic crisis: were it not for the timely introduction of computers, it was argued, not enough people could have been found to staff the banks, the ever increasingly complex communications and logistic problems of American armed forces spread all over the world could have not been met, and trading on the stock and commodity exchanges could not have been maintained. . . . Unprecedentedly large and complex computation tasks awaited American society at the end of the Second World War, and the computer, almost miraculously it would seem, arrived just in time to handle them.

Weizenbaum, however, is quick to conclude that this "just in time miracle" was nothing but a collective mental construct—a zeitgeist—of all the parties interested in the introduction of computers in mainstream America, given that the future that unfolded afterward was by no means the only one possible at the time. He supported this view by saying that most of the war effort, including the Manhattan Project, which led to the implementation of the atomic bomb, was executed successfully without the widespread availability of computers. Instead, human brain power was used to perform all the needed computations, from the most tedious to the most complex. Computers certainly sped up the process considerably, but they did not introduce any new basic understanding or knowledge of the processes—or science—that benefited from their introduction. In fact, Weizenbaum argues that although a growing number of early users began to see computers as indispensable tools, it did not follow that they really were. In those early days of digital computing, speed to yield the final result became the key variable supporting the immediate acceptance of computers in most aspects of American life. According to Weizenbaum, "The digital computer was not a prerequisite to the survival of modern society in the post-war period and beyond; its enthusiastic, uncritical embrace by the most 'progressive' elements of American government, business, and industry quickly made it a resource essential to society's survival *in the form* that the computer itself had been instrumental in shaping." This notion has been reinforced by other authors in the last decades. For example, Paul Edwards follows Weizenbaum in stating that "tools and their use form an integral part of human discourse and, through discourse, not only shape material reality directly but also mold the mental models, concepts, and theories that guide that shaping."

This implies that our continuous and growing interactions with computers are likely to change the demands we impose on our brains through a process that is by no means without risks. Take, for example, the issue of navigation. For millions of years, the exquisite ability to recognize the detailed natural features of the outside world has been literally engraved in the neuronal flesh of our brains. That is because brain structures, such as the hippocampus—and likely the motor cortex, as we saw in chapter 7—contain neuronal-based representations of space that allow us to design optimal navigation strategies to move around the world that surrounds us. Interestingly, brain-imaging studies conducted by researchers at the University College of London have shown that the hippocampus of experienced London taxi drivers is significantly larger than those of us who do not make a living driving through every single corner of England's complicated capital. The caveat, however, is that these studies were conducted on drivers who did not learn to drive using modern digital GPS devices. Because navigation through GPS stimulates completely different brain circuits than those involved in natural navigation, one can almost predict that an increase in hippocampus volume is not likely to materialize in a younger generation of London cab drivers. But could they actually show a reduction in hippocampus volume below the baseline of typical adults? That possibility has been raised by some neuroscientists who worry that if that happens, not only will natural navigation skills be compromised but so will all sorts of other cognitive skills that depend on the integrity of the human hippocampus. In a nutshell, this is the kind of general problem hundreds of millions of people may be facing in the next few decades as they adopt new digitally inspired strategies: a dismantling of the organic neural apparatus that was incorporated into our brains as a result of selective pressures that occurred hundreds of thousands or even millions of years ago. That could be a recipe for trouble. A lot of trouble.

Indeed, although the artificial intelligence movement has so far failed to achieve anything like superhuman intelligence, its rhetoric provides our brains with more trouble at a different dimension: our ability to differentiate what constitutes a real scientific advance from mere propaganda to sell a product. Successful demonstrations of artificial intelligence, such as defeating human chess players and Go world champions, have been widely disseminated by the artificial intelligence lobby, helping to create a widespread feeling that it may have, at long last, succeeded in displacing human intelligence from its pedestal. In reality, these new approaches recapitulate old algorithms and multivariate statistical ideas and, at most, enhance the ability of modern systems to perform pattern recognition functions. For instance, Deep Learning, despite the pompous name, is nothing but an artificial neural network of the sort invented in the 1970s in which a much larger number of computational steps—also known as hidden layers—has been added to the algorithm. Such a maneuver helps increase the pattern recognition performance of the artificial intelligence system, but it does not solve the main deficiencies always found in such software since its origins sixty years ago: artificial intelligence systems are prisoners of past information and the rules used to create their databases and training sets embedded in the system. They cannot create new knowledge. In that sense, artificial intelligence basically reflects the Laplacian dream of a fully predictable universe, one in which the future is totally predicted by the past. Thus, if such a system, meant to create music, receives as a training set only Mozart symphonies, it will never be able to create other styles of music like the ones composed by Bach, Beethoven, the Beatles, or Elton John. That is because artificial intelligence does not create anything; it does not understand anything; it does not generalize anything. It only spits out what it was fed—fed by human hands, mind you. If there is one thing such "intelligent systems" are not, it is intelligent, in the human definition of this word. Thus, if one uses a human benchmark of intelligence as our gold standard of performance, artificial intelligence systems will fail miserably every time.

The trouble is, artificial intelligence doesn't need to succeed at surpassing human intelligence now to become more powerful than we are in the future. Such a future can be reached through a much more expedient and feasible detour: overexposing the human brain to digital systems until, not having other options left, the True Creator of Everything finds no meaningful alternative to becoming one of them. As author Nicholas Carr masterfully put it, "As we come to rely on computers to mediate our understanding of the world, it is our own intelligence that flattens into artificial intelligence."

The opposite, as we saw, is impossible (see chapter 6). Therefore, we have only ourselves to blame if the worst comes to pass and future generations are deprived of experiencing the true range of humanity as we knew it until recently. As is often the case, scenarios like the singularity or even my digital chameleon hypothesis, before they become topics for academic discussion, tend to play themselves out and reach the public in the form of science fiction. In *How We Became Posthuman*, Hayles describes how the concept of a posthuman era has played a major role in several popular science fiction books. In one example, Hayles analyzes how in Neal Stephenson's neuroscientific thriller *Snow Crash*, the main plot revolves around the possibility that a virus could infect the minds of people all over the planet and transform them into mere biological automata, devoid of any trace of real consciousness, free will, agency, or individuality.

Such a horrible possibility makes sense if one accepts the premise of the cybernetics movement that the brain behaves as a simple Shannon information-processing device. Obviously, I don't think it does. But I do worry that our constant reciprocal interaction with digital logic, particularly when it leads to powerfully hedonic experiences, will result in the slow compromise or even elimination of some of the behaviors and cognitive aptitudes that represent the most exquisite and cherished attributes of the human condition. How can this happen if the human brain is not a Turing machine and does not rely on Shannon information for its computations? At the most basic level, numerous genes in the human genome, selected by a multitude of evolutionary events, interact as part of a "genetic program" responsible for assembling the brain's natural three-dimensional structure during prenatal and early postnatal life. This genetic programming guarantees that our brain's initial physical config-uration reflects the evolutionary process that took place over millions of years until the present layout of the human central nervous system coalesced into the basic neural architecture that evolved in anatomically modern humans about one hundred thousand years ago. Once we are born, the brain's pro-gramming continues as a result of its bidirectional interaction with the body that houses it and our surrounding environment. Continuous immersion in human culture and its cornucopia of social interactions further "programs" the central nervous system. Of course that is not the only way to alter our brain. We can also assimilate mechanical, electronic, and digital tools into the workings of our brains, as my work on brain-machine interfaces proves. I think it is also possible for the brain not merely to assimilate a digital device, but to become one.

In the 1970s, Joseph Weizenbaum was already impressed by the stunning results obtained when people began to use his program ELIZA. In Weizen-baum's view, digital computers were the latest addition to a long sequence of intellectual technologies, such as maps and clocks, which influenced in a decisive way how we perceive and experience reality. Once they penetrate our lives, these technologies are assimilated as "the very stuff out of which man builds his world." As such, he warned that "the introduction of computers into some complex human activities may constitute an irreversible commitment." According to Weizenbaum, "an intellectual technology [like the computer] be-comes an indispensable component of any structure, once it is so thoroughly integrated with the structure, so enmeshed in various vital substructures, that it can no longer be factored out without fatally impairing the whole structure."

It is no wonder that, with ideas such as these, Joseph Weizenbaum became an outcast, a heretic in the very field he helped to found with his own research. Yet, four decades later, the deep questions Weizenbaum raised continue to haunt us. Over the past two decades more observational and experimental evi-dence has accumulated to back up the notion that our interactions with digital systems are not innocuous. Instead, they may affect some of our most regular mental functions. That means that for each specific benefit in brain function gained by interacting with digital logic reported—which some are quick to herald every time any objection is raised to the digital onslaught that our ana-log brains are suffering—profound and unexpected changes in the way our organic computers operate can be documented. Indeed, Patricia Greenfield argued that the evidence from a large group of studies on the impact of different forms of media on intelligence and learning indicates that human interactions with any type of new media leads to cognitive gains that manifest themselves at the expense of other mental skills. In the case of our interactions with the internet and screen-based technologies, Greenfield shows that the "widespread and sophisticated development of visual-spatial skills" is paralleled by impairment in our ability to carry out the kind of "deep [mental] processing" underlying "mindful knowledge acquisition, inductive analysis, critical thinking, imagination, and reflection."

In *The Glass Cage: Automation and Us*, Nicholas Carr reviews an extensive list of studies showing that continuous exposure to digital systems can have profound effects on human performance, from the flying skills of airplane pilots to the pattern recognition ability of radiologists to the broad sense of creativity of architects. In all these very diverse conditions and contexts, the result is always the same: the moment humans assume a subaltern position in relation to digital systems, meaning that they are not in control of the main action anymore but serve only as a sidekick of a master computer, which is really taking over the main brunt of the task at hand—like flying a plane, interpreting radiological images, or designing houses—human skills begin to degrade to the point in which errors that were not common before rise to the foreground.

In figure 12.3, I provide a graphic display of what I think may be going on in the human brain in most of the circumstances in which digital systems begin to dictate the way humans conduct their routine. According to this digital chameleon hypothesis, continuous passive immersion in the digital systems of modern airplanes (in the case of pilots), digital imaging diagnostics (radiologists), and computer-assistive design (architects) may gradually reduce the range of cognitive human brain functions by assigning more relevance or even priority to process Shannon rather than Gödelian information. Basically, this would happen because once the external world begins rewarding individuals for behaving like digital machines in their jobs, at school, at home, or in any other type of social interaction, our brains will adapt quickly to the "new rules of the game" and radically change the way they routinely operate. This plastic reorganization and the changes in human behavior it would trigger, once again, would be driven by the brain's attempt to maximize hedonic sensations generated by the release of dopamine and other chemicals from neural circuits that mediate reward. Thus, if the outside world begins to award

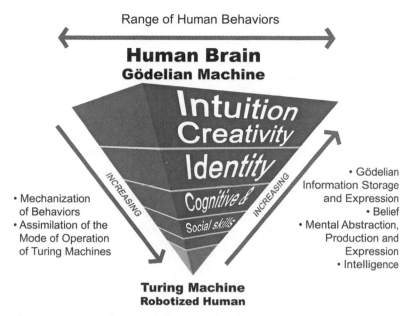

Figure 12.3. Inverted pyramid illustrates the clear contrast between the properties of Gödelian and Shannon information. (Image credit to Custódio Rosa.)

significant material or social gains to digital machine–like behaviors, human creativity and intuition may surrender to fixed protocol, ingenuity may succumb to rigid algorithmic procedures, critical thinking may be totally overcome by blind obedience to imposed rules, and novel artistic and scientific thinking may be obliterated by dogma. The longer this feedback loop continues to be enforced, the more the operation and behaviors of the brain will resemble those of a digital machine. Ultimately, this trend may lead to the compromising or pruning of a large variety of human attributes that depend on the expression of Gödelian information.

The neuroscientist Michael Merzenich, one of the pioneers in the investigation of adult brain plasticity, has this to say about the potential impact of the internet on the human brain: "When culture drives changes in the ways we engage our brains, it creates different brains." Merzenich's stark warning has been corroborated by several imaging studies that detected structural alterations in both the gray and white matter of the brains of adolescents diagnosed as suffering from internet addiction disorder. Although further studies based on larger samples will be required to determine the validity of such claims, these preliminary findings should not be taken lightly.

But one does not need to rely only on extreme cases of internet addiction to detect neurological or behavioral changes associated with our digital indulgence. Betsy Sparrow and colleagues showed that when people believe that a series of statements they have been asked to remember will be stored online, they perform worse than a control group that relies only on their own biological memory to remember the statements. This suggests that subcontracting some simple mental searches to Google may, after all, reduce our own brain's ability to store and recall memories reliably. These findings support an idea that Ronald Cicurel and I have debated for a while: when the brain is overwhelmed, either by information overload or by the need to engage in levels of multitasking it is not prepared to handle, one of its first reactions is to "forget"—either making it more difficult to access stored memories or, at the limit, simply erasing some information already stored. We believe this could be almost a defense mechanism of our brain to counter situations in which it is overtaxed beyond its processing limits.

Such an information overload can be clearly recognized in modern times by looking at the way people use the internet to keep in touch with family and friends. Not surprisingly, the impact of online social media on our natural social skills is another area in which we may be able to measure the true effects of digital systems on human behavior. For example, in *Alone Together*, Sherry Turkle describes her long experience interviewing teenagers and adults who are heavy users of texting, social media, and other online virtual environments. Social media and virtual reality environments can induce significant levels of anxiety; a profound lack of development in social skills, which invariably leads to a withdrawal from real social interactions; reduction in human empathy; and difficulties in handling solitude. Moreover, symptoms and signs of addiction to virtual life are often reported almost casually in some of these interviews.

After reading Sherry's book, I began wondering whether the new "always connected" routine is overtaxing the cerebral cortex by expanding dramatically the number of people with whom we can communicate, almost instantaneously, via the multitude of social media outlets available on the internet. Instead of respecting the group size limit (about 150 individuals) afforded by the volume of cortical tissue allocated to us by evolution, we are now in continuous contact with a much larger number of people constituting a virtual social group that can far exceed that neurobiological limit. Since the maturation of the human brain's white matter unfolds over the first few decades of postnatal life and does not reach a final level of maturity until the fourth decade of life, cortical overtaxing could be even more of a problem in teenagers and young adults, who have yet to reach a full, mature level of cortical connectivity. That could explain the high levels of anxiety and deficits of attention, cognitive ability, and even memory observed in heavy users of social media who fall into this segment of the population.

The compulsion so many of us feel to interact with digital systems such as the internet broadly and social media specifically also has an explanation in the digital chameleon hypothesis. Studies with young adults who have been diagnosed as being addicted to internet-based activities reveal clear disruptions in brain reward circuits. Once again, the key culprit involved is the neurotransmitter dopamine. These studies suggest that we engage more and more in online activities simply because they drive our brains to generate intense feelings of pleasure and reward. In this context, the interactive software we came to know as social media, such as Facebook, Twitter, WhatsApp, and WeChat, has become some kind of social glue, or, to use the type of language I have been employing throughout this book, the primary synchronizer of human brainets formed by thousands or even millions of people avid for an instantaneous satisfaction of our enormous craving for social bonding that emanates from within our brains. Call it virtual grooming; certainly the pleasures of both real grooming and internet surfing share the same neurochemical basis. The crucial involvement of dopaminergic circuits also explains why internet addiction exhibits clear parallels with compulsive gambling and drug dependence.

Is this something we should pay attention to? I think so. Not only because of the potential impact on the mental health of this and future generations but also because of the far-reaching consequences of our increasing interaction with digital systems. At the far limit, I can foresee that this staggering expansion in our online use and virtual social connectivity seems capable of providing a completely new type of selective pressure that may, eventually, bias the evolutionary future of our species. Based on that, one wonders whether the dawn of *Homo digital* is upon us or, more alarmingly, whether that species is already here, texting and tweeting without being noticed.

Even if that is not the case, it is interesting to ponder that after all the explosive growth in communication technology created and experienced by our species during the past century alone, which resulted in taking us one step closer to fulfilling Marshall McLuhan's prophecy of using artificial means to extend our central nervous systems to the point of almost linking every one of us to each other at light speed, the main by-product to emerge from this process seems to be an extreme fragmentation of the whole of humankind into a multitude of virtual tribes, each tightly linked by a specific set of beliefs, demands, concerns, likes and dislikes, moral and ethical values. Ironically, it seems that despite our push toward a high-tech society, what we have harvested out of this digital crop is a return to the essential mode of tribal social organization that gave rise to the True Creator of Everything millions of years ago. The only difference is that instead of spreading our bands of compatriots over the great forests and plains of the real world, we seem more and more committed to becoming mere hunters and gatherers of dispersed, dopamine-laced bits and bytes of cyberspace. That is all fine, provided we realize that the price we pay for this choice could be the loss of most of what has come to be known as the unique features of the human mind.

A few decades ago, Joseph Weizenbaum already conceived that something like this could happen in the future. To him, the only recipe to avoid the very fate we now face head-on was to adamantly refuse to subcontract "tasks that demand wisdom" to our own creations, things like digital computers and software. In his view, these should remain the sole prerogative of the True Creator of Everything.

Given all that I have seen, read, and experienced in recent years, I sincerely believe that the moment for taking a stand that follows Weizenbaum's wise recommendation is fast approaching since we are likely reaching an imminent point of no return in our overindulging love affair with digital machines. In this context, it seems only proper to finish this brief account on the dangers that the True Creator of Everything faces today by quoting one of the great poets of the twentieth century, T. S. Eliot, who in his 1934 "Choruses from *The Rock*" presciently pinpointed, in a mere three verses, the central predicament of our times:

> Where is the life that we lost in living?
> Where is the wisdom we have lost in knowledge?
> Where is the knowledge we have lost in information?

13 • Self-Annihilation or Immortality?

The Ultimate Choice of the True Creator of Everything

In the final years of the second decade of the twenty-first century, humanity, as a whole, finds itself dragging its collective feet to the edge of an existential bifurcation—or an evolutionary abyss, if you prefer—whose outcome, still nebulous, may well settle the future, or the lack thereof, of our embattled species *Homo sapiens* has a major collective decision to make. After hundreds of thousands years of an epic and intensively creative journey that yielded as its main mental edifice a whole new view of reality, the human universe, the True Creator of Everything finds itself embroiled in, mystified and, more often than not, misled by a couple of dominant mental abstractions that, despite some unequivocal benefits, carry hidden within themselves the potential to fully eradicate our human way of living and, at the limit, determine the complete and thorough obliteration of our kin from the face of the Earth. That this imminent cataclysmic threat has germinated within the depths of the human mind for the past few centuries, although ironic, cannot be considered a surprise at all. Once the human brain acquired the neurophysiological attributes to generate powerful mental abstractions and, later on, produced the technological means to induce and further enhance the synchronization of millions of minds into brainets capable of exponentially expanding the reach of our human social skills, one of the undesirable side effects to emerge from this process was the ability to self-destruct at the ultimate scale.

Even though the risk of total nuclear war has subsided somewhat in the past decades, today there is more than a terminal nuclear cataclysm to worry about. In fact, the time is fast approaching in which the True Creator will eventually have to make a decision: either succumb to the asphyxiating embrace of a pair of mental abstractions that threaten almost all modern human societies or, in a surprise move, make an unexpected U-turn that reasserts the central role played by our human brains in the making of our own universe. And here lies the existential dilemma I am referring to: choose wisely and the future, if not the immortality, of the entire human race will be ensured, or select a misguided course based on the mirages generated by mental abstractions running amok, and the prognosis for sealing self-annihilation may become irreversible.

Surprising as it may sound, the original climb toward the edges of this life-or-death precipice we all face today has its origins in the emergence of two intertwined mental abstractions that, by merging into a dominant and widely accepted worldview, can only be described today as a true new religious cult that aims at ruling and controlling every aspect of human life. Together they form a formidable and almost invincible opponent of the notion that humans should continue to exert full control over their own future. Merged into a unique and almost unbeatable symbiotic entity, these two mental abstractions clearly pose the most significant threat that the True Creator has engendered regarding the survival of our species. I am referring to the financial view of the human universe, which proposes to monetize every single aspect of human life, and the cult of the machine, a concept first described by Lewis Mumford that encompasses our species' capacity for becoming deeply bewitched by the tools and technologies we are able to develop to enhance our reach into the world. Over the past seventy-five years, this cult can be typified best by the views proposed by cybernetics and its most well-known offspring, artificial intelligence, since both movements share the mystifying belief that humans and their brains are nothing more than automata or Turing machines.

Although one could argue that the fusion of these two mental abstractions has led to an undeniable improvement in material development and the standard of living of humanity, one would have to immediately qualify this claim by the findings that most of these gains have been enjoyed in very unequal terms, given the extremely lopsided distribution of their benefits across the entirety of the human race. Moreover, once they fused into a single-minded framework, these two mental constructs immediately began to conspire, in a variety of ways, to create scenarios that threaten not only the future but also the viability of our species' way of living.

Essentially, my point is that if the ongoing fusion of ideologies that promote the full mechanization and monetization of human life into a single dominant worldwide operational construct continues at the current accelerating pace, or even speeds up, as some predict, there is a concrete chance that these developments may devour key aspects of human culture with such unprecedented levels of voracity that recovery may prove utterly impossible.

According to the dominant view today, every object and every aspect of our lives, including life itself, has a finite monetary value. For those who profess this belief, the only value one can assign to human life and human endeavors is the one determined by "the market." Yet proponents of this view seem to ignore that the market is nothing but an abstract entity, a brain-generated creature, which over the past centuries has acquired almost as mystical a status as the different types of gods created by the True Creator during the eons of human history. As the newly enthroned god of humanity, the market, despite being an offspring of the neurobiological principles that govern the human brain, has now turned against its own creator, like Zeus turned against Cronos, with the intent of gaining the complete surrender and submission of humanity to its moral and ethical values, or lack thereof. Indeed, the ethical values of this new human-created deity can be summarized simply as the relentless pursuit of infinite profits and unlimited greed at any cost. Consequently, both the followers and the cardinals of this Church of the Market seem to share and profess the same type of religious fervor that characterized the militants of the Catholic or the Protestant Church, or any other form of organized religion, for that matter. Yet nothing in the universe out there but the human mind has endowed the market with the power it currently exerts on every single aspect of our lives.

According to the braincentric view, the true roots of the Church of the Market's very successful proselytization, which incentivizes the exercise of irresponsible, unlimited greed by seeking the maximum possible financial return in all possible human activities, including the relentless mining of every one of our behaviors and opinions, sprout from the same basic neurobiological mechanisms that account for how brainets are formed and employed to dictate large-scale human social behaviors. Essentially, it all boils down to the extreme power that the neurotransmitter dopamine and other reward-related chemicals have in consolidating and amplifying, with the fundamental assistance of information viruses and different communication media, the spread of mental abstractions among human societies. As in the case of the deadly brainets discussed in chapter 11, today's financial mental abstractions, spread by powerful and tightly coupled brainets, are dictating illogical economic and fiscal policies as well as misguided moral and ethical values that tend to go directly against the best interests of most of humanity while favoring a tiny economic elite. As we saw in previous chapters, this happens because dopamine contributes decisively to the amalgamation of human brainets that propagate mental abstractions that, appealing to our most primitive instincts and archetypes, compete for assuming a dominant role in a human society.

By all recent accounts, particularly those made public during the 2008 U.S. banking crisis, dopamine-mediated, reward-seeking behaviors similar to those observed in drug, sex, or gambling addiction seem to have become ubiquitously embedded in the decision-making process employed by a significant number of market operators, big and small.

Financial gain at any cost: this seems to be the dominant motto of our times. Again, one only needs to recall the catastrophic events of the 2008 financial crash, an event that brought the entire planet to the brink of an unprecedented economic collapse, to realize how dangerous the unregulated operation of such financial brainets has become for the future of humanity. This is why I disagree with the social constructionism view that to understand phenomena like the financial markets we can simply focus on the study of the dynamics of human social behaviors, culture, and language. For starters, these are only second-order emergent phenomena generated by a large number of interacting human brains. Therefore, to fully understand how these second-order phenomena are generated, and how they can either be controlled or moderated, we need to dive into the neurobiological principles of how human brains operate, in isolation and as part of huge human social groups, in their pursuit of power and unlimited reward. Otherwise, we would be behaving like someone who claims that the turning of the key in the ignition explains how an automobile engine works.

Engaging in the debate about the true primary origins of complex human social constructs like financial, economic, and political systems and ideologies is essential because, as we saw, human brains are very plastic throughout our lives. This means that through education, one can demystify these mental abstractions by demonstrating that they are man-made, not the product of some divine intervention. That alone may pave the way for our education system to instill a much more grounded and relevant humanistic view in the minds of society's future decision makers regarding the wisdom of compromising the well-being of hundreds of millions of people in favor of following a mental mirage. Put in other words, by demonstrating that market-based ideology is neither a god nor part of a divine plan, we have a better chance of promoting an economic and political agenda that aims to improve the quality of life for people worldwide, while preserving Earth's natural environment for future generations.

Unlimited fairness, education, and opportunities, instead of irresponsible greed, should be the real motivators that drive the human universe.

As the Church of the Market's main tangible medium of exchange and repository of value, money has ascended to become the epicenter of this financial cosmological view of the human universe. One just needs to look at figure 13.1, where I depict a simplified historical description of the different mediums human societies have chosen to employ over time in order to acquire or exchange goods among themselves, to see how a representation of money, after a few decades into the digital revolution, allowed the financial cosmological view to seamlessly merge with its twin, the mechanized view of the human universe. From cacao seeds in the Aztec Empire to gold nuggets to metal coins, to the letters of credit issued by the Florentine and Venetian financiers to merchants and explorers to paper bills, credit cards, and all sorts of bonds and financial instruments all the way to the latest digital representation of money, depicted in a series of zeroes and ones, and even to the ever-growing spectrum of cryptocurrencies like the bitcoin, there is just one common thing that unifies all these mediums: their value has always been arbitrarily set by human trade around the world, through a consensual mental abstraction, sealed by an almost silent social contract, signed by all of those who will accept money as payment for goods and services. People all over the world are prepared to sell their labor, skills, creativity, thoughts, and ideas—not to mention fool, kill, enslave, and exploit others—in order to gather a collection of otherwise worthless pieces of printed paper or, more recently, a particular binary sequence in their digital banking accounts. That happens not because the paper or the bits in the bank account are worth anything real but because, in our times, the global finance system, materialized as the Church of the Market, has the monopoly in endowing those bills with a particular purchase value. The other face of this revelation is that at any moment that value can be totally wiped out, meaning that a $20 bill may become worthless, in terms of its real purchasing power. That is precisely what happened during the hyperinflation years of the Weimer Republic in the 1920s in Germany, a crucial development leading to the explosion of World War II. Sadly, that scenario could repeat itself any day now, as the 2008 banking crisis that was triggered in the United States and spread all over the world taught us.

Obviously, I am well aware that, given the complexity that economies acquired in human history, a medium like money had to be invented and disseminated widely to allow for trade to materialize on a large scale, to enable large human economies to produce and distribute the vital goods and services

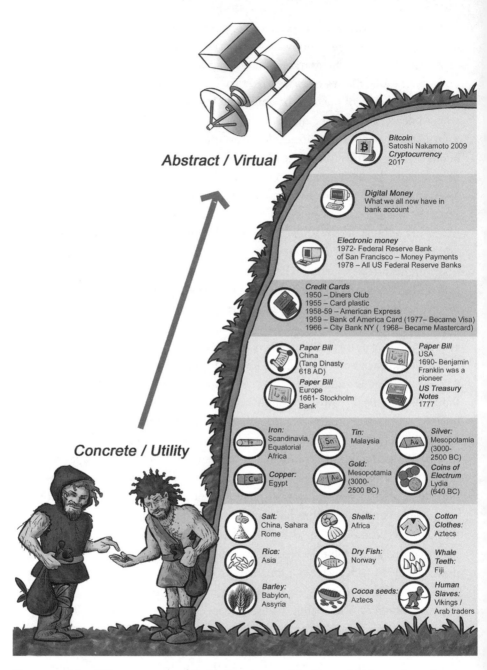

Figure 13.1. The different representations of money used by humankind over time. (Image credit to Custódio Rosa.)

needed to sustain about 7 billion people's needs for food, clothing, and shelter. Yet in the past seven hundred years, particularly since the emergence of banking during the Italian Renaissance, and later during the industrial revolution, the medium of money, pushed by the Church of the Market, morphed into more complex—and often incomprehensible to the layperson—financial abstractions that over time have run totally amok. As we have seen in the recent Greek debt crisis in Europe, money clearly occupies a much higher status in the minds of decision makers than does the well-being of large human societies. Indeed, at this point, a very significant portion of the developed world economy has nothing to do with the production and distribution of goods. Instead, a large percentage of economic activity depends solely on the issuing and trade of financial assets that have little relationship with concrete economic activity. Call it the grand casino of world finances if you will. The name is appropriate since, as of now, the dynamics of the world financial system have all but escaped human control, being driven, millisecond by millisecond, by the continuous virtual struggle of a number of supercomputers that dispute the supremacy of the markets on behalf of their human overlords, who now watch the game at a distance, having lost any tangible comprehension of the economic ecosystem, and nervously cross their fingers and hope for the best.

The ascent of the Church of the Market and the Money God to the summit of modern society explains why, when confronted with the choice of either guaranteeing payment to the European banks that had loaned money to fuel an insane real estate boom in Greece, or making sure that the Greeks could maintain some minimum standard of living, not to mention their dignity, the European Union's economic and political authorities did not hesitate a minute: no matter what excruciating level of human sacrifice it brought to the Greeks, the loans had to be repaid under the original conditions stipulated by financial institutions.

Ultimately, the Greek debt crisis made explicit what had already been a widely recognized fact in the financial world for many decades: in the financial cosmological view of the human universe, the Church of the Market holds greater value than the existence of nations, societies, and the livelihood of billions of people. From the perspective of a financially centered universe, every other human construct melts away as insignificant and irrelevant before the true holder of ominous power of our times—the Church of the Market—and its main agent of domination, the Money God.

The historian Eric Hobsbawm used the expression "the age of extremes" to encapsulate his views on how the history of the twentieth century could be best described and comprehended from a historical perspective. Hobsbawm suggests that the advent of modernity in the first few decades of the twentieth century resulted from the combination of three main forces: the acceleration of the process of the total surrender of political institutions and programs to narrow economic agendas, dictated by the desire of a small global elite to acquire the highest possible financial gain; the consolidation of the process of economic globalization without an equivalent globalization of the processes of political governance and human mobility; and the dramatic contraction of both temporal and spatial constraints in global human interactions due to a revolution in communication technologies. Altogether, these factors contributed to never before seen technological progress and record-breaking economic growth in the world's economic output. Yet the hefty price paid for these outcomes was a profound destabilization of political institutions, like nations and their sovereignty. Consequently, in the second decade of the twenty-first century, one could argue that the traditional mental concept of the national state, as well as its abstract borders, has been overrun by the dominant values and objectives derived from the mental abstractions favored by multinational corporations and the international financial system.

Ultimately, the process that enthroned the Church of the Market contributed to the virtual disintegration of the traditional way of living of many human societies, not only those that could not follow the rate of change but also those living in the leading economies involved in this process, like the United States and western Europe. Today, we live immersed in this potential global hecatomb, one in which neither institutions nor human societies—nor the human brain, for that matter—can really keep pace and cope with either the scope or the speed of change that results from such momentous transformations. Because the central emphasis and priorities of corporations and nations are primarily focused on achieving financial goals and enhancing productivity, nothing in our living routine seems to stand a chance to resist and survive the tsunami of continuous change imposed on most of humanity by the need to fulfill those objectives. Nothing seems to be immune to the ominous reach of the Church of the Market's greed. This may explain the tremendous level of anxiety and fear that is pervasive all over the world: nobody can be assured of having a permanent job, decent shelter, health care, education, or even a plan for the immediate future, since everything seems to be in a perpetual state of flux.

This overwhelming perception of total unpredictability experienced by most of humanity led the Polish sociologist and philosopher Zygmunt Bauman to describe the moment in which we live thus: "What was some time ago dubbed (erroneously) 'post-modernity,' and what I've chosen to call, more to the point, 'liquid modernity,' is the growing conviction that change is the only permanence, and uncertainty the only certainty. A hundred years ago 'to be modern' meant to chase 'the final state of perfection'—now it means an infinity of improvement, with no 'final state' in sight and none desired." Bauman diagnosed the problems we face:

> I am increasingly inclined to surmise that we presently find ourselves in a time of "interregnum"—when the old ways of doing things no longer work, the old learned or inherited modes of life are no longer suitable for the current *conditio humana*, but when the new ways of tackling the challenges and new modes of life better suited to the new conditions have not as yet been invented, put in place and set in operation.
>
> Forms of modern life may differ in quite a few respects—but what unites them all is precisely their fragility, temporariness, vulnerability and inclination to constant change. To "be modern" means to modernize—compulsively, obsessively; not so much just "to be," let alone to keep its identity intact, but forever "becoming," avoiding completion, staying under defined.

Bauman concludes: "Living under liquid modern conditions can be compared to walking in a minefield: everyone knows an explosion might happen at any moment and in any place, but no one knows when the moment will come and where the place will be. On a globalized planet, that condition is universal—no one is exempt and no one is insured against its consequences."

In Marshall McLuhan's prophetic words: "Now that man has extended his central nervous system by electric technology, the field of battle has shifted to mental image-making-and-breaking, both in war and in business."

Lost in the fray of this state of permanent flux, nobody seems to have stopped to reflect how the human brain would react to these new living conditions and how it would cope when immersed in a scenario where there is no solid ground but only a permanently fluid interface between its organic circuits and the external rules of social and economic engagement imposed by a new dominant and ruthless religion on the entirety of humanity.

From the braincentric view I introduced in this book, Hobsbawm's age of excess can be described as the period in human history in which a mental abstraction—capitalism—became powerful enough to reshape the dynamics of human interactions on a global scale, crossing a dangerous threshold that,

at the limit, may usher humanity into a black hole from which it cannot extri-
cate itself ever again. Basically, the market/money mental constructs and their
infinite derivatives acquired such a vital role in dictating all aspects of human
life and survival that, as they took off, spread, and ran away at speeds never
before witnessed or experienced by the human brain, those abstractions acquired a life of their own and surreptitiously began to threaten the survival of
a variety of key aspects of human culture. These manifestations included not
only wars and genocides but also economic and political proposals that began
to promote egregious levels of inequality and poverty, joblessness, and social
strife, as well as environmental damage at such a level that the threat of human self-inflicted mass extinction cannot be ignored any longer. Such a threat
could come from many sources and directions, from climate change—due
to the blind resistance of big business and governments to move away from
fossil fuels because of their focus on short-term financial gains—to a global
pandemic, allowed to spread by the continuous degradation of public funding
for preventive health care, basic research, and the lack of primary health cover-
age for billions of people worldwide.

Under the dominating mental construct of our times, financial cost is the
key—and in most cases the only—variable involved in all political, social, and
strategic decisions, including those determining which basic needs human
beings are entitled to and who should have access to the resources, including
new technologies, needed to fulfill these needs. How ironic it is that, on behalf
of a mental abstraction, modern governments, usually with the approval of
misled constituencies, continue to undermine the food security and education of our children, the health of our communities, decent housing for all
families, and the level of opportunities offered to people who, more and more,
do not have the minimal means required to fulfill their aspirations and their
search for integral fulfillment as human beings. At the limit, how can we still
be so naïve as to refer to a political system driven by special-interest groups
and the global financial lobby's agenda, freely disseminated and lauded by
corporate mass media, as truly democratic?

In 1949, Albert Einstein published a little article in which he described
his impressions of the impact of capitalism on human life at the time. In
what one today could call "Einstein's one-hundred-year progress report" on
the capitalist utopia, the great man wrote:

> Private capital tends to become concentrated in few hands, partly be-
> cause of competition among capitalists, and partly because technologi-

cal developments and the increasing division of labor encourage the formation of larger units of production at the expense of smaller ones. The result of these developments is an oligarchy of private capital the enormous power of which cannot be effectively checked even by a democratically organized political society. This is true since the members of legislative bodies are selected by political parties, largely financed or otherwise influenced by private capitalists who, for all practical purposes, separate the electorate from the legislature. The consequence is that the representatives of the people do not in fact sufficiently protect the interests of the underprivileged sections of the population. Moreover, under existing conditions, private capitalists inevitably control, directly or indirectly, the main sources of information (press, radio, education). It is thus extremely difficult, and indeed in most cases quite impossible, for the individual citizen to come to objective conclusions and to make intelligent use of his political rights.

The money-centric view of the human universe represents just half of the threat faced by humanity in the near future. The second mental abstraction responsible for brewing the perfect storm hovering around the future of human nature, the Cult of the Machine, can be as dangerous as the first since, ultimately, it holds as its holy grail, its supreme objective, the total elimination of the human labor force from the global economy. In our times, the Cult of the Machine professes that by combining modern techniques in artificial intelligence and robotics, eventually most jobs performed today by humans will be transferred to a new generation of smart machines and expert systems, all based on digital logic. In this human dystopia, the goal is even to replace the human brain through some sort of digital simulation running in a very powerful supercomputer that, in the end, will be able to mimic and reproduce all elements and attributes that define the human condition—even if, as I've argued in the previous chapter, the likelier reality is not that digital computers can simulate the human mind, but that the human brain will come to emulate digital computers.

The evangelists of artificial intelligence are excited to claim that this replacement of humans by machines will unleash a paradise on Earth. When the whole process of machine replacement is completed, they argue, all of us billions of human beings will have a lot of free time on our hands to simply exploit the limits of our creativity and pursue all sorts of intellectual and leisure activities. Of course, this new lifestyle would come at the price of massive levels of human unemployment, and the little detail of how each of us living in this machine-controlled paradise on Earth would be able to earn the means to enjoy our new lifestyle and still eat, dress, commute, pay our rent or mortgage, and send our kids to school seems to have escaped the creative minds proposing to replace us with their computer code, expert systems, and humanlike robots. Some have suggested that once machines take over most jobs, each of us should be entitled to a minimum income to cover our living necessities. Interestingly, not much thought has been given to who would be in charge of setting this minimum income and what "minimum living necessities" might mean and for whom. It does not take a rocket scientist, nor a neuroscientist, to realize that in the mind of the artificial intelligence prophets this job could be assigned only to the greatest oracle of modern times: the Church of the Market! Yet judging by what the very same oracle offered as advice regarding how to set the minimal living necessities the Greeks were entitled to, I would be really skeptical about, not to mention radically opposed to, leaving such a life-or-death decision in the hands of a mental abstraction that long ago ran out of any human empathy.

At this point, it is important to ask, why has such a delirium of playing God and striving to produce machines that attempt to replace human beings, and even the human brain, contaminated so many brilliant scientific minds? And why has artificial intelligence climbed to the top of the agenda of the business community as the potential solver of all problems, the fixer of all troubles that haunt humanity at this stage of its development, despite its well-known flagrant shortcomings?

I believe that all the excitement surrounding artificial intelligence research and its current and future applications derives from the seamless merging of the two dominant mental abstractions of our times, the Church of the Market and the Cult of the Machine, into a single entity. As a result of this marriage, I believe, the enormous drive toward the implementation of artificial intelligence applications in many industries has its origins in the misguided notion that, by replacing or drastically diminishing the use of human labor, these businesses can dramatically reduce their production expenses—including their most nagging annoyance, human labor costs—to a bare minimum, while taking their profits to sky-high levels never before seen. As such, the economic rationale that seems to be behind the new push for artificial intelligence could be described as follows: if a company can show that a piece of code or a smart robot can do the job of an experienced worker, the leverage it gains in negotiating wages and benefits with its workforce become enormous and almost impossible for that workforce to counter. Accordingly, debasing humans and their physical and mental skills by saying—and supposedly demonstrating—that a piece of metal or a few thousand lines of code can perform jobs better than the workers who used to perform these tasks is, in my opinion, part of a well-thought-out strategy of modern capitalists and big business to drive their profits close to the infinite. The only problem is that they seem to have forgotten to make an agreement with economists and scientists, who are not ready to conceal their opinions—and their data proving those beliefs—that this modern capitalist premise is false and utterly immoral. I say that because most of these "high-tech entrepreneurs" seem to either ignore or show no care for the potential social consequences of obliterating millions of jobs with their creations. Moreover, they appear to set aside serious concerns that the devastation of the workforce will also lead to a massive contraction in both the size and the purchasing power of their own consumer base.

Money-driven minds are not the only ones behind the push for artificial intelligence. Ultimately, the contemporary "artificial intelligence gold rush" can also be explained by an expanded and pretty terrifying Orwellian slogan that is driving most of the modern agenda and plans for the future: "Total control to ensure total security!"

By buying into this dystopia, a few governments have sold to their constituencies the false notion that to ensure their total security against all possible enemies, real or imaginary, people need to accept the surrender of their privacy and allow official surveillance to obtain the means for total control of their citizenries. As part of the brewing nightmare, some governments may hope to use artificial intelligence technology to acquire the capability of anticipating every decision and move of each of its citizens; at the limit, the extraordinary push for artificial intelligence underwritten by the defense and intelligence communities around the world aim at nothing short of the establishment of the "total surveillance state," a new type of totalitarian regime in which this technology is employed by governments to anticipate the behavior and, if they get their wishes, even the thoughts of every individual. In this Orwellian world, potential "crimes against the state" would be detected even as they germinate inside an individual's mind, before any action has materialized in the external world. Although some may argue that such a capability could be very useful to reduce crime worldwide, it is again important to emphasize that such a technology, if it can ever be developed, has a tremendous potential for being abused by governments as a way to create the ultimate means of political censorship, at a scale never before witnessed in human societies. In comparison to this apparatus for potential political persecution, Stalin's NKVD and Hitler's Gestapo could be considered child's play.

Curiously, the golden dream of dictatorships, intelligence services, and military and civilian tyrants was not implemented by institutions that belong to the deep state apparatus of modern nations. Rather, it was pioneered and first translated into action by a new business plan developed by one of the most emblematic companies of Silicon Valley, by those entrepreneurs who, just a few years ago, promised emphatically not to take advantage of their growing internet monopoly to do evil. After being incubated and launched with explosive success by Google, this new business practice was transferred by executives to another giant internet company, Facebook. This history was reconstructed by Shoshana Zuboff, a professor emerita of Harvard Business School, in *The Age of Surveillance Capitalism: The Fight for a Human Future at the New Frontier of Power*, published in 2018 when I was carrying out the last revision of this chapter. Professor Zuboff describes the same scenario of privacy intrusion I outlined a few paragraphs above as part of the new business model pioneered by Google. Zuboff baptized the marriage of convenience between the Church of the Market and the Cult of the Machine as the emergence of "surveillance capitalism." In her definition, surveillance capitalism is "a new economic order that claims human experience as free raw material for hidden commercial practices of extraction, prediction, and sales." In total agreement with my own assessment, Professor Zuboff believes that surveillance capitalism represents "a significant threat to human nature in the twenty-first century as industrial capitalism was to the natural world in the nineteenth and twentieth." She goes further to state that this "rogue mutation of capitalism" has allowed the emergence of "a new instrumentarian power that asserts dominance over society and presents startling challenges to market democracy."

Although there is an abundance of implicit and explicit signs that governments in many countries, including the United States, would be more than willing to adopt even more elaborate surveillance technologies if they ever become available, the only reason I can still sleep at night is that I am fully aware that limitations in the digital approach of artificial intelligence will not allow this plan to unfold in the near future, if ever. Yet these shortcomings will not prevent the same agents from continuing to pursue ways to harness the power of the human brain to create new surveillance tools or even a new generation of brain-controlled weapons, leading to an era in which the human brain becomes fully integrated into new warfare mediums. Given the recent significant interest demonstrated by the U.S. defense and intelligence communities in becoming full partners of the U.S. Brain Initiative established by President Barack Obama, the full "weaponization" of the human brain, seen before as a subject only for science fiction movies, should be taken very seriously by both the neuroscientific community and society at large. In this new reality, neuroscientists in particular have to reflect deeply before they make the decision to accept funds from military and intelligence sources, since the danger that the product of their intellectual and experimental work could be misappropriated to harm people has never been so high and concrete. For the first time in its short history, neuroscience has a central role to play as one of the gatekeepers and safeguards of the common good of society. It will be up to neuroscientists as a community to work as a protective shield, constantly warning society of the risks of any present and future attacks on elementary human rights, such as the rights of privacy and freedom of speech, or any attempt to compromise the fullest possible expression of human behaviors by invading the sacrosanct refuge of our own minds.

From this viewpoint, the dramatic existential risks imposed on the future of humanity by the growing worship of the Church of the Market and the Cult of the Machine clearly demonstrate why adopting a braincentric cosmology is so fundamental for ensuring that we, as a species, can regain our collective footing as the center of our own human universe. For starters, this braincentric view demystifies the origins of the dominant forces of modern life—markets, money, and machines—showing them to be nothing but by-products of the human brain, mental mirages built inside ourselves that, after centuries of evolution and trial and error, have acquired a life of their own, defining priorities, strategies, and practices that tend to relegate humankind's interventions, needs, and aspirations to a much diminished and devaluated secondary role.

As such, the braincentric view of the universe exposes in a very explicit way the sad reality that, for millennia, human societies have been driven to make decisions that impact decisively the future of human culture and, ultimately, the survival of our species, based on mental constructs that are not grounded in the true best interests of the vast majority of the living and still to be born members of our kin. Implacable religious dogmas, prejudices of any sort, economic systems based on vast inequality, and other distorted worldviews should not dictate human actions and behavior. That is why I insist on repeating that by knowing their true origins—our own brains—we may be able to convince more and more people why these mental abstractions should not dominate our ways of living.

By the same token, as we saw in chapter 10, the braincentric view shows that even science and the scientific method are limited, by the very constraints imposed by the neurobiological properties of our central nervous system, in what they can offer as a description of the universe that is out there. Because of these undeniable limitations, manifested, for instance, by a series of unsolved enigmas in quantum physics, science and scientists have the duty to inform society that, notwithstanding the magnificent wonders they have achieved in the past few centuries and will continue to produce in the future, they cannot promise to deliver the ultimate truth. In this context, concepts like the theory of everything, the chimera that a single mathematical formulation can be found to describe the entirety of the universe, or the belief that a machine can be invented that reproduces human brains, are not only untenable fantasies—they are complete fallacies that have contributed to misguiding and misleading millions of people into believing in a fairy tale. Science does not need to resort to this type of shallow propaganda because what it is truly capable of achieving is more than enough cause to validate all efforts to disseminate and democratize its practice among us. As Niels Bohr explained so eloquently about a century ago, science is not the search for the ultimate truth about what reality is—that is beyond us—instead, science offers our best opportunity to obtain the greatest possible understanding of what is out there so we can use this knowledge first to enlighten us and eventually to manipulate the world around us in order to improve the lives of humankind. And despite misleading attempts by some to classify Bohr's view as a mere expression of metaphysical solipsism, the same cosmological viewpoint has been supported and shared over the past hundred years by a large number of major intellectuals, philosophers, mathematicians, and physicists alike.

Following Bohr's philosophy, the braincentric view positions human reasoning right at the center of its own human universe, since this is the only universe we can truly speak about: the one sculptured by the mental constructs of the more than 100 billion human beings who have set foot on this beautiful blue planet over the past one hundred thousand years. As such, the change in framework suggested by a braincentric cosmology makes explicit the urgent need for a radical change in the priorities of our current economic and political systems, not to mention our postmodern culture, in order to refocus their actions to better provide for the basic necessities and legitimate wishes and existential rights of all living human beings. Essentially, what I am trying to say is that the broad range of human necessities, which are already considered inalienable human rights, should take precedence over goals artificially generated as mental abstractions, which are spiraling out of control and conspiring against our collective well-being and the survival of our species.

The braincentric view also categorically refutes the contemporary thesis, professed enthusiastically by the worshipers of the artificial intelligence myth, that we and our brains can be reduced to biological machines or automata whose actions and thoughts can be replicated and simulated by mathematical algorithms and digital hardware or software, no matter how elaborate and complex. Unless humanity as a whole decides to take one more crucial step toward self-annihilation by renouncing its birthright as knowledge gatherers and universe creators, the future scenario proposed by some radical artificial intelligence researchers is yet another example of the type of hollow mental fantasy that will take us nowhere. Moving in the totally opposite direction, the braincentric view proposes that human beings should assert their collective claim as creators of their own universe and never relinquish control of their own destinies to what amounts to a bunch of glorified machines.

But what is the alternative to this dominant worldview of our times? My answer is pretty simple. By continuing to fulfill its almost holy duty—to dissipate energy to accumulate knowledge and use it to build a more complete description of the human universe and provide for the betterment of its kin—the True Creator of Everything can wisely choose the only meaningful alternative for the future, one that ensures the enduring survival and blossoming of the human condition, arguably our best and only ticket for the fulfillment of the so eagerly sought dream of human immortality.

I say that because I truly believe that in this universe of ours, there is nothing that even comes close to the beauty, the elegance, and the eloquence of the mental monuments that the True Creator of Everything has erected, since the beginning of time, out of its tiny neural electromagnetic storms, to leave an enduring legacy that defines, for better or for worse, the essence of what it is to be human.

EPILOGUE

Over millions and millions of years of a mostly random journey, natural evolution on Earth carefully sewed a three-dimensional mesh composed of copious bundles, sheets, and coils of white neurobiological matter. By conducting and accelerating mundane electrobiological sparks generated by tens of billions of neurons, this organic scaffold gave birth to a unique type of noncomputable electromagnetic interaction, one that endowed a relativistic primate brain with a precious gift: its own point of view.

This almost miraculous event happened because tiny electromagnetic waves, acting like an invisible glue, induced the same tens of billions of neurons to coalesce into a seamless neuronal space-time continuum. From the unpredictable recursive processing of this electromagnetic analog-to-digital organic computer, the True Creator of Everything emerged about one hundred thousand years ago. And in fewer than five thousand generations, it mastered the keystone biological mechanism of life, which consists of dissipating excess entropy to embedded semantic-rich Gödelian information. Using this recipe of the living, the True Creator erected a human universe from the soup of potential information generously made available by the cosmos. It did that by using its growing stocks of Gödelian information to dissipate further entropy into knowledge, tool making, language, social bonding, and reality building.

To fulfill its ultimate ambition, the True Creator took advantage of how its inner connecting core also provided optimal conditions for the emergence of tight interbrain synchronization, involving millions or even billions of individual brains across the confines of time and space. Through these brainets, the True Creator gave rise to the most creative, resilient, prosperous, and dangerous animal social groups ever assembled on Earth.

From the onset, human brainets obsessively engaged in the task of trying to explain everything that exists in the vast cosmos around them. For that, they resorted to a unique universe-making mental toolbox that included art, myths, religion, time and space, mathematics, technology, and science. By stitching the by-products of those mental tools and all other individual experiences of more than 100 billion people into its private mental tapestry, the True Creator at last achieved its final masterpiece: the creation of the human universe, the only possible account of material reality available to us.

And then, in what can only be described as an ironic twist of fate, as ever more powerful brainets created a never-ending succession of mental abstractions that grew increasingly more seductive, more enticing, and even more bewitching than human life itself, some of the True Creator's offspring eventually rose to conspire and threaten, in a perfect Shakespearean plot, the very existence of their own master.

What will the future bring for the True Creator? Self-annihilation, a new human species made of biological digital zombies, or the long-expected ultimate triumph of the human condition? At this point, nobody can answer this question with certainty. Whichever destiny is reserved for us, there will be no human-made machine capable of overcoming the True Creator's most intimate skills.

Nor the wondrous braincentric human universe it built.

BIBLIOGRAPHY

Al-Khalili, Jim. *The House of Wisdom: How Arabic Science Saved Ancient Knowledge and Gave Us the Renaissance.* New York: Penguin, 2011.

Anastassiou, Costas A., Sean M. Montgomery, Mauricio Barahona, György Buzsáki, and Christof Koch. "The Effect of Spatially Inhomogeneous Extracellular Electric Fields on Neurons." *Journal of Neuroscience* 30, no. 5 (February 2010): 1925–36.

Anfinsen, Christian B. "Principles That Govern the Folding of Protein Chains." *Science* 181, no. 4096 (July 1973): 223–30.

Annese, Jacopo, Natalie M. Schenker-Ahmed, Hauke Bartsch, Paul Maechler, Colleen Sheh, Natasha Thomas, Junya Kayano, Alexander Ghatan, Noah Bresler, Matthew P. Frosch, Ruth Klaming, and Suzanne Corkin. "Postmortem Examination of Patient H.M.'s Brain Based on Histological Sectioning and Digital 3D Reconstruction." *Nature Communications* 5 (2014): 3122.

Arendt, Hannah. *The Human Condition.* Chicago: University of Chicago Press, 1998.

Arii, Yoshihar, Yuki Sawada, Kazuyuki Kawamura, Sayaka Miyake, Yasuo Taichi, Yuishin Izumi, Yukiko Kuroda, et al. "Immediate Effect of Spinal Magnetic Stimulation on Camptocormia in Parkinson's Disease." *Journal of Neurology, Neurosurgery & Psychiatry* 85 no. 11 (November 2014): 1221–26.

Arvanitaki, A. "Effects Evoked in an Axon by the Activity of a Contiguous One." *Journal of Neurophysiology* 5 (1942): 89–108.

Bailly, Francis, and Giuseppe Longo. *Mathematics and the Natural Sciences: The Physical Singularity of Life.* London: Imperial College Press, 2011.

Bakhtiari, Reyhaneh, Nicole R. Zürcher, Ophélie Rogier, Britt Russo, Loyse Hippolyte, Cristina Granziera, Babak Nadjar Araabi, Majid Nili Ahmadabadi, and Nouchine Hadjikhani. "Differences in White Matter Reflect Atypical Developmental Trajectory in Autism: A Tract-Based Spatial Statistics Study." *Neuroimage: Clinical* 1, no. 1 (September 2012): 48–56.

Barbour, Julian B. *The End of Time: The Next Revolution in Physics*. Oxford: Oxford University Press, 2000.

Barra, Allen. "Moneyball: Was the Book That Changed Baseball Built on a False Premise?" *Guardian*, April 21, 2017. https://www.theguardian.com/sport/2017/apr/21/moneyball-baseball-oakland-book-billy-beane.

Barrow, John D. *New Theories of Everything: The Quest for Ultimate Explanation*. Oxford: Oxford University Press, 2007.

Bauman, Zygmunt. *Liquid Love: On the Frailty of Human Bonds*. Cambridge: Polity, 2003.

———. *Liquid Modernity*. Cambridge: Polity, 2000.

———. *Liquid Times: Living in an Age of Uncertainty*. Cambridge: Polity, 2007.

Beane, Samuel C. *The Religion of Man-Culture: A Sermon Preached in the Unitarian Church, Concord, N.H., January 29, 1882*. Concord: Republican Press Association, 1882.

Bennett, C. H. "Logical Reversibility of Computation." *IBM Journal of Research and Development* 17, no. 6 (1973): 525–32.

Bentley, Peter J. "Methods for Improving Simulations of Biological Systems: Systemic Computation and Fractal Proteins." *Journal of the Royal Society Interface* 6, supplement 4 (August 2009): S451–66.

Berger, Hans. "Electroencephalogram in Humans." *Archiv für Psychiatrie und Nervenkrankheiten* 87 (1929): 527–70.

Berger, Lee, and John Hawks. *Almost Human: The Astonishing Tale of Homo Naledi and the Discovery That Changed Our Human Story*. New York: National Geographic, 2017.

Bickerton, Derek. *Adam's Tongue: How Humans Made Language, How Language Made Humans*. New York: Hill and Wang, 2009.

Boardman, John, Jasper Griffin, and Oswyn Murray. *The Oxford History of Greece and the Hellenistic World*. Oxford: Oxford University Press, 1991.

Born, H. A. "Seizures in Alzheimer's Disease." *Neuroscience* 286 (February 2015): 251–63.

Botvinick, Matthew, and Jonathan Cohen. "Rubber Hands 'Feel' Touch That Eyes See." *Nature* 391, no. 6669 (February 1998): 756.

Bringsjord, Selmer, and Konstantine Arkoudas. "The Modal Argument for Hypercomputing Minds." *Theoretical Computer Science* 317, nos. 1–3 (June 2004): 167–90.

Bringsjord, Selmer, and Michael Zenzen. "Cognition Is Not Computation: The Argument from Irreversibility." *Synthese* 113, no. 2 (November 1997): 285–320.

Brooks, Rosa. *How Everything Became War and the Military Became Everything: Tales from the Pentagon*. New York: Simon and Schuster, 2016.

Burgelman, Robert A. "Prigogine's Theory of the Dynamics of Far-from-Equilibrium Systems Informs the Role of Strategy Making in Organizational Evolution." *Stanford University, Graduate School of Business, Research Papers* (2009).

Caminiti, Roberto, Hassan Ghaziri, Ralf Galuske, Patrick R. Hof, and Giorgio M. Inno-
centie. "Evolution Amplified Processing with Temporally Dispersed Slow Neuronal
Connectivity in Primates." *Proceedings of the National Academy of Sciences USA* 106,
no. 46 (November 2009): 19551–56.

Campbell, Joseph. *Myths to Live By.* New York: Viking, 1972.

Campbell, Joseph, and Bill D. Moyers. *The Power of Myth.* New York: Doubleday, 1988.

Carmena, Jose M., Mikhail A. Lebedev, Roy E. Crist, Joseph E. O'Doherty, David M.
Santucci, Dragan F. Dimitrov, Parag G. Patil, Craig S. Henriquez, and Miguel A. L.
Nicolelis. "Learning to Control a Brain-Machine Interface for Reaching and Grasp-
ing by Primates." *Public Library of Science Biology* 1, no. 2 (November 2003): E42.

Carmena, Jose M., Mikhail A. Lebedev, Craig S. Henriquez, and Miguel A. Nicolelis. "Stable Ensemble Performance with Single-Neuron Variability during Reaching Movements in Primates." *Journal of Neuroscience* 25, no. 46 (November 2005):
10712–16.

Carr, Nicholas G. *The Glass Cage: Automation and Us.* New York: Norton, 2014.

———. *The Shallows: What the Internet Is Doing to Our Brains.* New York: Norton,
2010.

Carroll, Sean M. *The Big Picture: On the Origins of Life, Meaning, and the Universe Itself.*
New York: Dutton, 2016.

Castells, Manuel. *Communication Power.* Oxford: Oxford University Press, 2013.

———. *Networks of Outrage and Hope: Social Movements in the Internet Age.* 2nd ed.
Cambridge: Polity, 2015.

———. *The Rise of the Network Society.* The Information Age: Economy, Society, and
Culture. Chichester, UK: Wiley-Blackwell, 2010.

Casti, John L., and Werner DePauli. *Gödel: A Life of Logic, the Mind, and Mathematics.*
Cambridge, MA: Perseus, 2000.

Ceruzzi, Paul E. *Computing: A Concise History.* The MIT Press Essential Knowledge
Series. Cambridge, MA: MIT Press, 2012.

Chaitin, Gregory J. *The Limits of Mathematics.* London: Springer-Verlag, 2003.

———. *Meta Math! The Quest for Omega.* New York: Pantheon Books, 2005.

Chaitin, Gregory, Newton C. da Costa, and Francisco A. Dória. *Goedel's Way: Exploits
into an Undecided World.* London: CRC, 2011.

Chalmers, David John. *The Conscious Mind: In Search of a Fundamental Theory.* Philosophy of Mind Series. New York: Oxford University Press, 1996.

Chapin, John K., Karen A. Moxon, Ronald S. Markowitz, and Miguel A. Nicolelis.
"Real-Time Control of a Robot Arm Using Simultaneously Recorded Neurons in
the Motor Cortex." *Nature Neuroscience* 2, no. 7 (July 1999): 664–70.

Chervyakov, Alexander V., Andrey Y. Chernyavsky, Dmitry O. Sinitsyn, and Michael A.
Piradov. "Possible Mechanisms Underlying the Therapeutic Effects of Transcranial
Magnetic Stimulation." *Frontiers in Human Neuroscience* 9 (June 2015): 303.

Chiang, Chia-Chu, Rajat S. Shivacharan, Xile Wei, Luis E. Gonzalez-Reyes, and Dominique M. Durand. "Slow Periodic Activity in the Longitudinal Hippocampal Slice Can Self-Propagate Non-synaptically by a Mechanism Consistent with Ephaptic Coupling." *Journal of Physiology* 597, no. 1 (January 2019): 249–69.

Christensen, Mark Schram, Jesper Lundbye-Jensen, Michael James Grey, Alexandra Damgaard Vejlby, Bo Belhage, and Jens Bo Nielsen. "Illusory Sensation of Movement Induced by Repetitive Transcranial Magnetic Stimulation." *Public Library of Science One* 5, no. 10 (October 2010): e13301.

Cicurel, Ronald. *L'ordinateur ne digérera pas le cerveau: Sciences et cerveaux artificiels; Essai sur la nature du réel.* Lausanne: CreateSpace, 2013.

Cicurel, Ronald, and Miguel A. L. Nicolelis. *The Relativistic Brain: How It Works and Why It Cannot be Simulated by a Turing Machine.* Lausanne: Kios, 2015.

Clottes, Jean. *Cave Art.* London: Phaidon, 2008.

Cohen, Leonardo G., Pablo Celnik, Alvaro Pascual-Leone, Brian Corwell, Lala Faiz, James Dambrosia, Manabu Honda, Norihiro Sadato, Christian Gerloff, M. Dolores Catalá, and Mark Hallett. "Functional Relevance of Cross-Modal Plasticity in Blind Humans." *Nature* 389, no. 6647 (September 1997): 180–83.

Copeland, B. Jack. "Hypercomputation." *Minds and Machines* 12, no. 4 (November 2002): 461–502.

———. "Turing's O-Machines, Searle, Penrose and the Brain (Human Mentality and Computation)." *Analysis* 58, no. 2 (April 1998): 128–38.

Copeland, B. Jack, Carl J. Posy, and Oron Shagrir, eds. *Computability: Turing, Gödel, and Beyond.* Cambridge, MA: MIT Press, 2013.

Costa, Rui M., Ranier Gutierrez, Ivan E. de Araujo, Monica R. Coelho, Alexander D. Kloth, Raul R. Gainetdinov, Marc G. Caron, Miguel A. Nicolelis, and Sidney A. Simon. "Dopamine Levels Modulate the Updating of Tastant Values." *Genes, Brain and Behavior* 6, no. 4 (June 2007): 314–20.

Curtis, Gregory. *The Cave Painters: Probing the Mysteries of the World's First Artists.* New York: Knopf, 2006.

Dawkins, Richard. *The Selfish Gene.* Oxford: Oxford University Press, 1976.

Debener, Stefan, Markus Ullsperger, Markus Siegel, Katja Fiehler, D. Yves von Cramon, and Andreas K. Engel. "Trial-by-Trial Coupling of Concurrent Electroencephalogram and Functional Magnetic Resonance Imaging Identifies the Dynamics of Performance Monitoring." *Journal of Neuroscience* 25, no. 50 (December 2005): 11730–37.

Dennett, Daniel C. *Consciousness Explained.* Boston: Little, Brown, 1991.

Derbyshire, John. *Unknown Quantity: A Real and Imaginary History of Algebra.* Washington, DC: Joseph Henry, 2006.

de Souza, Carolina Pinto, Maria Gabriela Ghilardi dos Santos, Clement Hamani, and Erich Talamoni Fonoff. "Spinal Cord Stimulation for Gait Dysfunction in Parkinson's Disease: Essential Questions to Discuss." *Movement Disorders* 32, no. 2 (November 2018): 1828–29.

Deutsch, David. *The Beginning of Infinity: Explanations That Transform the World.* New York: Viking, 2011.

———. *The Fabric of Reality.* Harmondsworth, UK: Allen Lane, 1997.

Devlin, Keith. *The Man of Numbers: Fibonacci's Arithmetic Revolution.* New York: Bloomsbury USA, 2011.

Dikker, Suzanne, Lu Wan, Ido Davidesco, Lisa Kaggen, Matthias Oostrik, James McClintock, Jess Rowland, Georgios Michalareas, Jay J. Van Bavel, Mingzhou Ding, and David Poeppel. "Brain-to-Brain Synchrony Tracks Real-World Dynamic Group Interactions in the Classroom." *Current Biology* 27, no. 9 (May 2017): 1375–80.

di Pellegrino, G., L. Fadiga, L. Fogassi, V. Gallese, and G. Rizzolatti. "Understanding Motor Events: A Neurophysiological Study." *Experimental Brain Research* 91, no. 1 (1992): 176–80.

Domingos, Pedro. *The Master Algorithm: How the Quest for the Ultimate Learning Machine Will Remake Our World.* New York: Basic Books, 2015.

Donati, Anna R., Solaiman Shokur, Edgard Morya, Debora S. Campos, Renan C. Moioli, Claudia M. Gitti, Patricia B. Augusto, et al. "Long-Term Training with a Brain-Machine Interface-Based Gait Protocol Induces Partial Neurological Recovery in Paraplegic Patients." *Scientific Reports* 6 (August 2016): 30383.

Dreyfus, Hubert L. *What Computers Still Can't Do: A Critique of Artificial Reason.* Cambridge, MA: MIT Press, 1992.

Dunbar, R. I. M. *Grooming, Gossip, and the Evolution of Language.* Cambridge, MA: Harvard University Press, 1996.

———. "Neocortex Size as a Constraint on Group Size in Primates." *Journal of Human Evolution* 20 (1992): 469–93.

———. *The Trouble with Science.* Cambridge, MA: Harvard University Press, 1996.

Dunbar, R. I. M., and Suzanne Shultz. "Evolution in the Social Brain." *Science* 317, no. 5843 (September 2007): 1344–47.

Dyson, Freeman J. *Origins of Life.* Cambridge: Cambridge University Press, 1999.

Dzirasa, Kafui, Laurent Coque, Michelle M. Sidor, Sunil Kumar, Elizabeth A. Dancy, Joseph S. Takahashi, Colleen A. McClung, and Miguel A. L. Nicolelis. "Lithium Ameliorates Nucleus Accumbens Phase-Signaling Dysfunction in a Genetic Mouse Model of Mania." *Journal of Neuroscience* 30, no. 48 (December 2010): 16314–23.

Dzirasa, Kafui, Romulo Fuentes, Sunil Kumar, Juan M. Potes, and Miguel A. Nicolelis. "Chronic in Vivo Multi-circuit Neurophysiological Recordings in Mice." *Journal of Neuroscience Methods* 195, no. 1 (January 2011): 36–46.

Dzirasa, Kafui, Sunil Kumar, Benjamin D. Sachs, Marc G. Caron, and Miguel A. Nicolelis. "Cortical-Amygdalar Circuit Dysfunction in a Genetic Mouse Model of Serotonin Deficiency." *Journal of Neuroscience* 33, no. 10 (March 2013): 4505–13.

Dzirasa, Kafui, DeAnna L. McGarity, Anirban Bhattacharya, Sunil Kumar, Joseph S. Takahashi, David Dunson, Colleen A. McClung, and Miguel A. Nicolelis. "Impaired Limbic Gamma Oscillatory Synchrony during Anxiety-Related Behavior in a Genetic Mouse Model of Bipolar Mania." *Journal of Neuroscience* 31, no. 17 (April 2011): 6449–56.

Dzirasa, Kafui, H. Westley Phillips, Tatyana D. Sotnikova, Ali Salahpour, Sunil Kumar, Raul R. Gainetdinov, Marc G. Caron, and Miguel A. Nicolelis. "Noradrenergic Control of Cortico-Striato-Thalamic and Mesolimbic Cross-Structural Synchrony." *Journal of Neuroscience* 30, no. 18 (May 2010): 6387–97.

Dzirasa, Kafui, Amy J. Ramsey, Daniel Y. Takahashi, Jennifer Stapleton, Juan M. Potes, Jamila K. Williams, Raul R. Gainetdinov, et al. "Hyperdopaminergia and Nmda Receptor Hypofunction Disrupt Neural Phase Signaling." *Journal of Neuroscience* 29, no. 25 (June 2009): 8215–24.

Dzirasa, Kafui, Sidarta Ribeiro, Rui Costa, Lucas M. Santos, Shieh-Chi Lin, Andre Grosmark, Tatyana D. Sotnikova, et al. "Dopaminergic Control of Sleep-Wake States." *Journal of Neuroscience* 26, no. 41 (October 2006): 10577–89.

Dzirasa, K., L. M. Santos, S. Ribeiro, J. Stapleton, R. R. Gainetdinov, M. G. Caron, and M. A. Nicolelis. "Persistent Hyperdopaminergia Decreases the Peak Frequency of Hippocampal Theta Oscillations during Quiet Waking and Rem Sleep." *Public Library of Science One* 4, no. 4 (2009): e5238.

Eddington, Arthur Stanley. *The Nature of the Physical World.* Cambridge: Macmillan/Cambridge University Press, 1928.

Edwards, Paul N. *The Closed World: Computers and the Politics of Discourse in Cold War America.* Inside Technology. Cambridge, MA: MIT Press, 1996.

Ehrenzweig, Anton. *The Hidden Order of Art: A Study in the Psychology of Artistic Imagination.* London: Weidenfeld and Nicolson, 1967.

Einstein, Albert. *Relativity: The Special and the General Theory.* 1954. Reprint, London: Routledge, 2001.

———. "Why Socialism?" *Monthly Review* 1, no. 1 (1949).

Engel, Andreas K., Pascal Fries, and Wolf Singer. "Dynamic Predictions: Oscillations and Synchrony in Top-Down Processing." *Nature Reviews Neuroscience* 2, no. 10 (October 2001): 704–16.

Englander, Zoe A., Carolyn E. Pizoli, Anastasia Batrachenko, Jessica Sun, Gordon Worley, Mohamad A. Mikati, Joanne Kurtzberg, and Allen W. Song. "Diffuse Reduction of White Matter Connectivity in Cerebral Palsy with Specific Vulnerability of Long Range Fiber Tracts." *Neuroimage: Clinical* 2 (March 2013): 440–47.

Fagan, Brian M. *Cro-Magnon: How the Ice Age Gave Birth to the First Modern Humans.* New York: Bloomsbury, 2010.

Fanselow, Erica E., and Miguel A. Nicolelis. "Behavioral Modulation of Tactile Responses in the Rat Somatosensory System." *Journal of Neuroscience* 19, no. 17 (September 1999): 7603–16.

Fanselow, Erica E., Ashlan P. Reid, and Miguel A. Nicolelis. "Reduction of Pentylene-tetrazole-Induced Seizure Activity in Awake Rats by Seizure-Triggered Trigeminal Nerve Stimulation." *Journal of Neuroscience* 20, no. 21 (November 2000): 8160–68.

Ferguson, Niall. *The Ascent of Money: A Financial History of the World.* New York: Penguin, 2008.

———. *The House of Rothschild.* Vol. 1: *Money's Prophets.* New York: Penguin, 1998.

Ferrari, Pier Francesco, and Giacomo Rizzolatti. *New Frontiers in Mirror Neurons Research.* Oxford: Oxford University Press, 2015.

Fingelkurts, Andrew A. "Timing in Cognition and EEG Brain Dynamics: Discreteness versus Continuity." *Cognitive Processing* 7, no. 3 (September 2006): 135–62.

Fitzsimmons, Nathan A., Weying Drake, Timothy L. Hanson, Mikhail A. Lebedev, and Miguel A. Nicolelis. "Primate Reaching Cued by Multichannel Spatiotemporal Cortical Microstimulation." *Journal of Neuroscience* 27, no. 21 (May 2007): 5593–602.

Fitzsimmons, Nathan A., Mikhail A. Lebedev, Ian D. Peikon, and Miguel A. Nicolelis. "Extracting Kinematic Parameters for Monkey Bipedal Walking from Cortical Neuronal Ensemble Activity." *Frontiers in Integrative Neuroscience* 3 (March 2009): 3.

Flor, Herta, Lone Nikolajsen, and Troels Staehelin Jensen. "Phantom Limb Pain: A Case of Maladaptive CNS Plasticity?" *Nature Reviews Neuroscience* 7, no. 11 (November 2006): 873–81.

Fodor, Jerry. *The Language of Thought.* Cambridge, MA: MIT Press, 1975.

Ford, Martin. *Rise of the Robots: Technology and the Threat of a Jobless Future.* New York: Basic Books, 2015.

Foucault, Michel. *The Order of Things: An Archaeology of the Human Sciences.* World of Man. New York: Pantheon Books, 1971.

Freed-Brown, Grace, and David J. White. "Acoustic Mate Copying: Female Cowbirds Attend to Other Females' Vocalizations to Modify Their Song Preferences." *Proceedings of the Royal Society—Biological Sciences* 276, no. 1671 (September 2009): 3319–25.

Freeman, Charles. *The Closing of the Western Mind: The Rise of Faith and the Fall of Reason.* New York: Vintage Books, 2005.

Frenkel, Edward. *Love and Math: The Heart of Hidden Reality.* New York: Basic Books, 2013.

Frostig, Ron D., Cynthia H. Chen-Bee, Brett A. Johnson, and Nathan S. Jacobs. "Imaging Cajal's Neuronal Avalanche: How Wide-Field Optical Imaging of the Point-Spread Advanced the Understanding of Neocortical Structure-Function Relationship." *Neurophotonics* 4, no. 3 (July 2017): 031217.

Fuentes, Romulo, Per Petersson, and Miguel A. Nicolelis. "Restoration of Locomotive Function in Parkinson's Disease by Spinal Cord Stimulation: Mechanistic Approach." *European Journal of Neuroscience* 32, no. 7 (October 2010): 1100–8.

Fuentes, Romulo, Per Petersson, William B. Siesser, Marc G. Caron, and Miguel A. Nicolelis. "Spinal Cord Stimulation Restores Locomotion in Animal Models of Parkinson's Disease." *Science* 323, no. 5921 (March 2009): 1578–82.

Gallese, Vittori, Christian Keysers, and Giacomo Rizzolatti. "A Unifying View of the Basis of Social Cognition." *Trends in Cognitive Sciences* 8, no. 9 (September 2004): 396–403.

Gamble, Clive, John Gowlett, and R. I. M. Dunbar. *Thinking Big: How the Evolution of Social Life Shaped the Human Mind.* London: Thames and Hudson, 2014.

Gane, Simon, Dimitris Georganakis, Klio Maniati, Manolis Vamvakias, Nikitas Ragoussis, Efthimios M. Skoulakis, and Luca Turin. "Molecular Vibration-Sensing Component in Human Olfaction." *Public Library of Science One* 8, no. 1 (January 2013): e55780.

Gardner, Howard. *Multiple Intelligences: New Horizons.* New York: Basic Books, 2006.

Gertner, Jon. *The Idea Factory: Bell Labs and the Great Age of American Innovation.* New York: Penguin, 2012.

Ghazanfar, Asif A., and Charles E. Schroeder. "Is Neocortex Essentially Multisensory?" *Trends in Cognitive Sciences* 10, no. 6 (June 2006): 278–85.

Gleick, James. *The Information: A History, a Theory, a Flood.* New York: Pantheon Books, 2011.

Gleiser, Marcelo. *The Island of Knowledge: The Limits of Science and the Search for Meaning.* New York: Basic Books, 2014.

———. *A Tear at the Edge of Creation: A Radical New Vision for Life in an Imperfect Universe.* Hanover: Dartmouth College Press, 2013.

Gödel, Kurt. "Some Basic Theorems on the Foundations of Mathematics and Their Philosophical Implications." In *Collected Works*, vol. 3: *Unpublished Essays and Lectures*, edited by Solomon Feferman, John W. Dawson Jr., Warren Goldfarb, Charles Parsons, and Robert M. Solovay. New York: Oxford University Press, 1995.

———. "Über Formal Unentscheidbare Sätze der Principia Mathematica und Verwandter Systeme L." *Monatshefte für Mathematik und Physik* 38 (1931): 173–98.

Goff, Philip. "A Way Forward to Solve the Hard Problem of Consciousness." *Guardian*, January 28, 2015.

Gombrich, Ernst H. *The Story of Art.* Englewood Cliffs, NJ: Prentice-Hall, 1995.

Gosling, David L. *Science and the Indian Tradition: When Einstein Met Tagore.* India in the Modern World Series. London: Routledge, 2007.

Gould, Stephen Jay. *Wonderful Life: The Burgess Shale and the Nature of History.* New York: Norton, 1989.

Gray, Jeffrey. *Consciousness: Creeping up on the Hard Problem.* Oxford: Oxford University Press 2004.

Greene, Brian. *The Hidden Reality: Parallel Universes and the Deep Laws of the Cosmos.* New York: Knopf, 2011.

Greenfield, Patricia M. "Technology and Informal Education: What Is Taught, What Is Learned." *Science* 323, no. 5910 (January 2009): 69–71.

Halpern, Paul. *Einstein's Dice and Schrödinger's Cat: How Two Great Minds Battled Quantum Randomness to Create a Unified Theory of Physics.* New York: Basic Books, 2015.

Hamilton, Edith. *The Greek Way*. New York: Norton, 2017.

Hanson, Timothy L., Andrew M. Fuller, Mikhail A. Lebedev, Dennis A. Turner, and Miguel A. Nicolelis. "Subcortical Neuronal Ensembles: An Analysis of Motor Task Association, Tremor, Oscillations, and Synchrony in Human Patients." *Journal of Neuroscience* 32, no. 25 (June 2012): 8620–32.

Harari, Yuval N. *Homo Deus: A Brief History of Tomorrow*. New York: Harper, 2017.

———. *Sapiens: A Brief History of Humankind*. New York: Harper, 2015.

Haroutunian, V., P. Katsel, P. Roussos, K. L. Davis, L. L. Altshuler, and G. Bartzokis. "Myelination, Oligodendrocytes, and Serious Mental Illness." *Glia* 62, no. 11 (November 2014): 1856–77.

Harris, Tristan. *How a Handful of Tech Companies Control Billions of Minds Every Day*. TED 2017, session 11, April 2017. https://www.ted.com/talks/tristan_harris_the _manipulative_tricks_tech_companies_use_to_capture_your_attention.

Hart, Peter. *The Somme: The Darkest Hour on the Western Front*. New York: Pegasus Books, 2008.

Hartmann, Konstantin, Eric E. Thomson, Ivan Zea, Richy Yun, Peter Mullen, Jay Canarick, Albert Huh, and Miguel A. Nicolelis. "Embedding a Panoramic Representation of Infrared Light in the Adult Rat Somatosensory Cortex through a Sensory Neuroprosthesis." *Journal of Neuroscience* 36, no. 8 (February 2016): 2406–24.

Hartt, Frederick, and David G. Wilkins. *History of Italian Renaissance Art: Painting, Sculpture, Architecture*. New York: H. N. Abrams, 1994.

Hartwig, Valentina, Giulio Giovannetti, Nicola Vanello, Massimo Lombardi, Luigi Landini, and Silvana Simi. "Biological Effects and Safety in Magnetic Resonance Imaging: A Review." *International Journal of Environmental Research and Public Health* 6, no. 6 (June 2009): 1778–98.

Harvey, David. *The Enigma of Capital: And the Crises of Capitalism*. Oxford: Oxford University Press, 2010.

Hasson, Uri, Asif A. Ghazanfar, Bruno Galantucci, Simon Garrod, and Christian Keysers. "Brain-to-Brain Coupling: A Mechanism for Creating and Sharing a Social World." *Trends in Cognitive Sciences* 16, no. 2 (February 2012): 114–21.

Hasson, Uri, Yuval Nir, Ifat Levy, Galit Fuhrmann, and Rafael Malach. "Intersubject Synchronization of Cortical Activity during Natural Vision." *Science* 303, no. 5664 (March 2004): 1634–40.

Hawking, Stephen, and Leonard Mlodinow. *The Grand Design*. New York: Bantam Books, 2010.

Hayles, N. Katherine. *How We Became Posthuman: Virtual Bodies in Cybernetics, Literature, and Informatics*. Chicago: University of Chicago Press, 1999.

———. *How We Think: Digital Media and Contemporary Technogenesis*. Chicago: University of Chicago Press, 2012.

Hebb, Donald O. *The Organization of Behavior: A Neuropsychological Theory*. A Wiley Book in Clinical Psychology. New York: Wiley, 1949.

Hecht, Erin E., Lauren E. Murphy, David A. Gutman, John R. Votaw, David M. Schuster, Todd M. Preuss, Guy A. Orban, Dietrich Stout, and Lisa A. Parr. "Differences in Neural Activation for Object-Directed Grasping in Chimpanzees and Humans." *Journal of Neuroscience* 33, no. 35 (August 2013): 14117–34.

Hecht, Erin E. and Lisa Parr. "The Chimpanzee Mirror System and the Evolution of Frontoparietal Circuits for Action Observation and Social Learning." In *New Frontiers in Mirror Neurons Research,* edited by Pier Francesco Ferrari and Giacomo Rizzolatti. Oxford: Oxford University Press, 2015.

Henrich, Joseph Patrick. *The Secret of Our Success: How Culture Is Driving Human Evolution, Domesticating Our Species, and Making Us Smarter.* Princeton: Princeton University Press, 2016.

Henry, Richard Conn. "The Mental Universe." *Nature* 436, no. 29 (July 2005): 29.

Hey, Anthony J. G., and Patrick Walters. *The New Quantum Universe.* Cambridge: Cambridge University Press, 2003.

Hidalgo, César A. *Why Information Grows: The Evolution of Order, from Atoms to Economies.* New York: Basic Books, 2015.

Hobsbawm, Eric J. *The Age of Capital, 1848–1875.* New York: Vintage Books, 1996.

———. *The Age of Empire, 1875–1914.* New York: Vintage Books, 1989.

———. *The Age of Extremes: A History of the World, 1914–1991.* New York: Pantheon Books, 1994.

———. *The Age of Revolution, 1789–1848.* New York: Vintage Books, 1996.

Hobsbawm, Eric J., and Chris Wrigley. *Industry and Empire: From 1750 to the Present Day.* New York: New Press, 1999.

Hoffmann, D. L., C. D. Standish, M. Garcia-Diez, P. B. Pettitt, J. A. Milton, J. Zilhao, J. J. Alcolea-Gonzalez, et al. "U-Th Dating of Carbonate Crusts Reveals Neandertal Origin of Iberian Cave Art." *Science* 359, no. 6378 (February 2018): 912–15.

Hofstadter, Douglas R. *Gödel, Escher, Bach: An Eternal Golden Braid.* New York: Basic Books, 1999.

Hossenfelder, Sabine. *Lost in Math: How Beauty Leads Physics Astray.* New York: Basic Books, 2018.

Hubel, David H. *Eye, Brain, and Vision.* Scientific American Library Series. New York: Scientific American Library, 1995.

Huxley, Aldous. *The Doors of Perception and Heaven and Hell.* New York: Perennial Classics, 2004.

Ifft, Peter J., Solaiman Shokur, Zheng Li, Mikhail A. Lebedev, and Miguel A. Nicolelis. "A Brain-Machine Interface Enables Bimanual Arm Movements in Monkeys." *Science Translational Medicine* 5, no. 210 (November 2013): 210ra154.

Ingraham, Christopher. "Poetry Is Going Extinct, Government Data Show." *Washington Post,* April 24, 2015.

Jackson, Maggie. *Distracted: The Erosion of Attention and the Coming Dark Age.* Amherst, MA: Prometheus Books, 2008.

James, Steven R. "Hominid Use of Fire in the Lower and Middle Pleistocene: A Review of the Evidence." *Current Anthropology* 30, no. 1 (1989): 1–26.

Jameson, Fredric. *The Ancients and the Postmoderns*. London: Versos, 2015.

Janicak, Philip G., and Mehmet E. Dokucu. "Transcranial Magnetic Stimulation for the Treatment of Major Depression." *Neuropsychiatric Disease and Treatment* 11 (2015): 1549–60.

Jeans, James. *The Mysterious Universe*. Cambridge: Macmillan/Cambridge University Press, 1930.

Jefferys, J. G. "Nonsynaptic Modulation of Neuronal Activity in the Brain: Electric Currents and Extracellular Ions." *Physiological Reviews* 75, no. 4 (October 1995): 689–723.

Jibu, Mari, and Kunio Yasue. *Quantum Brain Dynamics and Consciousness: An Introduction*. Advances in Consciousness Research 3. Amsterdam: John Benjamins, 1995.

Johanson, Donald C., and Kate Wong. *Lucy's Legacy: The Quest for Human Origins*. New York: Harmony Books, 2009.

John, E. R. "A Field Theory of Consciousness." *Consciousness and Cognition* 10, no. 2 (June 2001): 184–213.

Jung, Carl G. *Archetypes and the Collective Unconscious*. In vol. 9 of *The Collected Works of C. G. Jung*. Bollingen Series. Princeton: Princeton University Press, 1980.

———. *Psychological Types*. In vol. 6 of *The Collected Works of C. G. Jung*. Bollingen Series. Princeton: Princeton University Press, 1976.

———. *Synchronicity: An Acausal Connecting Principle*. In vol. 8 of *The Collected Works of C. G. Jung*. Bollingen Series. Princeton: Princeton University Press, 2010.

———. *The Undiscovered Self*. New York: Signet, 2006.

Kaas, Jon H. "The Evolution of Neocortex in Primates." *Progress in Brain Research* 195 (2012): 91–102.

Kaspersky Lab. "The Rise and Impact of Digital Amnesia: Why We Need to Protect What We No Longer Remember," 2015. https://media.kasperskycontenthub.com/wp-content/uploads/sites/100/2017/03/10084613/Digital-Amnesia-Report.pdf.

Kauffman, Stuart A. *At Home in the Universe: The Search for Laws of Self-Organization and Complexity*. New York: Oxford University Press, 1995.

Keenan, Julian Paul, Gordon G. Gallup, and Dean Falk. *The Face in the Mirror: The Search for the Origins of Consciousness*. New York: Ecco, 2003.

Kennedy, Hugh. *When Baghdad Ruled the Muslim World: The Rise and Fall of Islam's Greatest Dynasty*. Cambridge, MA: Da Capo, 2005.

Keynes, John Maynard. *The General Theory of Employment, Interest and Money (Illustrated)*. Kindle ed. Green World, 2015.

Kieu, Tien D. "Quantum Algorithm for Hilbert's Tenth Problem." *International Journal of Theoretical Physics* 42, no. 7 (2003): 1461–78.

Kim, Sang H., Sang-Hyun Baik, Chang S. Park, Su J. Kim, Sung W. Choi, and Sang E. Kim. "Reduced Striatal Dopamine D2 Receptors in People with Internet Addiction." *Neuroreport* 22, no. 8 (June 2011): 407–11.

Kim, Yoo-H., Rong Yu, Sergei P. Kulik, Yanhua Shih, and Marian O. Scully. "Delayed 'Choice' Quantum Eraser." *Physical Review Letters* 84, no. 1 (January 2000): 1–5.

King, Ross. *Brunelleschi's Dome: How a Renaissance Genius Reinvented Architecture*. New York: Walker, 2000.

Klein, Richard G., and Blake Edgar. *The Dawn of Human Culture*. New York: Wiley, 2002.

Köhler, Wolfgang. *Dynamics in Psychology*. New York: Liveright, 1940.

———. *Gestalt Psychology: An Introduction to New Concepts in Modern Psychology*. New York: Liveright, 1992.

Korzybski, Alfred. *Selections from Science and Sanity: An Introduction to Non-Aristotelian Systems and General Semantics*. Fort Worth, TX: Institute of General Semantics, 2010.

Kreiter, A. K., and W. Singer. "Stimulus-Dependent Synchronization of Neuronal Responses in the Visual Cortex of the Awake Macaque Monkey." *Journal of Neuroscience* 16, no. 7 (April 1996): 2381–96.

Krupa, David J., Matthew S. Matell, Amy J. Brisben, Laura M. Oliveira, and Miguel A. Nicolelis. "Behavioral Properties of the Trigeminal Somatosensory System in Rats Performing Whisker-Dependent Tactile Discriminations." *Journal of Neuroscience* 21, no. 15 (August 2001): 5752–63.

Krupa, David J., Michael C. Wiest, Marshall G. Shuler, Mark Laubach, and Miguel A. Nicolelis. "Layer-Specific Somatosensory Cortical Activation during Active Tactile Discrimination." *Science* 304, no. 5679 (June 2004): 1989–92.

Kuhn, Thomas S. *The Structure of Scientific Revolutions*. Chicago: University of Chicago Press, 1996.

Kupers, R., M. Pappens, A. M. de Noordhout, J. Schoenen, M. Ptito, and A. Fumal. "rTMS of the Occipital Cortex Abolishes Braille Reading and Repetition Priming in Blind Subjects." *Neurology* 68, no. 9 (February 2007): 691–93.

Kurzweil, Ray. *In the Age of Spiritual Machines: When Computers Exceed Human Intelligence*. New York: Penguin Books, 2000.

———. *The Singularity Is Near: When Humans Transcend Biology*. New York: Viking, 2005.

Lakoff, George, and Rafael E. Núñez. *Where Mathematics Comes From: How the Embodied Mind Brings Mathematics into Being*. New York: Basic Books, 2000.

Lane, Nick. *The Vital Question: Energy, Evolution, and the Origins of Complex Life*. New York: Norton, 2015.

Lashley, K. S., K. L. Chow, and J. Semmes. "An Examination of the Electrical Field Theory of Cerebral Integration." *Psychological Review* 58, no. 2 (March 1951): 123–36.

Laubach, Mark, Johan Wessberg, and Miguel A. Nicolelis. "Cortical Ensemble Activity Increasingly Predicts Behaviour Outcomes during Learning of a Motor Task." *Nature* 405, no. 6786 (June 2000): 567–71.

Lebedev, Mikhail A., Jose M. Carmena, Joseph E. O'Doherty, Miriam Zacksenhouse, Craig S. Henriquez, Jose C. Principe, and Miguel A. Nicolelis. "Cortical Ensemble Adaptation to Represent Velocity of an Artificial Actuator Controlled by a Brain-Machine Interface." *Journal of Neuroscience* 25, no. 19 (May 2005): 4681–93.

Lebedev, Mikhail A., and Miguel A. Nicolelis. "Brain-Machine Interfaces: From Basic Science to Neuroprostheses and Neurorehabilitation." *Physiological Reviews* 97, no. 2 (April 2017): 767–837.

———. "Brain-Machine Interfaces: Past, Present and Future." *Trends in Neuroscience* 29, no. 9 (September 2006): 536–46.

———. "Toward a Whole-Body Neuroprosthetic." *Progress in Brain Research* 194 (2011): 47–60.

Lewis, Michael. *Moneyball: The Art of Winning an Unfair Game.* New York: Norton, 2003.

Lewis, Paul. "'Our Minds Can Be Hijacked': The Tech Insiders Who Fear a Smartphone Dystopia." *Guardian*, October 6, 2017.

Lewis-Williams, J. David. *Conceiving God: The Cognitive Origin and Evolution of Religion.* London: Thames and Hudson, 2010.

———. *The Mind in the Cave: Consciousness and the Origins of Art.* London: Thames and Hudson, 2002.

Lewis-Williams, J. David, and D. G. Pearce. *Inside the Neolithic Mind: Consciousness, Cosmos, and the Realm of the Gods.* London: Thames and Hudson, 2005.

Lin, Rick C., M. A. Nicolelis, H. L. Zhou, and J. K. Chapin. "Calbindin-Containing Nonspecific Thalamocortical Projecting Neurons in the Rat." *Brain Research* 711, nos. 1–2 (March 1996): 50–55.

Lind, Johan, Magnus Enquist, and Stefano Ghirlanda. "Animal Memory: A Review of Delayed Matching-to-Sample Data." *Behavioral Processes* 117 (August 2015): 52–58.

Liu, Min, and Jianghong Luo. "Relationship between Peripheral Blood Dopamine Level and Internet Addiction Disorder in Adolescents: A Pilot Study." *International Journal of Clinical and Experimental Medicine* 8, no. 6 (2015): 9943–48.

Livio, Mario. *Is God a Mathematician?* New York: Simon and Schuster, 2009.

Lloyd, Seth. *Programming the Universe: A Quantum Computer Scientist Takes on the Cosmos.* New York: Knopf, 2006.

Lorkowski, C. M. "David Hume: Causation." In *Internet Encyclopedia of Philosophy.* https://www.iep.utm.edu/hume-cau/.

Lucas, J. R. "Minds, Machines and Gödel." *Philosophy* 36, nos. 112–27 (1961): 43–59.

Mach, Ernst. *The Analysis of Sensations and the Relation of the Physical to the Psychical.* Translated by C. M. Williams and Sydney Waterlow. Chicago: Open Court, 1914.

Maguire, Eleanor A., David G. Gadian, Ingrid S. Johnsrude, Catriona D. Good, John Ashburner, Richard S. J. Frackowiak, Christopher D. Firth. "Navigation-Related Structural Change in the Hippocampi of Taxi Drivers." *Proceedings of the National Academy of Sciences USA* 97, no. 8 (April 2000): 4398–403.

Malavera, Alejandra, Federico A. Silva, Felipe Fregni, Sandra Carrillo, and Ronald G. Garcia. "Repetitive Transcranial Magnetic Stimulation for Phantom Limb Pain in Land Mine Victims: A Double-Blinded, Randomized, Sham-Controlled Trial." *Journal of Pain* 17, no. 8 (August 2016): 911–18.

Maravita, Angelo, Charles Spence, and Jon Driver. "Multisensory Integration and the Body Schema: Close to Hand and within Reach." *Current Biology* 13, no. 13 (July 2003): R531–39.

Martin, Thomas R. *Ancient Greece: From Prehistoric to Hellenistic Times*. New Haven: Yale University Press, 2013.

Mas-Herrero, Ernest, Alain Dagher, and Robert J. Zatorre. "Modulating Musical Reward Sensitivity Up and Down with Transcranial Magnetic Stimulation." *Nature Human Behaviour* 2, no. 1 (January 2018): 27–32.

Matell, Matthew S., Warren H. Meck, and Miguel A. Nicolelis. "Interval Timing and the Encoding of Signal Duration by Ensembles of Cortical and Striatal Neurons." *Behavioral Neuroscience* 117, no. 4 (August 2003): 760–73.

Maturana, Humberto R., and Francisco J. Varela. *The Tree of Knowledge: The Biological Roots of Human Understanding*. Boston: Shambhala, 1992.

McFadden, Johnjoe. "The Conscious Electromagnetic Information (Cemi) Field Theory—The Hard Problem Made Easy?" *Journal of Consciousness Studies* 9, no. 8 (August 2002): 45–60.

———. "Synchronous Firing and Its Influence on the Brain's Electromagnetic Field—Evidence for an Electromagnetic Field Theory of Consciousness." *Journal of Consciousness Studies* 9, no. 4 (April 2002): 23–50.

McLuhan, Marshall. *Understanding Media: The Extensions of Man*. Corte Madera, CA: Gingko, 2013.

McLuhan, Marshall, W. Terrence Gordon, Elena Lamberti, and Dominique Scheffel-Dunand. *The Gutenberg Galaxy: The Making of Typographic Man*. Toronto: University of Toronto Press, 2011.

Meldrum, D. Jeffrey, and Charles E. Hilton. *From Biped to Strider: The Emergence of Modern Human Walking, Running, and Resource Transport*. American Association of Physical Anthropologists Meeting. New York: Kluwer Academic/Plenum, 2004.

Melzack, Ronald. "From the Gate to the Neuromatrix." *Pain*, supplement 6 (August 1999): S121–26.

———. *The Puzzle of Pain*. New York: Basic Books, 1973.

Melzack, Ronald, and Patrick D. Wall. *Textbook of Pain*. Edinburgh: Churchill Livingstone, 1999.

Menocal, Maria Rosa. *The Ornament of the World: How Muslims, Jews, and Christians Created a Culture of Tolerance in Medieval Spain*. Boston: Little, Brown, 2002.

Meredith, M. Alex, and H. Ruth Clemo. "Corticocortical Connectivity Subserving Different Forms of Multisensory Convergence." In *Multisensory Object Perception in the Primate Brain*, edited by M. J. Naumer and J. Kaiser. New York: Springer, 2010.

Merzbach, Uta C., and Carl B. Boyer. *A History of Mathematics*. Hoboken, NJ: John Wiley, 2011.

Miller, Arthur I. *Einstein, Picasso: Space, Time, and Beauty That Causes Havoc*. New York: Basic Books, 2001.

Miller, Daniel J., Tetyana Duka, Cheryl D. Stimpson, Steven J. Schapiro, Wallace B. Baze, Mark J. McArthur, Archibald J. Fobbs, et al. "Prolonged Myelination in Human Neocortical Evolution." *Proceedings of the National Academy of Sciences USA* 109, no. 41 (October 2012): 16480–85.

Mitchell, Melanie. *Complexity: A Guided Tour*. Oxford: Oxford University Press, 2009.

Mithen, Steven J. *After the Ice: A Global Human History, 20,000–5000 BC*. Cambridge, MA: Harvard University Press, 2004.

———. *Creativity in Human Evolution and Prehistory*. London: Routledge, 1998.

———. *The Prehistory of the Mind: The Cognitive Origins of Art, Religion and Science*. London: Thames and Hudson, 1996.

———. *The Singing Neanderthals: The Origins of Music, Language, Mind, and Body*. Cambridge, MA: Harvard University Press, 2006.

Moosavi-Dezfooli, Seyed-M., Alhussein Fawzi, Omar Fawzi, and Pascal Frossard. "Universal Adversarial Perturbations." *IEEE Conference on Computer Vision and Pattern Recognition (CVPR)* (2017): 86–94.

Morgan, T. J., N. T. Uomini, L. E. Rendell, L. Chouinard-Thuly, S. E. Street, H. M. Lewis, C. P. Cross, et al. "Experimental Evidence for the Co-evolution of Hominin Tool-Making Teaching and Language." *Nature Communications* 6 (January 2015): 6029.

Moyle, Franny. *Turner: The Extraordinary Life and Momentous Times of J.M.W. Turner*. New York: Penguin, 2016.

Mumford, Lewis. *Art and Technics*. Bampton Lectures in America. New York: Columbia University Press, 2000.

———. *The City in History: Its Origins, Its Transformations, and Its Prospects*. New York: Harcourt, 1961.

———. *The Condition of Man*. New York: Harcourt Brace Jovanovich, 1973.

———. *The Human Way Out*. Pendle Hill Pamphlet. Wallingford, PA: Pendle Hill, 1958.

———. *The Myth of the Machine: Technics and Human Development*. London: Secker and Warburg, 1967.

———. *The Pentagon of Power*. Vol. 2 of *The Myth of the Machine*. New York: Harcourt Brace Jovanovich, 1974.

———. *The Story of Utopias*. Kindle ed. Amazon Digital Services LLC, 2011.

———. *Technics and Civilization*. Chicago: University of Chicago Press, 2010.

Newberg, Andrew B., Eugene G. D'Aquili, and Vince Rause. *Why God Won't Go Away: Brain Science and the Biology of Belief*. New York: Ballantine Books, 2001.

Nicolelis, Miguel A. "Actions from Thoughts." *Nature* 409, no. 6818 (January 2001): 403–7.

———, ed. *Advances in Neural Population Coding*. Amsterdam: Elsevier, 2001.

———. "Are We at Risk of Becoming Biological Digital Machines?" *Nature Human Behavior* 1, no. 8 (January 2017): 1–2.

———. *Beyond Boundaries: The New Neuroscience of Connecting Brains with Machines— And How It Will Change Our Lives*. New York: Times Books/Henry Holt, 2011.

———. "Brain-Machine Interfaces to Restore Motor Function and Probe Neural Circuits." *Nature Reviews Neuroscience* 4, no. 5 (May 2003): 417–22.

———. "Controlling Robots with the Mind." *Scientific American Reports* 18 (February 2008): 72–79.

———. "Living with Ghostly Limbs." *Scientific American Mind* 18 (December 2007): 53–59.

———. *Methods for Neural Ensemble Recordings*. Boca Raton: CRC, 2008.

———. "Mind in Motion." *Scientific American* 307, no. 3 (September 2012): 58–63.

———. "Mind out of Body." *Scientific American* 304, no. 2 (February 2011): 80–83.

Nicolelis, M. A., L. A. Baccala, R. C. Lin, and J. K. Chapin. "Sensorimotor Encoding by Synchronous Neural Ensemble Activity at Multiple Levels of the Somatosensory System." *Science* 268, no. 5215 (June 1995): 1353–58.

Nicolelis, Miguel A., and John K. Chapin. "Controlling Robots with the Mind." *Scientific American* 287, no. 4 (October 2002): 46–53.

Nicolelis, Miguel A., Dragan Dimitrov, Jose M. Carmena, Roy Crist, Gary Lehew, Jerald D. Kralik, and Steven P. Wise. "Chronic, Multisite, Multielectrode Recordings in Macaque Monkeys." *Proceedings of the National Academy of Sciences USA* 100, no. 19 (September 2003): 11041–46.

Nicolelis, Miguel A., and Erica E. Fanselow. "Thalamocortical Optimization of Tactile Processing according to Behavioral State." *Nature Neuroscience* 5, no. 6 (June 2002): 517–23.

Nicolelis, Miguel A., Erica E. Fanselow, and Asif A. Ghazanfar. "Hebb's Dream: The Resurgence of Cell Assemblies." *Neuron* 19, no. 2 (August 1997): 219–21.

Nicolelis, Miguel A., Asif A. Ghazanfar, Barbara M. Faggin, Scott Votaw, and Laura M. Oliveira. "Reconstructing the Engram: Simultaneous, Multisite, Many Single Neuron Recordings." *Neuron* 18, no. 4 (April 1997): 529–37.

Nicolelis, Miguel A., and Mikhail A. Lebedev. "Principles of Neural Ensemble Physiology Underlying the Operation of Brain-Machine Interfaces." *Nature Reviews Neuroscience* 10, no. 7 (July 2009): 530–40.

Nicolelis, Miguel A., Laura M. Oliveira, Rick C. Lin, and John K. Chapin. "Active Tactile Exploration Influences the Functional Maturation of the Somatosensory System." *Journal of Neurophysiology* 75, no. 5 (May 1996): 2192–96.

Nicolelis, Miguel A., and Sidarta Ribeiro. "Seeking the Neural Code." *Scientific American* 295, no. 6 (December 2006): 70–77.

Nijholt, Anton. "Competing and Collaborating Brains: Multi-Brain Computer Interfacing." In *Brain-Computer Interfaces: Current Trends and Applications*, edited by A. E. Hassanien and A. T. Azar. Cham: Springer International Publishing Switzerland, 2015.

Nishitani, Nobuyuki, and Ritta Hari. "Viewing Lip Forms: Cortical Dynamics." *Neuron* 36, no. 6 (December 2002): 1211–20.

Noebels, Jeffrey. "A Perfect Storm: Converging Paths of Epilepsy and Alzheimer's Dementia Intersect in the Hippocampal Formation." *Epilepsia* 52, supplement 1 (2011): 39–46.

Notter, D. R., J. R. Lucas, and F. S. McClaugherty. "Accuracy of Estimation of Testis Weight from in Situ Testis Measures in Ram Lambs." *Theriogenology* 15, no. 2 (1981): 227–34.

Numan, Michael. *Neurobiology of Social Behavior: Toward an Understanding of the Prosocial and Antisocial Brain.* London: Elsevier Academic Press, 2015.

Oberman, Lindsay M., and Vilayanur S. Ramachandran. "The Role of the Mirror Neuron System in the Pathophysiology of Autism Spectrum Disorder." In *New Frontiers in Mirror Neurons Research*, edited by Pier Francesco Ferrari and Giacomo Rizzolatti. Oxford: Oxford University Press, 2015.

O'Doherty, Joseph E., Mikhail A. Lebedev, Timothy L. Hanson, Nathan A. Fitzsimmons, and Miguel A. Nicolelis. "A Brain-Machine Interface Instructed by Direct Intracortical Microstimulation." *Frontiers in Integrative Neuroscience* 3 (September 2009): 20.

O'Doherty, Joseph E., Mikhail A. Lebedev, Peter J. Ifft, Katie Z. Zhuang, Solaiman Shokur, Hannes Bleuler, and Miguel A. Nicolelis. "Active Tactile Exploration Using a Brain-Machine-Brain Interface." *Nature* 479, no. 7372 (November 2011): 228–31.

O'Dowd, Matt. *How the Quantum Eraser Rewrites the Past.* PBS Digital Studios, Space Time, 2016. https://www.youtube.com/watch?v=8ORLN_KwAgs&app=desktop.

O'Neill, Kristie. "The Hutu and Tutsi Distinction." From *Advanced Topics in Sociology: The Sociology of Genocide—SOC445H5.* Ontario, Canada: University of Toronto— Mississauga, November 13, 2009. http://docplayer.net/33422656-The-distinction -between-hutu-and-tutsi-is-central-to-understanding-the-rwandan.html.

Pais-Vieira, Miguel, Gabriela Chiuffa, Mikhail Lebedev, Amol Yadav, and Miguel A. Nicolelis. "Building an Organic Computing Device with Multiple Interconnected Brains." *Scientific Reports* 5 (July 2015): 11869.

Pais-Vieira, Miguel, Carolina Kunicki, Po H. Tseng, Joel Martin, Mikhail Lebedev, and Miguel A. Nicolelis. "Cortical and Thalamic Contributions to Response Dynamics across Layers of the Primary Somatosensory Cortex during Tactile Discrimination." *Journal of Neurophysiology* 114, no. 3 (September 2015): 1652–76.

Pais-Vieira, Miguel, Mikhail Lebedev, Carolina Kunicki, Jing Wang, and Miguel A. Nicolelis. "A Brain-to-Brain Interface for Real-Time Sharing of Sensorimotor Information." *Scientific Reports* 3 (2013): 1319.

Pais-Vieira, Miguel, Mikhail A. Lebedev, Michael C. Wiest, and Miguel A. Nicolelis. "Simultaneous Top-Down Modulation of the Primary Somatosensory Cortex and Thalamic Nuclei during Active Tactile Discrimination." *Journal of Neuroscience* 33, no. 9 (February 2013): 4076–93.

Pais-Vieira, Miguel, Amol P. Yadav, Derek Moreira, David Guggenmos, Amilcar Santos, Mikhail Lebedev, and Miguel A. Nicolelis. "A Closed Loop Brain-Machine Interface for Epilepsy Control Using Dorsal Column Electrical Stimulation." *Scientific Reports* 6 (September 2016): 32814.

Pallasmaa, Juhani. *The Eyes of the Skin: Architecture and the Senses.* Chichester, UK: Wiley-Academy; Hoboken, NJ: John Wiley and Sons, 2012.

————. *The Thinking Hand: Existential and Embodied Wisdom in Architecture.* Chichester, UK: Wiley, 2009.

Papagianni, Dimitra, and Michael Morse. *Neanderthals Rediscovered: How Modern Science Is Rewriting Their Story.* New York: Thames and Hudson, 2013.

Papanicolaou, Andrew C. *Clinical Magnetoencephalography and Magnetic Source Imaging.* Cambridge: Cambridge University Press, 2009.

Papoušek, Hanus, and Mechthild Papoušek. "Mirror Image and Self-Recognition in Young Human Infants: I. A New Method of Experimental Analysis." *Developmental Psychobiology* 7, no. 2 (March 1974): 149–57.

Patil, Parag G., Jose M. Carmena, Miguel A. Nicolelis, and Dennis A. Turner. "Ensemble Recordings of Human Subcortical Neurons as a Source of Motor Control Signals for a Brain-Machine Interface." *Neurosurgery* 55, no. 1 (July 2004): 27–38.

Pedrosa, Mário. *Primary Documents.* Edited by Glória Ferreira and Paulo Herkenhoff. New York: Museum of Modern Art, 2015.

Pedrosa, Mário. *Arte ensaios.* São Paulo: Cosac Naify, 2015.

Penrose, Roger. *The Emperor's New Mind: Concerning Computers, Minds, and the Laws of Physics.* New York: Penguin Books, 1991.

————. *Fashion, Faith, and Fantasy in the New Physics of the Universe.* Princeton: Princeton University Press, 2016.

————. *Shadows of the Mind: A Search for the Missing Science of Consciousness.* Oxford: Oxford University Press, 1994.

Petersen, A. "The Philosophy of Niels Bohr." *Bulletin of the Atomic Scientists* 19, no. 7 (1963): 8–9.

Petrides, Michael. *Neuroanatomy of Language Regions of the Human Brain*. Amsterdam: Elsevier Academic Press, 2014.

Piccinini, Gualtiero. "Computationalism in the Philosophy of Mind." *Philosophy Compass* 4, no. 3 (2009): 515–32.

Pickering, Andrew. *The Cybernetic Brain: Sketches of Another Future*. Chicago: University of Chicago Press, 2010.

Piketty, Thomas, and Arthur Goldhammer. *Capital in the Twenty-First Century*. Cambridge, MA: Belknap Press of Harvard University Press, 2014.

Pockett, Susan. "Field Theories of Consciousness." *Scholarpedia* (2013). doi:10.4249/scholarpedia.4951, http://www.scholarpedia.org/article/Field_theories_of_consciousness.

———. *The Nature of Consciousness: A Hypothesis*. Lincoln, NE: iUniverse, 2000.

Poincaré, Henri. *Leçons de mécanique celeste*. Paris: Gauthier-Villars, 1905.

———. *La science e l'hypothèse*. Paris: Flammarion, 1902.

———. *The Value of Science: Essential Writings of Henri Poincaré*. Modern Library Science Series. New York: Modern Library, 2001.

Pollard, Justin, and Howard Reid. *The Rise and Fall of Alexandria: Birthplace of the Modern Mind*. New York: Penguin, 2007.

Popper, Karl R. *The Logic of Scientific Discovery*. London: Routledge, 1992.

Pour-El, Marian B., and J. Ian Richards. *Computability in Analysis and Physics*. Berlin: Springer-Verlag, 1989.

Prigogine, Ilya. *The End of Certainty*. New York: Free Press, 1996.

Prigogine, Ilya, and Isabelle Stengers. *The End of Certainty: Time, Chaos, and the New Laws of Nature*. New York: Free Press, 1997.

———. *Order out of Chaos: Man's New Dialogue with Nature*. Toronto: Bantam Books, 1984.

Puchner, Martin. *The Written World: The Power of Stories to Shape People, History, Civilization*. New York: Random House, 2017.

Putnam, Hilary. "Brains and Behavior." In *Analytical Philosophy: Second Series*, edited by Ronald J. Butler. Oxford: Blackwell, 1963.

———. *The Many Faces of Realism*. The Paul Carus Lectures. La Salle, IL: Open Court, 1987.

———. *Mathematics, Matter, and Method*. Cambridge: Cambridge University Press, 1979.

Radman, Thomas, Yuzho Su, Je H. An, Lucas C. Parra, and Marom Bikson. "Spike Timing Amplifies the Effect of Electric Fields on Neurons: Implications for Endogenous Field Effects." *Journal of Neuroscience* 27, no. 11 (March 2007): 3030–36.

Rajangam, Sankaranarayani, Po H. Tseng, Allen Yin, Gary Lehew, David Schwarz, Mikhail A. Lebedev, and Miguel A. Nicolelis. "Wireless Cortical Brain-Machine Interface for Whole-Body Navigation in Primates." *Scientific Reports* 6 (March 2016): 22170.

Ramakrishnan, Arjun, Yoon W. Byun, Kyle Rand, Christian E. Pedersen, Mikhail A. Lebedev, and Miguel A. L. Nicolelis. "Cortical Neurons Multiplex Reward-Related Signals along with Sensory and Motor Information." *Proceedings of the National Academy of Sciences USA* 114, no. 24 (June 2017): E4841–50.

Ramakrishnan, Arjun, Peter J. Ifft, Miguel Pais-Vieira, Yoon W. Byun, Katie Z. Zhuang, Mikhail A. Lebedev, and Miguel A. Nicolelis. "Computing Arm Movements with a Monkey Brainet." *Scientific Reports* 5 (July 2015): 10767.

Raphael, Max. *Prehistoric Cave Paintings*. Translated by Norbert Guterman. Bollingen Series. New York: Pantheon Books, 1945.

Rasch, Bjorn, and Jan Born. "About Sleep's Role in Memory." *Physiological Reviews* 93, no. 2 (April 2013): 681–766.

Reimann, Michael W., Costas A. Anastassiou, Rodrigo Perin, Sean L. Hill, Henry Markram, and Christof Koch. "A Biophysically Detailed Model of Neocortical Local Field Potentials Predicts the Critical Role of Active Membrane Currents." *Neuron* 79, no. 2 (July 2013): 375–90.

Renfrew, Colin, Christopher D. Frith, and Lambros Malafouris. *The Sapient Mind: Archaeology Meets Neuroscience*. Oxford: Oxford University Press, 2009.

Rilling, James K. "Comparative Primate Neuroimaging: Insights into Human Brain Evolution." *Trends in Cognitive Sciences* 18, no. 1 (January 2014): 46–55.

Robb, L. P., J. M. Cooney, and C. R. McCrory. "Evaluation of Spinal Cord Stimulation on the Symptoms of Anxiety and Depression and Pain Intensity in Patients with Failed Back Surgery Syndrome." *Irish Journal of Medical Science* 186, no. 3 (August 2017): 767–71.

Robinson, Andrew. *The Last Man Who Knew Everything: Thomas Young, the Anonymous Polymath Who Proved Newton Wrong, Explained How We See, Cured the Sick, and Deciphered the Rosetta Stone, among Other Feats of Genius*. New York: Pi, 2006.

Rogawski, Michael A., and Wolfgang Loscher. "The Neurobiology of Antiepileptic Drugs for the Treatment of Nonepileptic Conditions." *Nature Medicine* 10, no. 7 (July 2004): 685–92.

Ronen, Itamar, Matthew Budde, Ece Ercan, Jacopo Annese, Aranee Techawiboonwong, and Andrew Webb. "Microstructural Organization of Axons in the Human Corpus Callosum Quantified by Diffusion-Weighted Magnetic Resonance Spectroscopy of N-Acetylaspartate and Post-mortem Histology." *Brain Structure and Function* 219, no. 5 (September 2014): 1773–85.

Rothbard, Murray Newton. *A History of Money and Banking in the United States: The Colonial Era to World War II*. Auburn, AL: Ludwig von Mises Institute, 2002.

Rovelli, Carlo. "Relational Quantum Mechanics." *arXiv:quant-ph/9609002v2*, February 24, 1997. https://arxiv.org/abs/quant-ph/9609002v2.

Rozzi, Stefano. "The Neuroanatomy of the Mirror Neuron System." In *New Frontiers in Mirror Neurons Research*, edited by Pier Francesco Ferrari and Giacomo Rizzolatti. Oxford: Oxford University Press, 2015.

Rubino, Giulia, Lee A. Rozema, Adrien Feix, Mateus Araujo, Jonas M. Zeuner, Lorenzo M. Procopio, Caslav Brukner, and Philip Walther. "Experimental Verification of an Indefinite Causal Order." *Science Advances* 3, no. 3 (March 2017): e1602589.

Russell, Bertrand. *A History of Western Philosophy, and Its Connection with Political and Social Circumstances from the Earliest Times to the Present Day.* New York: Simon and Schuster, 1945.

———. *An Inquiry into Meaning and Truth.* New York: Norton, 1940.

Sacks, Oliver. *Hallucinations.* Waterville, ME: Thorndike, 2013.

Sadato, Norihiro, Alvaro Pascual-Leone, Jordan Grafman, Vicente Ibanez, Marie-P. Deiber, George Dold, and Mark Hallett. "Activation of the Primary Visual Cortex by Braille Reading in Blind Subjects." *Nature* 380, no. 6574 (April 1996): 526–28.

Saliba, George. *Islamic Science and the Making of the European Renaissance.* Transformations. Cambridge, MA: MIT Press, 2007.

Samotus, Olivia, Andrew Parrent, and Mandar Jog. "Spinal Cord Stimulation Therapy for Gait Dysfunction in Advanced Parkinson's Disease Patients." *Movement Disorders* 33, no. 5 (2018): 783–92.

Santana, Maxwell B., Pär Halje, Hougelle Simplicio, Ulrike Richter, Marco A. M. Freire, Per Petersson, Romulo Fuentes, and Miguel A. L. Nicolelis. "Spinal Cord Stimulation Alleviates Motor Deficits in a Primate Model of Parkinson Disease." *Neuron* 84, no. 4 (November 2014): 716–22.

Scharf, Caleb A. *The Copernicus Complex: Our Cosmic Significance in a Universe of Planets and Probabilities.* New York: Scientific American/Farrar, Straus and Giroux, 2014.

Schneider, Michael L., Christine A. Donnelly, Stephen E. Russek, Burm Baek, Matthew R. Pufall, Peter F. Hopkins, Paul D. Dresselhaus, Samuel P. Benz, and William H. Rippard. "Ultralow Power Artificial Synapses Using Nanotextured Magnetic Josephson Junctions." *Science Advances* 4, no. 1 (January 2018): e1701329.

Schrödinger, Erwin. *What Is Life? The Physical Aspect of the Living Cell.* Cambridge: Cambridge University Press, 1944.

———. *What Is Life? With Mind and Matter and Autobiographical Sketches.* Canto Classics. Cambridge: Cambridge University Press, 1992.

Schwarz, David A., Mikhail A. Lebedev, Timothy L. Hanson, Dragan F. Dimitrov, Gary Lehew, Jim Meloy, Sankaranarayani Rajangam, et al. "Chronic, Wireless Recordings of Large-Scale Brain Activity in Freely Moving Rhesus Monkeys." *Nature Methods* 11, no. 6 (June 2014): 670–76.

Searle, John R. *The Construction of Social Reality.* New York: Free Press, 1995.

———. *Freedom and Neurobiology.* New York: Columbia University Press, 2007.

———. *Making the Social World: The Structure of Human Civilization.* Oxford: Oxford University Press, 2010.

———. *Seeing Things as They Are: A Theory of Perception.* Oxford: Oxford University Press, 2015.

Seddon, Christopher. *Humans: From the Beginning; From the First Apes to the First Cities.* London: Glanville, 2014.

Selfslagh, A., S. Shokur, D. S. Campos, A. R. Donati, S. Almeida, M. Bouri, and M. A. Nicolelis. "Non-invasive, Brain-Controlled Functional Electrical Stimulation for Locomotion Rehabilitation in Paraplegic Patients." *In Review* (2019).

Shannon, Claude. "A Mathematical Theory of Communication." *Bell System Technical Journal* 47, no. 3 (1948): 379–423.

Sherwood, Chet C., Cheryl D. Stimpson, Mary A. Raghanti, Derek E. Wildman, Monica Uddin, Lawrence I. Grossman, Morris Goodman, et al. "Evolution of Increased Glia-Neuron Ratios in the Human Frontal Cortex." *Proceedings of the National Academy of Sciences USA* 103, no. 37 (September 2006): 13606–11.

Shlain, Leonard. *Art & Physics: Parallel Visions in Space, Time, and Light.* New York: Quill/W. Morrow, 1993.

Shokur, Solaiman, Ana R. C. Donati, Debora S. Campos, Claudia Gitti, Guillaume Bao, Dora Fischer, Sabrina Almeida, Vania A. S. Braga, et al. "Training with Brain-Machine Interfaces, Visuo-Tactile Feedback and Assisted Locomotion Improves Sensorimotor, Visceral, and Psychological Signs in Chronic Paraplegic Patients." *Public Library of Science One* 13, no. 11 (2018): e0206464.

Shokur, Solaiman, Simone Gallo, Renan C. Moioli, Ana R. Donati, Edgard Morya, Hannes Bleuler, and Miguel A. Nicolelis. "Assimilation of Virtual Legs and Perception of Floor Texture by Complete Paraplegic Patients Receiving Artificial Tactile Feedback." *Scientific Reports* 6 (September 2016): 32293.

Shokur, Solaiman, Joseph E. O'Doherty, Jesse. A. Winans, Hannes Bleuler, Mikhail A. Lebedev, and Miguel A. Nicolelis. "Expanding the Primate Body Schema in Sensorimotor Cortex by Virtual Touches of an Avatar." *Proceedings of the National Academy of Sciences USA* 110, no. 37 (September 2013): 15121–26.

Siegelmann, Hava T. "Computation beyond the Turing Limit." *Science* 268, no. 5210 (April 1995): 545–48.

Sigmund, Karl. *Exact Thinking in Demented Times: The Vienna Circle and the Epic Quest for the Foundations of Science.* New York: Basic Books, 2017.

Sivakumar, Siddharth S., Amalia G. Namath, and Roberto F. Galan. "Spherical Harmonics Reveal Standing EEG Waves and Long-Range Neural Synchronization during Non-REM Sleep." *Frontiers in Computational Neuroscience* 10 (2016): 59.

Smaers, Jeroen B., Axel Schleicher, Karl Zilles, and Lucio Vinicius. "Frontal White Matter Volume Is Associated with Brain Enlargement and Higher Structural Connectivity in Anthropoid Primates." *Public Library of Science One* 5, no. 2 (February 2010): e9123.

Smolin, Lee. *Time Reborn: From the Crisis in Physics to the Future of the Universe.* New York: Houghton Mifflin Harcourt, 2013.

———. *The Trouble with Physics: The Rise of String Theory, the Fall of a Science, and What Comes Next.* New York: Houghton Mifflin Harcourt, 2006.

Snow, C. P., and Stefan Collini. *The Two Cultures*. Canto Classics. Cambridge: Cambridge University Press, 2012.

Sparrow, Betsy, Jenny Liu, and Daniel M. Wegner. "Google Effects on Memory: Cognitive Consequences of Having Information at Our Fingertips." *Science* 333, no. 6043 (August 2011): 776–78.

Sperry, R. W., N. Miner, and R. E. Myers. "Visual Pattern Perception Following Subpial Slicing and Tantalum Wire Implantations in the Visual Cortex." *Journal of Comparative and Physiological Psychology* 48, no. 1 (February 1955): 50–58.

Sproul, Barbara C. *Primal Myths: Creation Myths around the World*. New York: Harper Collins, 1979.

Starr, S. Frederick. *Lost Enlightenment: Central Asia's Golden Age from the Arab Conquest to Tamerlane*. Princeton: Princeton University Press, 2013.

Stephens, Greg J., Lauren J. Silbert, and Uri Hasson. "Speaker-Listener Neural Coupling Underlies Successful Communication." *Proceedings of the National Academy of Sciences USA* 107, no. 32 (August 2010): 14425–30.

Stiefel, Klaus M., Benjamin Torben-Nielsen, and Jay S. Coggan. "Proposed Evolutionary Changes in the Role of Myelin." *Frontiers in Neuroscience* 7 (2013): 202.

"The Story of Us." Special issue, *Scientific American* 25, no. 4S (2016).

Stout, Dietrich. "Tales of a Stone Age Neuroscientist." *Scientific American* 314, no. 4 (April 2016): 28–35.

Stout, Dietrich, Erin Hecht, Nada Khreisheh, Bruce Bradley, and Thierry Chaminade. "Cognitive Demands of Lower Paleolithic Toolmaking." *Public Library of Science One* 10, no. 4 (2015): e0121804.

Strathern, Paul. *The Medici: Power, Money, and Ambition in the Italian Renaissance*. New York: Pegasus Books, 2016.

Sumpter, David J. T. *Collective Animal Behavior*. Princeton: Princeton University Press, 2010.

Sypeck, Jeff. *Becoming Charlemagne: Europe, Baghdad, and the Empires of A.D. 800*. New York: Ecco, 2006.

Tagore, Rabindranath. *The Collected Works of Rabindranath Tagore (Illustrated Edition)*. New Delhi: General Press, 2017.

———. *The Religion of Man: Rabindranath Tagore*. Kolkata, India: Rupa, 2005.

Taylor, Timothy. *The Artificial Ape: How Technology Changed the Course of Human Evolution*. New York: Palgrave Macmillan, 2010.

Temin, Peter. *The Vanishing Middle Class: Prejudice and Power in a Dual Economy*. Cambridge, MA: MIT Press, 2017.

Temkin, Owsei. *The Falling Sickness: A History of Epilepsy from the Greeks to the Beginnings of Modern Neurology*. Baltimore: Johns Hopkins University Press, 1971.

Thomson, Eric E., Rafael Carra, and Miguel A. Nicolelis. "Perceiving Invisible Light through a Somatosensory Cortical Prosthesis." *Nature Communications* 4 (2013): 1482.

Thomson, Eric E., Ivan Zea, William Windham, Yohann Thenaisie, Cameron Walker, Jason Pedowitz, Wendy Franca, Ana L. Graneiro, and Miguel A. L. Nicolelis. "Cortical Neuroprosthesis Merges Visible and Invisible Light without Impairing Native Sensory Function." *eNeuro* 4, no. 6 (November–December 2017).

Tononi, Giulio. *Phi: A Voyage from the Brain to the Soul.* Singapore: Pantheon Books, 2012.

Toynbee, Arnold. *A Study of History.* Abridgement of Volumes I–VI by D. C. Somervell. New York: Oxford University Press, 1946.

———. *A Study of History.* Abridgement of Volumes VII–X by D. C. Somervell. New York: Oxford University Press, 1946.

Travers, Brittany G., Nagesh Adluru, Chad Ennis, Do P. M. Tromp, Dan Destiche, Sam Doran, Erin D. Bigler, et al. "Diffusion Tensor Imaging in Autism Spectrum Disorder: A Review." *Autism Research* 5, no. 5 (October 2012): 289–313.

Tsakiris, Manos, Marcello Costantini, and Patrick Haggard. "The Role of the Right Temporo-Parietal Junction in Maintaining a Coherent Sense of One's Body." *Neuropsychologia* 46, no. 12 (October 2008): 3014–18.

Tseng, Po H., Sankaranarayani Rajangam, Gary Lehew, Mikhail A. Lebedev, and Miguel A. L. Nicolelis. "Interbrain Cortical Synchronization Encodes Multiple Aspects of Social Interactions in Monkey Pairs." *Scientific Reports* 8, no. 1 (March 2018): 4699.

Tuchman, Roberto, and Isabelle Rapin. "Epilepsy in Autism." *Lancet Neurology* 1, no. 6 (October 2002): 352–58.

Tulving, Endel, and Fergus I. M. Craik. *The Oxford Handbook of Memory.* Oxford: Oxford University Press, 2000.

Turing, Alan M. "Computing Machinery and Intelligence." *Mind* (1950): 433–60.

———. "On Computable Numbers, with an Application to the Entscheidungsproblem." *Proceedings of the London Mathematical Society* 2, no. 42 (1936): 230–65.

———. "Systems of Logic Based on Ordinals." PhD diss., Princeton University, 1939.

Turkle, Sherry. *Alone Together: Why We Expect More from Technology and Less from Each Other.* New York: Basic Books, 2011.

———. *Reclaiming Conversation: The Power of Talk in a Digital Age.* New York: Penguin, 2015.

———. *The Second Self: Computers and the Human Spirit.* Cambridge, MA: MIT Press, 2005.

Uttal, William R. *Neural Theories of Mind: Why the Mind-Brain Problem May Never Be Solved.* Mahwah, NJ: Lawrence Erlbaum Associates, 2005.

van der Knaap, Lisette J., and Ineke J. van der Ham. "How Does the Corpus Callosum Mediate Interhemispheric Transfer? A Review." *Behavioural Brain Research* 223, no. 1 (September 2011): 211–21.

Varela, Francisco J., Evan Thompson, and Eleanor Rosch. *The Embodied Mind: Cognitive Science and Human Experience.* Cambridge, MA: MIT Press, 1991.

Varoufakis, Yanis. *Adults in the Room: My Battle with Europe's Deep Establishment.* London: Bodley Head, 2017.

Verhulst, Ferdinand. *Henri Poincaré: Impatient Genius.* New York: Springer, 2012.

Verschuur, Gerrit L. *Hidden Attraction: The History and Mystery of Magnetism.* New York: Oxford University Press, 1993.

Vigneswaran, Ganesh, Roland Philipp, Roger N. Lemon, and Alexander Kraskov. "M1 Corticospinal Mirror Neurons and Their Role in Movement Suppression during Action Observation." *Current Biology* 23, no. 3 (February 2013): 236–43.

von der Malsburg, Christoph. "Binding in Models of Perception and Brain Function." *Current Opinion in Neurobiology* 5, no. 4 (August 1995): 520–26.

von Foerster, Heinz, ed. *Cybernetics: Circular Causal and Feedback Mechanisms in Biological and Social Systems.* Vols. 6–10. New York: Josiah Macy Jr. Foundation, 1949–55.

Vossel, Keith A., Maria C. Tartaglia, Haakon B. Nygaard, Adam Z. Zeman, and Bruce L. Miller. "Epileptic Activity in Alzheimer's Disease: Causes and Clinical Relevance." *Lancet Neurology* 16, no. 4 (April 2017): 268.

Wallace, Alan. *The Nature of Reality: A Dialogue between a Buddhist Scholar and a Theoretical Physicist.* Institute for Cross Disciplinary Engagement, Dartmouth College, 2017. https://www.youtube.com/watch?t=195s&v=pLbSlCoPucw&app=desktop.

Wang, Jun, Jamie Barstein, Lauren E. Ethridge, Matthew W. Mosconi, Yukari Takarae, and John A. Sweeney. "Resting State EEG Abnormalities in Autism Spectrum Disorders." *Journal of Neurodevelopmental Disorders* 5, no. 1 (September 2013): 24.

Wawro, Geoffrey. *A Mad Catastrophe: The Outbreak of World War I and the Collapse of the Habsburg Empire.* New York: Basic Books, 2014.

Weatherford, Jack. *The History of Money: From Sandstone to Cyberspace.* New York: Crown, 1997.

Weinberg, Steven. *To Explain the World: The Discovery of Modern Science.* New York: Harper, 2015.

Weizenbaum, Joseph. *Computer Power and Human Reason: From Judgment to Calculation.* San Francisco: W. H. Freeman, 1976.

Weizenbaum, Joseph, and Gunna Wendt. *Islands in the Cyberstream: Seeking Havens of Reason in a Programmed Society.* Sacramento: Litwin Books, 2015.

Wessberg, Johan, Christopher R. Stambaugh, Jerald D. Kralik, Pam D. Beck, Mark Laubach, John K. Chapin, Jung Kim, et al. "Real-Time Prediction of Hand Trajectory by Ensembles of Cortical Neurons in Primates." *Nature* 408, no. 6810 (November 2000): 361–65.

West, Meredith J., and Andrew P. King. "Female Visual Displays Affect the Development of Male Song in the Cowbird." *Nature* 334, no. 6179 (July 1988): 244–46.

West, Meredith J., Andrew P. King, David J. White, Julie Gros-Louis, and Grace Freed-Brown. "The Development of Local Song Preferences in Female Cowbirds (Molothrus Ater): Flock Living Stimulates Learning." *Ethology* 112, no. 11 (2006): 1095–107.

Wheeler, John Archibald. "Information, Physics, Quantum: The Search for Links." In *Proceedings of the 3rd International Symposium on Foundations of Quantum Mechanics in the Light of New Technology*, edited by S. Kobayashi, H. Ezawa, Y. Murayama, and S. Nomura, 354–68. Tokyo: Physical Society of Japan, 1990.

Wigner, Eugene. "Remarks on the Mind-Body Question: Symmetries and Reflections." In *Philosophical Reflections and Syntheses: The Collected Works of Eugene Paul Wigner (Part B, Historical, Philosophical, and Socio-political Papers)*, edited by J. Mehra. Berlin: Springer, 1995.

Wilson, Frank R. *The Hand: How Its Use Shapes the Brain, Language, and Human Culture*. New York: Pantheon Books, 1998.

Wittgenstein, Ludwig. *Philosophical Investigations*. Translated by G. E. M. Anscombe, P. M. S. Hacker, and Joachim Schulte. Edited by P.M.S. Hacker and Joachim Schulte. Chichester, UK: Wiley-Blackwell, 2009.

———. *Tractatus Logico-Philosophicus*. Routledge Great Minds. London: Routledge, 2014.

Witthaut, Dirk, Sandro Wimberger, Raffaella Burioni, and Marc Timme. "Classical Synchronization Indicates Persistent Entanglement in Isolated Quantum Systems." *Nature Communications* 8 (April 2017): 14829.

Wong, Julie Carrie. "Former Facebook Executive: Social Media Is Ripping Society Apart." *Guardian*, December 12, 2017.

Wrangham, Richard W. *Catching Fire: How Cooking Made Us Human*. New York: Basic Books, 2009.

Yadav, Amol P., Romulo Fuentes, Hao Zhang, Thais Vinholo, Chi-H. Wang, Marco A. Freire, and Miguel A. Nicolelis. "Chronic Spinal Cord Electrical Stimulation Protects against 6-Hydroxydopamine Lesions." *Scientific Reports* 4 (2014): 3839.

Yadav, Amol P., and Miguel A. L. Nicolelis. "Electrical Stimulation of the Dorsal Columns of the Spinal Cord for Parkinson's Disease." *Movement Disorders* 32, no. 6 (June 2017): 820–32.

Yin, Allen, Po H. Tseng, Sankaranarayani Rajangam, Mikhail A. Lebedev, and Miguel A. L. Nicolelis. "Place Cell-like Activity in the Primary Sensorimotor and Premotor Cortex during Monkey Whole-Body Navigation." *Scientific Reports* 8, no. 1 (June 2018): 9184.

Zajonc, Arthur. *Catching the Light: The Entwined History of Light and Mind*. New York: Oxford University Press, 1995.

Zhang, Kechen, and Terrence J. Sejnowski. "A Universal Scaling Law between Gray Matter and White Matter of Cerebral Cortex." *Proceedings of the National Academy of Sciences USA* 97, no. 10 (May 2000): 5621–26.

Zuboff, Shoshana. *The Age of Surveillance Capitalism: The Fight for a Human Future at the New Frontier of Power*. New York: Public Affairs, 2018.

Index

Page numbers in *italics* refer to figures.